Statistics for Biology and Health

Series Editors
K. Dietz, M. Gail, K. Krickeberg, J. Samet, A. Tsiatis

Springer
New York
Berlin
Heidelberg
Barcelona
Hong Kong
London
Milan
Paris
Singapore
Tokyo

Statistics for Biology and Health

Warren J. Ewens   Gregory R. Grant

Statistical Methods in Bioinformatics: An Introduction

With 30 Illustrations

 Springer

Warren J. Ewens
Department of Biology
University of Pennsylvania
Philadelphia, PA 19104-6018
USA
wewens@sas.upenn.edu

Gregory R. Grant
Penn Center for Computational Biology
University of Pennsylvania
Philadelphia, PA 19104
USA
ggrant@pcbi.upenn.edu

Series Editors

K. Dietz
Institut für Medizinische Biometrie
Universität Tübingen
Westbahnhofstrasse 55
D-72070 Tübingen
Germany

M. Gail
National Center Institute
Rockville, MD 20892
USA

K. Krickeberg
Le Chatelet
F-63270 Manglieu
France

J. Samet
School of Public Health
Department of Epidemiology
Johns Hopkins University
615 Wolfe Street
Baltimore, MD 21205-2103
USA

A. Tsiatis
Department of Statistics
North Carolina State University
Raleigh, NC 27695
USA

R 858
. E 986
2002 ×

Library of Congress Cataloging-in-Publication Data
Ewens, W.J. (Warren John), 1937–
    Statistical methods in bioinformatics : an introduction / Warren Ewens, Gregory Grant.
        p. cm. — (Statistics for biology and health)
    Includes bibliographical references and index.
    ISBN 0-387-95229-2 (alk. paper)
    1. Bioinformatics. 2. Statistical Methods. I. Grant, Gregory, 1965– II. Title. III. Series.
R858 .E986 2001
570′.1′5195—dc21

00-067923

Printed on acid-free paper.

Production managed by MaryAnn Brickner; manufacturing supervised by Jerome Basma.
Camera-ready copy prepared from the authors' LaTeX files.
Printed and bound by Edwards Brothers, Inc., Ann Arbor, MI.
Printed in the United States of America.

9 8 7 6 5 4 3 2 (Corrected second printing, 2002)

ISBN 0-387-95229-2   SPIN 10850423

Springer-Verlag   New York Berlin Heidelberg
A member of BertelsmannSpringer Science+Business Media GmbH

For Kathy and Elisabetta

Preface

We take *bioinformatics* to mean the emerging field of science growing from the application of mathematics, statistics, and information technology, including computers and the theory surrounding them, to the study and analysis of very large biological, and particularly genetic, data sets. The field has been fueled by the increase in DNA data generation leading to the massive data sets already generated, and yet to be generated, in particular the data from the human genome project, as well as other genome projects.

Bioinformatics does not aim to lay down fundamental mathematical laws that govern biological systems parallel to those laid down in physics. Such laws, if they exist, are a long way from being determined for biological systems. Instead, at this stage the main utility of mathematics in the field is in the creation of tools that investigators can use to analyze data. For example, biologists need tools for the statistical assessment of the similarity between two or more DNA or protein sequences, for finding genes in genomic DNA, and for estimating differences in how genes are expressed in different tissues. Such tools involve statistical modeling of biological systems, and it is our belief that there is a need for a book that introduces probability, statistics, and stochastic processes in the context of bioinformatics. We hope to fill that need here.

The material in this text assumes little or no background in biology. The basic notions of biology that one needs in order to understand the material are outlined in Appendix A. Some further details that are necessary to understand particular applications are given in the context of those applications. The necessary background in mathematics is introductory courses in calculus and linear algebra. In order to be clear about notation and terminology, as well as to organize several results that are needed in the text, a review of basic notions in mathematics is given in Appendix B. No computer science knowledge is assumed and no programming is necessary to understand the material.

Why are probability and statistics so important in bioinformatics? Bioinformatics involves the analysis of biological data. Many chance mechanisms are involved in the creation of these data, most importantly the many random processes inherent in biological evolution and the randomness inherent in any sampling process. Stochastic process theory involves the description of the evolution of random processes occurring over time or space. Biological evolution over eons has provided the outcome of one of the most complex stochastic processes imaginable, and one that requires complex stochastic process theory for its description and analysis.

Our aim is to give an introductory account of some of the probability theory, statistics, and stochastic process theory appropriate to computational biology and bioinformatics. This is not a "how-to" book, of which there are several in the literature, but it aims to fill a gap in the literature in the statistical and probabilistic aspects of bioinformatics. The earlier chapters in this book contain standard introductory material suitable for any statistics course. Even here, however, we have departed somewhat from well-trodden paths, beginning to focus on material of interest in bioinformatics, for example the theory of the maximum of several random variables, moment-generating functions, geometric random variables and their various generalizations, together with information theory and entropy. We have also provided, in Appendix B, some standard mathematical results that are needed as background for this and other material.

This text is by no means comprehensive. There are several books that cover some of the topics we consider at a more advanced level. The reader should approach this text in part as an introduction and a means of assessing his/her interest in and ability to pursue this field further. Thus we have not tried to cover a comprehensive list of topics, nor to cover those topics discussed in complete detail. No book can ever fulfill the task of providing a complete introduction to this subject, since bioinformatics is evolving too quickly. To learn this subject as it evolves one must ultimately turn to the literature of published articles in journals. We hope that this book will provide a first stepping stone leading the reader in this direction.

We also wish to appeal to trained statisticians and to give them an introduction to bioinformatics, since their contribution will be vital to the analysis of the biological data already at hand and, more important, to developing analyses for new forms of data to arrive in the future. Such readers should be able to read the latter chapters directly.

The statistical procedures currently used in this subject are often ad hoc, with different methods being used in different parts of the subject. We have tried to provide as many threads running through the book as possible in order to overcome this problem and to integrate the material.

One such thread is provided by aspects of the material on stochastic processes. BLAST is one of the most frequently used algorithms in applied statistics, one BLAST search being made every few seconds on average by bioinformatics researchers around the world. However, the stochastic process theory behind the statistical calculations used in this algorithm is not widely understood. We approach this theory by starting with random walks, and through these to sequential analysis theory and to Markov chains, and ultimately to BLAST. This sequence also leads to the theory of hidden Markov models and to evolutionary analyses.

We have chosen this thread for three reasons. The first is that BLAST theory is intrinsically important. The second, as just mentioned, is that this provides a coherent thread to the often unconnected aspects of stochastic process theory used in various areas of bioinformatics. The final reason is that, with the human genome data and the genomes of other important species complete at least in first draft, we wish to emphasize procedures that lead to the analysis of these data. The analysis of these data will require new and currently unpredictable statistical analyses, and in particular, the theory for the most recent and sophisticated versions of BLAST, and for its further developments, will require new advanced theory.

So far as more practical matters are concerned, we are well aware of the need for precision in presenting any mathematically based topic. However, we are also aware, for an applied field, of the perils of a too mathematically precise approach to probability, perhaps through measure theory. Our approach has tended to be less rather than more pedantic, and detailed qualifications that interrupt the flow and might annoy the reader have been omitted. As one example, we assume throughout that all random variables we consider have finite moments of all orders. This assumption enables us to avoid many minor (and in practice unimportant) qualifications to the analysis we present.

So far as statistical theory is concerned, the focus in this book is on *discrete* as opposed to *continuous* random variables, since (especially with DNA and protein sequences) discrete random variables are more relevant to bioinformatics. However, some aspects of the theory of discrete random variables are difficult, with no limiting distribution theory available for the maximum of these random variables. In this case progress is made by using theory from continuous random variables to provide bounds and approximations. Thus continuous random variables are also discussed in some detail in the early chapters.

The focus in this book is, as stated above, on probability, statistics, and stochastic processes. We do, however, discuss aspects of the important algorithmic side of bioinformatics, especially when relevant to these probabilistic topics. In particular, the dynamic programming algorithm is introduced because of its use in various probability applications, especially in hidden Markov models. Several books are already available that are devoted to algorithmic aspects of the subject.

In a broad interpretation of the word "bioinformatics" there are several areas of the application of statistics to bioinformatics that we do not develop. Thus we do not cover aspects of the statistical theory in genetics associated with disease finding and linkage analysis. This subject deserves an entire book on its own. Nor do we discuss the increasingly important applications of bioinformatics in the stochastic theory of evolutionary population genetics. Again, all these topics deserve a complete treatment of their own.

This book is based on lectures given to students in the two-semester course in bioinformatics and computational biology at the University of Pennsylvania given each year during the period 1995–2000. We are most indebted to Elisabetta Manduchi, from PCBI/CBIL, who helped at every stage in revising the material. We are also grateful to the late Christian Overton for guidance, inspiration, and friendship. We thank all other members of PCBI/CBIL who supported us in this task and patiently answered many questions, in particular Brian Brunk, Jonathan Schug, Chris Stoeckert, Jonathan Crabtree, Angel Pizarro, Deborah Pinney, Shannon McWeeney, Joan Mazzarelli, and Eugene Buehler. We also thank Warren Gish for his help on BLAST, and for letting us reproduce his BLAST printout examples. We thank Chris Burge, Sandrine Dudoit, Terry Speed, Matt Werner, Alessandra Gallinari, Sam Sokolovski, Helen Murphy, Ethan Fingerman, Aaron Shaver, and Sue Wilson for their help. Finally, we thank students in the computational biology courses we have taught for their comments on the material, which we have often incorporated into this book. Any errors or omissions are, of course, our own responsibility. An archive of errata will be maintained at http://www.textbook-errata.org.

Warren J. Ewens
Gregory R. Grant
Philadelphia, Pennsylvania
February, 2001

Contents

1

Probability Theory (i): One Random Variable

1.1 Introduction

The DNA in an organism consists of very long sequences from an alphabet of four letters (nucleotides), a, g, c, and t (for adenine, guanine, cytosine, and thymine, respectively).[1] These sequences undergo change within any population over the course of many generations, as random mutations arise and become fixed in the population. Therefore, two rather different sequences may well derive from a common ancestor. Suppose we have two small DNA sequences such as those in (1.1) below, perhaps from two different species, where the arrows indicate paired nucleotides that are the same in both sequences.

$$
\begin{array}{c}
\downarrow \quad \downarrow \qquad \downarrow \quad \downarrow\downarrow\downarrow \quad \downarrow\downarrow \qquad\qquad\qquad \downarrow\downarrow\downarrow \\
g\ g\ a\ g\ a\ c\ t\ g\ t\ a\ g\ a\ c\ a\ g\ c\ t\ a\ a\ t\ g\ c\ t\ a\ t\ a \qquad (1.1) \\
g\ a\ a\ c\ g\ c\ c\ c\ t\ a\ g\ c\ c\ a\ c\ g\ a\ g\ c\ c\ c\ t\ t\ a\ t\ c
\end{array}
$$

We wish to gauge whether the two sequences show significant similarity, to assess, for example, whether they have a remote common ancestor.

If the sequences were each generated at random, with the four letters a, g, c, and t having equal probabilities of occurring at any position, then the two sequences should tend to agree at about one quarter of the positions. The two sequences above agree at 11 out of 26 positions. How unlikely is

---

[1] The reader unfamiliar with the basic notions of biology should refer to Appendix A for the necessary background.

this outcome if the sequences were generated at random? We cannot answer this question until we understand the properties of random sequences. Probability theory shows that under the assumptions of equal probabilities for a, g, c, and t at any site, and independence of all nucleotides involved, the probability that there will be 11 or more matches in a sequence comparison of length 26 is approximately 0.04. Therefore, our observation of 11 matches gives evidence that something other than chance is at work.

We have just carried out a statistical operation. That is, we have observed some data, in this case the number of matches between two sequences of length 26, and on the basis of some probability calculation and using some hypothetical value for some *parameter* (the unknown probability of a match), we made a statement about our level of belief in the value of that parameter.

This statement depended on some probability calculation. No valid statistical operation can be carried out without first making the probability calculation appropriate to that operation. Thus a study of probability theory is essential for an understanding of statistics. Probability theory is important also on its own account, quite apart from its underpinning of statistics. This is particularly true in bioinformatics. Thus because of the intrinsic importance of probability theory, and because of its relevance to statistics, this chapter provides an introduction to the probability theory relating to a single random variable. In the next chapter we extend this theory to the case of many random variables.

Statistics concerns the *optimal* methods of analyzing data generated from some chance mechanism. A significant component of this optimality requirement is the appropriate choice of what is to be computed from the data in order to carry out the statistical analysis. Should we focus on the total number of matches (in this case, 11) between the two sequences? Should we focus on the size of the longest observed run of matches (here 3, occurring in positions 9–11 and also positions 23–25)? We will see later that the most frequently used statistical method for assessing the similarity of two sequences uses neither of these quantities. The question of what should be computed from complex data in order to make a statistical inference leads to sometimes difficult statistical and probabilistic calculations. We address these matters in Chapter 8.

The probability calculations that led to our conclusion were based on assumptions about the nature of the data (equal probabilities and independence). The accuracy of such conclusions depends on the accuracy of the assumptions made. Methods to test for the accuracy of assumptions lie in the realm of statistics, and will be discussed at length later. The necessity of having to make simplifying assumptions, even when they do not hold, brings up one of the most important issues in the application of statistics to bioinformatics. This issue is discussed in Section 4.9.

1.2 Discrete Random Variables, Definitions

1.2.1 Probability Distributions and Parameters

In line with the comments made in the Preface, we give informal definitions in this book for random variables, probability distributions, and parameters, rather than the formal definitions often found in statistics textbooks.

A *discrete random variable* is a numerical quantity that in some experiment that involves some degree of randomness takes one value from some discrete set of possible values. For example, the experiment might be the rolling of two six-sided dice, and the random variable might be the sum of the two numbers showing on the dice. In this case the possible values of the random variable are $2, 3, \ldots, 12$. Often in practice the possible values of a discrete random variable consist of some subset of the integers $\{\ldots, -2, -1, 0, 1, 2, \ldots\}$, but the theory given below allows a general discrete set of possible values. In some cases there may be an infinite number of possible values for a random variable; for example, the random number of tosses of a coin until the first head appears can take any value $1, 2, 3, \ldots$.

By convention, random variables are written as uppercase symbols, often X, Y, and Z, while the eventually observed values of a random variable are written in lowercase, for example x, y, and z. Thus Y might be the conceptual (random) number of matches between two DNA sequences of length 26 before they are actually obtained, and in the observed comparison given in (1.1) the observed value y of this random variable is 11.

We discuss both discrete and continuous random variables and will also discuss the relation between them. In order to clarify the distinction between the two, we frequently use the notation Y for a discrete random variable and X for a continuous random variable. If some discussion applies for both discrete and continuous random variables, we will use the notation X for both.

The *probability distribution* of a discrete random variable Y is the set of values that this random variable can take, together with their associated probabilities. Probabilities are numbers between zero and one inclusive that always add to one when summed over all possible values of the random variable. An example is given in (1.2).

The probability distribution is often presented, as in (1.2), in the form of a table, listing the possible values that the random variable can take together with the probabilities of each value. For example, if we plan to toss a fair coin twice, and the random variable Y is the number of heads that eventually turn up, the probability distribution of this random variable can be presented as follows:

Possible values of the random variable Y	0	1	2
Associated probabilities	.25	.50	.25

$$(1.2)$$

We show how these probabilities are calculated in the next section. In practice, the probabilities associated with the possible values of the random variable of interest are often unknown. For example, if a coin is biased, and the probability of a head on each toss is unknown to us, then the probabilities for 0, 1, or 2 heads when this coin is tossed twice are unknown.

There are two other frequently used methods of presenting a probability distribution. The first is by using a "chart," or "diagram," in which the possible values of the random variable are indicated on the horizontal axis, and their corresponding probabilities by the heights of vertical lines, or bars, above each respective possible value (see Figure 1.1). When the possible values of the random variable are integers there is an appealing geometric interpretation. In this case the rectangles all have width one, and so the probability of a set of values is equal to the total area of the rectangles above them. This representation bears a strong analogy to that for the continuous case, to be discussed in Section 1.8.

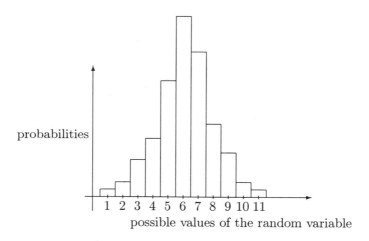

possible values of the random variable

FIGURE 1.1.

However, a third method of presentation is more appropriate in theoretical work, namely through a mathematical function. In this functional approach we denote the probability that discrete random variable Y takes the value y by $P_Y(y)$. Thus $P_Y(4)$ is the probability that the random variable Y takes the value 4. The suffix "Y" indicates that the probability relates to the random variable Y. This is necessary since we often discuss probabilities associated with several random variables simultaneously. In the functional approach $P_Y(y)$ is written in the form

$$P_Y(y) = g(y), \quad y = y_1, y_2, y_3, \ldots,$$
(1.3)

where $g(y)$ is some specific mathematical function of y. Apart from specifying the mathematical function $g(y)$, it is also necessary to indicate the *range* of the random variable, that is, the set of values that it can take. In (1.3) above, $y_1, y_2, y_3, \ldots$ are the possible values of Y. The range of Y should not be confused with the range of $g(y)$. The range of Y is in fact *the domain* of $g(y)$.

As an example, suppose that the possible values of the random variable Y are 1, 2, and 3, and that

$$P_Y(y) = y^2/14, \quad y = 1, 2, 3.$$

Here $g(y) = y^2/14$. Of course, in this simple case we could rewrite the probabilities explicitly in tabular form:

Possible values of the random variable Y	1	2	3
Associated probabilities	1/14	4/14	9/14

(1.4)

However, the functional form is often more convenient, and further, for random variables with an infinite range, listing all possible values and their probabilities in tabular form is impossible.

As another example, suppose that Y can take the possible values 1, 2, and 3, and that now

$$P_Y(y) = \frac{\theta^{2y}}{\theta^2 + \theta^4 + \theta^6}, \quad y = 1, 2, 3,$$

where θ is some fixed nonzero real number. Whatever the value of θ, $P_Y(y) > 0$ for $y = 1, 2, 3$, and $P_Y(1) + P_Y(2) + P_Y(3) = 1$. Therefore, Y is a well-defined random variable. This is true even though the value of θ might be unknown to us. In such cases we refer to θ as a *parameter:* a parameter is some constant, usually unknown, involved in a probability distribution. The important thing to note is that Y is a well-defined random variable even though we may not know any of the probabilities of any of its possible values.

Another important function is the *distribution function* $F_Y(y)$ of the discrete random variable Y. This is the probability that the random variable Y takes a value y or less, so that

$$F_Y(y) = \sum_{y' \leq y} P_Y(y').$$ (1.5)

An example of the functions $P_Y(y)$ and $F_Y(y)$ is given in Figure 1.2. Notice that the rightmost rectangle of the distribution function has height one. We reserve the notation $P_Y(y)$ throughout this book to denote the probability that the discrete random variable Y takes the value y, and the notation $F_Y(y)$ as defined in (1.5) to denote the distribution function of this discrete random variable.

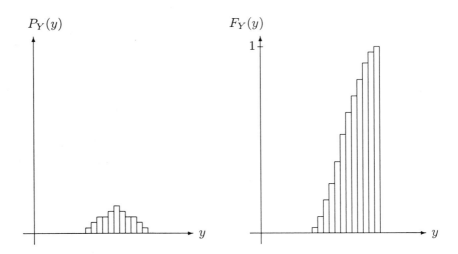

FIGURE 1.2. An example of a probability distribution (left) and its distribution function (right).

1.2.2   Independence

The concept of independence is central in probability and statistics. We formally define independence of events in Section 1.12.5 and independence of discrete random variables in Section 2.1.1. It is, however, necessary to give an informal definition here.

For the moment we take the word "independent" to have its common everyday meaning: Two or more events are independent if the outcome of one event does not affect in any way the outcome of any other event. Discrete random variables are independent if the value of one does not affect in any way the probabilities associated with the possible values of any other random variable.

A frequently occurring case of independent events is typified by the outcomes of different rolls of a die. If a die is rolled any number of times, the number turning up on any one roll is independent of the number turning up on any other roll.

As another example, suppose that a fair die is to be rolled. Let one event be that the number is even, and the other event that the number is greater than or equal to 3. The probability of the former event is $\frac{1}{2}$ and of the latter event is $\frac{2}{3}$. These events are independent: If it is known that it is even, then we know that it is 2, 4, or 6, and two of these three possibilities are greater than or equal to 3. Therefore, the probability the number is greater than or equal to 3 given that it is even is $\frac{2}{3}$. Thus the probability that it is greater than or equal to 3 is the same whether or not we know that the number is even. Similarly, given that the number is a 3, 4, 5, or 6, the probability that it is even remains at $\frac{1}{2}$. On the other hand, if the second event is that the number is 4, 5, or 6, the two events are not independent: The probability of the second event is $\frac{1}{2}$, but given that the number is even, we know it must be 2, 4, or 6, so the probability of the second event, that it is either 4, 5, or 6, becomes $\frac{2}{3}$.

1.3   Six Important Discrete Probability Distributions

In this section we describe the six discrete probability distributions that occur most frequently in bioinformatics and computational biology. They are all presented in the functional form (1.3).

1.3.1   One Bernoulli Trial

A *Bernoulli trial* is a single trial with two possible outcomes, often called "success" and "failure." The probability of success is denoted by p and the probability of failure, $1 - p$, is sometimes denoted by q.

A Bernoulli trial is so simple that it seems unnecessary to introduce a random variable associated with it. Nevertheless, it is useful to do so, and indeed there are two random variables that we will find it useful to associate with a Bernoulli trial.

The first of these, the Bernoulli random variable, is the number of successes Y obtained on this trial. Clearly, $Y = 0$ with probability $1 - p$, and $Y = 1$ with probability p. The probability distribution of Y can then be written in the mathematical form (1.3) as

$$P_Y(y) = p^y (1 - p)^{1-y}, \quad y = 0, 1. \tag{1.6}$$

The second random variable, denoted here by S, takes the value -1 if the trial results in failure and $+1$ if it results in success. Here

$$P_S(s) = p^{(1+s)/2} (1 - p)^{(1-s)/2}, \quad s = -1, +1. \tag{1.7}$$

1.3.2   The Binomial Distribution

A binomial random variable is the number of successes in a fixed number n of independent Bernoulli trials with the same probability of success for each trial. The number of heads in some fixed number of tosses of a coin is an example of a binomial random variable.

More precisely, the binomial distribution arises if all four of the following requirements hold. First, each trial must result in one of two possible outcomes, often called (as with a Bernoulli trial) "success" and "failure."

Second, the various trials must be independent. Third, the probability of success must be the same on all trials. Finally, the number n of trials must be fixed in advance, not determined by the outcomes of the trials as they occur. The probability of success is denoted, as in the Bernoulli case, by p. We call p the parameter, and n the index, of this distribution.

The random variable of interest is the total number of successes in the n trials, denoted by Y. The probability distribution of Y is given by the formula

$$P_Y(y) = \binom{n}{y} p^y (1-p)^{n-y}, \quad y = 0, 1, 2, \ldots, n. \tag{1.8}$$

Section B.4 provides the definition of $\binom{n}{y}$. We will derive this mathematical form for the binomial distribution when discussing the combinatorial term $\binom{n}{y}$ in Section B.6.

Figure 1.3 shows the probability distribution for two different binomials, the first for $n = 10$ and $p = \frac{1}{2}$, the second for $n = 20$ and $p = \frac{1}{4}$. The first is symmetric, while the second is not. The probabilities in the second distribution are nonzero up to and including $y = 20$, but the probabilities for values of y exceeding 12 are too small to be visible on the graph.

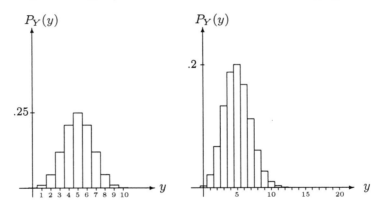

FIGURE 1.3. Two binomial distributions, the first with $n = 10$ and $p = \frac{1}{2}$, the second with $n = 20$ and $p = \frac{1}{4}$.

There are several important comments to make concerning Bernoulli trials and the binomial distribution.

First, the single Bernoulli trial distribution (1.6) is the special case ($n = 1$) of the binomial distribution.

Second, the quantity p in (1.8) is often an unknown parameter.

Third, there is no simple formula for the distribution function $F_Y(y)$ (defined in (1.5)) for the binomial distribution. Probabilities associated with this function are often approximated by the "normal approximation to the binomial," discussed in Section 1.10.3, or are calculated numerically by summation, using equations (1.5) and (1.8). Current computing power makes direct calculation possible even for quite large values of n. Some perhaps unexpected aspects of the computation of this distribution function are given in Appendix C.

Fourth, one must be careful when using a binomial distribution that the four defining conditions above all hold. For example, this is relevant to the comparison of the two DNA sequences given in (1.1). Our assumptions leading to the likelihood of observing 11 or more matches were based on the assumption that the number of matches follows a binomial distribution. Suppose that a "success" is the event that the two nucleotides in corresponding positions in the two sequences match. It is not necessarily true that the probability of success is the same at all sites. Nor is it necessarily true that independence holds: It is a result of population genetics theory, for example, that the nucleotide frequencies at very close sites tend to evolve in a dependent fashion, leading to a potential nonindependence of observing a success at very close sites. Thus two of the central requirements for a binomial distribution might not hold for this sequence comparison. Depending on our purposes, however, it might still be desirable to make these assumptions as an approximation. When we construct models, we invariably must make simplifying assumptions about the processes leading to the data. This issue is discussed further in Section 4.9.

The final point illustrates a concept that will arise again in Chapter 4 and it is instructive to illustrate it with an example. A fair 400-sided die is rolled once. The probability that the number 1 turns up on the die is $p = 1/400$. Intuition might lead one to assume that the probability of seeing a 1 is approximately doubled if the die is rolled twice, is tripled if the die is rolled three times, and so on. The binomial distribution shows, more precisely, that the probability that the number 1 turns up at least once in two rolls is $2p - p^2$, not $2p$. With three rolls the probability is $3p - 3p^2 + p^3$, not $3p$. The intuition, however, is correct. When p is small, the probability of rolling a 1 at least once in n rolls is $np + o(np)$ (as $np \to 0$), and so it is very nearly np as long as np is small. (The "o" notation is defined in Section B.8.) The graph in Figure 1.4 shows the probability that the number 1 turns up at least once in n rolls. The graph is almost perfectly linear in n up to about $n = 15$, and only after that does it significantly deviate from the straight line. This kind of approximation will be important when we discuss Poisson processes in Chapter 4.

1.3.3   The Uniform Distribution

Perhaps the simplest discrete probability distribution is the uniform distribution. In the case of most interest to us, a random variable Y has the *uniform distribution* if the possible values of Y are $a, a+1, \ldots, a+b-1$, for two integer constants a and b, and the probability that Y takes any

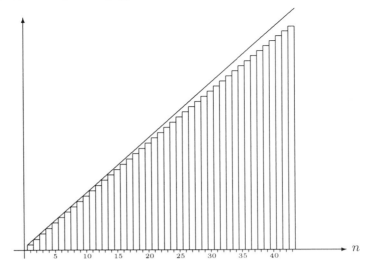

FIGURE 1.4. The probability of at least one success in n binomial trials, with probability $p = 1/400$ for success on each trial, as a function of n.

specified one of these b possible values is b^{-1}. That is,

$$P_Y(y) = b^{-1}, \quad y = a, a+1, \ldots, a+b-1. \tag{1.9}$$

Properties of a random variable having this uniform distribution are discussed later.

1.3.4   The Geometric Distribution

The geometric distribution arises in a situation similar to that of the binomial. Suppose that a sequence of independent Bernoulli trials is conducted, each trial having probability p of success. The random variable of interest is the number Y of trials before but not including the first failure. The possible values of Y are $0, 1, 2, \ldots$. As will be shown in Section 1.12.5, the probability that several independent events all occur is the product of the probabilities of the individual events. Since if $Y = y$ there must have been y successes followed by one failure, it follows that

$$P_Y(y) = (1-p)p^y, \quad y = 0, 1, 2, \ldots. \tag{1.10}$$

The distribution function $F_Y(y)$ can be calculated from (1.10) as

$$F_Y(y) = \text{Prob}(Y \le y) = 1 - p^{y+1}, \quad y = 0, 1, 2, \ldots. \tag{1.11}$$

The geometric distribution with $p = .7$ is shown in Figure 1.5.

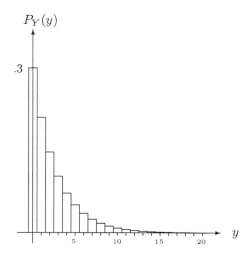

FIGURE 1.5. The geometric distribution with $p = .7$.

We also call Y the length of a "success run." It is often of interest to consider the length of a success run immediately following a failure. The length of such a run also has the distribution (1.10). Note that a possible value of Y is 0, and this value will arise if another failure occurs immediately after the initial failure considered.

One example of the use of success runs occurs in the comparison of the two sequences in (1.1), where the length of the longest success run in this comparison is three. We might wish to use geometric distribution probabilities to assess whether this is a significantly long run. "Significance" in this sense will be defined formally in Section 3.4, and aspects of this test will be discussed in Section 6.3.

Geometric-Like Random Variables

Suppose Y is a random variable with range $0, 1, 2, \ldots$ and that, as $y \to \infty$,

$$1 - F_Y(y - 1) = \text{Prob}(Y \geq y) \sim Cp^y, \qquad (1.12)$$

for some fixed constant C, $0 < C < 1$. In this case we say that the random variable is *geometric-like*. The symbol "$\sim$" in (1.12) is the asymptotic symbol defined in Section B.8. From equation (1.11), the geometric distribution behaves as in (1.12) if we allow $C = 1$ and the asymptotic relation is replaced by an equality.

Geometric-like distributions are not typically part of a standard treatment of elementary probability theory. However, we introduce them here, since they are central to BLAST theory.

1.3.5   The Negative Binomial and the Generalized Geometric Distributions

Suppose that a sequence of independent Bernoulli trials is conducted, each with a probability p of success. The binomial distribution arises when the number of trials n is fixed in advance, and the random variable is the number of successes in these n trials. In some applications the role of the number of trials and the number of successes is reversed, in that the number of successes is fixed in advance (at the value m), and the random variable N is the number of trials up to and including this mth success. The random variable N is then said to have the *negative binomial* distribution. The probability distribution of N is found as follows.

The probability $P_N(n)$ that $N = n$ is the probability that the first $n-1$ trials result in exactly $m-1$ successes and $n-m$ failures (in some order) and that trial n results in success. The former probability is found from (1.8) to be

$$\binom{n-1}{m-1} p^{m-1}(1-p)^{(n-1)-(m-1)}, \quad n = m, m+1, m+2, \ldots,$$

while the latter probability is simply p. Thus

$$P_N(n) = \binom{n-1}{m-1} p^{m}(1-p)^{n-m}, \quad n = m, m+1, m+2, \ldots. \tag{1.13}$$

An example for $p = .75$ and $m = 10$ is shown in Figure 1.6.

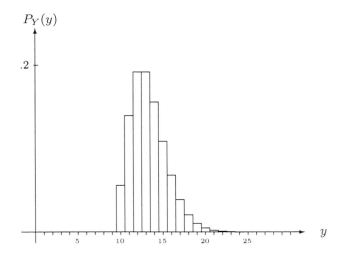

FIGURE 1.6. The negative binomial distribution with $p = .75$ and $m = 10$.

We will later be interested in the case where we fix in advance the number of failures (at the value $k+1$), and consider the random number Y of trials preceding but not including this $(k+1)$th failure. An argument similar to that giving (1.13) shows that

$$P_Y(y) = \binom{y}{k} p^{y-k}(1-p)^{k+1}, \quad y = k, k+1, k+2, \ldots. \tag{1.14}$$

Although this equation is derived in the same way as a negative binomial distribution, it is also the generalization of the geometric distribution equation (1.10) to the case of general k, reducing to the geometric distribution when $k = 0$. For this reason we will refer to the distribution given in equation (1.14) as the *generalized geometric* distribution, with parameters p and k. (This distribution should not be confused with the geometric-like distribution defined in Section 1.3.4).

The distribution functions for these distributions do not admit a simple mathematical form. We discuss aspects of these distribution functions in Section 6.3 and Appendix C.

1.3.6 The Poisson Distribution

A random variable Y has a Poisson distribution (with parameter $\lambda > 0$) if

$$P_Y(y) = \frac{e^{-\lambda}\lambda^y}{y!}, \quad y = 0, 1, 2, \ldots. \tag{1.15}$$

An example with $\lambda = 5$ is given in Figure 1.7. An expression for the distribution function of the Poisson distribution is given by the right-hand side in equation (4.14), replacing k by $k+1$.

The Poisson distribution arises as a limiting form of the binomial distribution. If the number of trials n in a binomial distribution is large, the probability of success p on each trial is small, and the product $np = \lambda$ is moderate, then the binomial probability of y successes is very close to the probability that a Poisson random variable with parameter λ takes the value y. A more precise statement, with a formal mathematical derivation, is given in Chapter 4. This limiting property makes the Poisson distribution particularly useful, since the condition "n large, p small, np moderate" arises often in applications of the binomial distribution in bioinformatics.

The Poisson distribution also arises in its own right in certain biological processes; these are described in Section 4.1.

1.4   The Mean of a Discrete Random Variable

The mean of a random variable is often confused with the concept of an average, and it is important to keep the distinction between the two concepts

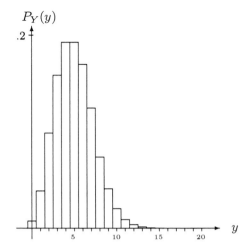

FIGURE 1.7. The Poisson distribution with $\lambda = 5$.

clear. The mean of a discrete random variable Y is defined as

$$\sum_y y P_Y(y), \tag{1.16}$$

the summation being over all possible values that the random variable Y can take (i.e., over its range). As an example, the mean of a random variable having the binomial distribution (1.8) is

$$\sum_{y=0}^n y \binom{n}{y} p^y (1-p)^{n-y}, \tag{1.17}$$

and this can be shown (see Problem 1.1) to be np.

If the range of the random variable is infinite, the mean might not exist (that is, the sum in (1.16) might diverge). In all cases we consider, the mean exists (i.e., is some finite number), and since qualifications to the theory are needed when the mean might be infinite, we assume throughout, without further discussion, that all means of interest are finite.

There are several remarks to make regarding the mean of a discrete random variable.

(i) The notation μ is often used for a mean. An alternative name for the mean of a random variable is the "expected value" of that random variable, and this leads to a second frequently used notation, namely $E(Y)$, for the expected value, or mean, of the random variable Y.

(ii) The word "average" is *not* an alternative for the word "mean," and has a quite different interpretation from that of "mean." This distinction will be discussed in detail in Section 2.10.1.

(iii) In many practical situations the mean μ of a discrete random variable Y is unknown to us, because we do not know the numerical values of the probabilities $P_Y(y)$. That is to say, μ is a parameter. Of all the estimation and hypothesis testing procedures discussed in detail in Chapters 3 and 8, estimation of, and testing hypotheses about, a mean are among the most important.

(iv) The mean is not necessarily a realizable value of a discrete random variable. For example, the distribution in (1.4) has mean

$$1 \times \frac{1}{14} + 2 \times \frac{4}{14} + 3 \times \frac{9}{14} = \frac{18}{7},$$

yet 18/7 is not a realizable value of the random variable.

(v) The expressions "the mean of the random variable Y" and "the mean of the probability distribution of the random variable Y" are equivalent and are used interchangeably, depending on which is more natural in the context.

(vi) The concept of the mean of a discrete random variable Y can be generalized to the mean, or expected value, of any function $g(Y)$ of Y. The function $g(Y)$ is itself a random variable, and the expected value of $g(Y)$, denoted by $E(g(Y))$, is defined (using (1.3)) by

$$E(g(Y)) = \sum_z z P_{g(Y)}(z), \tag{1.18}$$

where the sum is over all values z of $g(Y)$. This is also equal to

$$E(g(Y)) = \sum_y g(y) P_Y(y), \tag{1.19}$$

where the sum is over all values y of Y, as can be seen from the equality

$$P_{g(Y)}(z) = \sum_{\substack{y \text{ such} \\ \text{that } g(y)=z}} P_Y(y).$$

Equation (1.19) is more convenient than (1.18) to use in practice, since it does not require us to find the probability distribution of $g(Y)$. As an example of the use of equation (1.19), if Y is a random variable having the distribution (1.2), $E(Y^2) = 0^2 \cdot \frac{1}{4} + 1^2 \cdot \frac{1}{2} + 2^2 \cdot \frac{1}{4} = 1.5$. Note that this is different from $\left(E(Y)\right)^2$, which in this example takes the value 1.

A linearity property of the mean follows from equation (1.19): If Y is a random variable with mean μ, and if α and β are constants, then the random variable $\alpha + \beta Y$ has mean

$$E(\alpha+\beta Y) = \sum(\alpha+\beta y)P_Y(y) = \alpha \sum P_Y(y)+\beta \sum yP_Y(y) = \alpha+\beta\mu.$$

(vii) If the graph of a probability distribution function is symmetric about some vertical line $y = a$, the mean is at this point of symmetry a.

Various problems at the end of this chapter ask for calculation of the means of the distributions described above. A summary of these means is given in Table 1.1 on page 18.

1.5   The Variance of a Discrete Random Variable

A quantity of importance equal to that of the mean of a random variable is its *variance*. The variance (denoted by σ^2) of the discrete random variable Y is defined by

$$\sigma^2 = \sum_y (y - \mu)^2 P_Y(y), \tag{1.20}$$

the summation being taken over all possible values of Y.

There are several points to note concerning the variance of a discrete random variable.

(i) The variance has the standard notation σ^2, anticipated above.

(ii) The variance is a measure of the dispersion of the probability distribution of the random variable around its mean (see Figure 1.8).

(iii) The expressions "the variance of the random variable Y" and "the variance of the probability distribution of the random variable Y" are equivalent and are used interchangeably.

(iv) A quantity that is often more useful than the variance of a probability distribution is the *standard deviation*. This is defined as the positive square root of the variance, and (naturally enough) is denoted by σ.

(v) The variance, like the mean, is often unknown to us.

(vi) If Y is a random variable with variance σ^2, and if α and β are constants, then the variance of the random variable $\alpha + \beta Y$ is $\beta^2\sigma^2$.

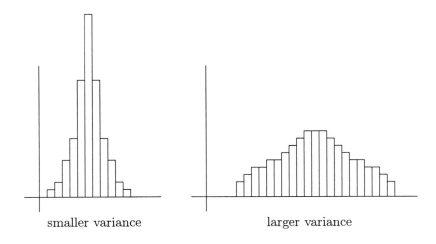

smaller variance   larger variance

FIGURE 1.8.

(vii) An equivalent and often more convenient formula for the variance is

$$\sigma^2 = \left(\sum_y y^2 P_Y(y) \right) - \mu^2 = E(Y^2) - (E(Y))^2, \qquad (1.21)$$

from which it follows that

$$E(Y^2) = \sigma^2 + \mu^2. \qquad (1.22)$$

(viii) A further formula, useful for random variables that take nonnegative integer values only, is

$$\sigma^2 = \left(\sum_y y(y-1) P_Y(y) \right) + \mu - \mu^2. \qquad (1.23)$$

Various problems at the end of this chapter ask for the calculation of the variances of the distributions described above, and Problem 1.3 asks for verification of equations (1.21) and (1.23).

As with the mean, the variance of any random variable we consider in this book is finite, so in order to avoid pedantic qualifications we assume from now on that every variance of interest to us is finite.

The means and variances of the distributions discussed above are listed in Table 1.1.

distribution	mean	variance
Bernoulli	p	$p(1-p)$
Binomial	np	$np(1-p)$
Uniform	$a + (b-1)/2$	$(b^2 - 1)/12$
Geometric	$p/(1-p)$	$p/(1-p)^2$
Negative Binomial	m/p	$m(1-p)/p^2$
Poisson	λ	λ

TABLE 1.1. Means and variances of discrete random variables.

1.6   General Moments of a Probability Distribution

The mean and variance are special cases of *moments* of a discrete probability distribution, and in this section we explore these moments in more detail.

If r is any positive integer, then $E(Y^r)$ is called the rth *moment of the probability distribution about zero*, and is usually denoted by μ'_r. From (1.19),

$$E(Y^r) = \sum_y y^r P_Y(y). \tag{1.24}$$

The mean is thus the first moment of the distribution about zero. Throughout this book we assume that for any random variable considered and for any r, the rth moment exists (i.e., the sum in (1.24) converges). This assumption holds for all of the random variables we consider. Note that $E(Y^r)$ is not the same as $\left(E(Y)\right)^r$, and the numerical values of the two are equal only in certain special cases.

Of almost equal importance for discrete integer-valued random variables is the rth *factorial moment* of a probability distribution. This is denoted by $\mu'_{[r]}$ and is defined by

$$\begin{aligned} \mu'_{[r]} &= E\left(Y(Y-1)\cdots(Y-r+1)\right) \\ &= \sum_y \left(y(y-1)\cdots(y-r+1)\right) P_Y(y). \end{aligned} \tag{1.25}$$

As an example, suppose that the random variable Y has the binomial distribution (1.8). Then the rth factorial moment is

$$\begin{aligned} E\left(Y(Y-1)\cdots(Y-r+1)\right) \\ = \sum_{y=0}^{n} y(y-1)\cdots(y-r+1)\binom{n}{y}p^y(1-p)^{n-y}. \end{aligned} \tag{1.26}$$

We evaluate this expression for the case $r \leq n$. (For $r > n$ the expression is necessarily 0.) The first r terms in the sum are zero (because of the factor $y(y-1)\cdots(y-r+1)$), so that the summation can start at $y = r$. Next,

$$y(y-1)\cdots(y-r+1)\binom{n}{y} = n(n-1)\cdots(n-r+1)\binom{n-r}{y-r}.$$

From this, equation (1.26) reduces to

$$E\big(Y(Y-1)\cdots(Y-r+1)\big)$$

$$= n(n-1)\cdots(n-r+1)p^r \sum_{y=r}^{n}\binom{n-r}{y-r}p^{y-r}(1-p)^{n-y}. \quad (1.27)$$

The sum in this expression can be simplified by writing $k = y-r, m = n-r$ to give

$$\sum_{k=0}^{m}\binom{m}{k}p^k(1-p)^{m-k}.$$

This sum is 1, since it is the sum of all the values of a binomial probability distribution. Thus, finally,

$$E\big(Y(Y-1)\cdots(Y-r+1)\big) = n(n-1)\cdots(n-r+1)p^r. \quad (1.28)$$

The rth *moment about the mean* of the probability distribution, denoted by μ_r, is defined by

$$\mu_r = \sum_y (y-\mu)^r P_Y(y). \quad (1.29)$$

The first such moment, μ_1, is identically zero, and is thus not of interest. The second, μ_2, is the variance σ^2 of the random variable. The third, μ_3, is related to the *skewness* of the probability distribution. The skewness γ_1 of a distribution is normally defined as

$$\gamma_1 = \mu_3/\sigma^3. \quad (1.30)$$

Distributions with positive skewness have long tails to the right, and distributions with negative skewness have long tails to the left.

1.7   The Probability-Generating Function

An important quantity for any discrete random variable, or equivalently its probability distribution, is its probability-generating function, or pgf for short. Any pgf will be denoted by $\mathrm{p}(t)$, or sometimes $\mathrm{q}(t)$, and for a discrete random variable Y, the pgf $\mathrm{p}(t)$ is defined by the equation

$$\mathrm{p}(t) = E(t^Y) = \sum_y P_Y(y)\,t^y, \quad (1.31)$$

the summation being taken over all possible values of Y. This sum always converges for $t = 1$ (to 1). If the sum only converges for $t = 1$, then $\mathbb{p}(t)$ is not useful. For all cases of interest to us $\mathbb{p}(t)$ exists in some interval $(1-a, 1+a)$ of values of t surrounding 1, where $a > 0$. The variable t is sometimes called a "dummy" variable, since it has no meaning or interpretation of its own. Instead, it is a mathematical tool that can be used, in conjunction with power series properties such as those given in Section B.12, to arrive efficiently at various desired results.

It is important to note that when these possible values are not all non-negative integers, the sum above is not a Taylor series. In our applications probability-generating functions will sometimes include negative powers of t. For the relevant background on such series, see Sections B.12 through B.14.

As a straightforward example, the pgf of the Bernoulli random variable defined in (1.6) is

$$1 - p + pt. \tag{1.32}$$

Similarly, the pgf of the Bernoulli random variable defined in (1.7) is

$$(1 - p)t^{-1} + pt. \tag{1.33}$$

As a less straightforward example, the pgf of the binomial distribution (1.8) is

$$\mathbb{p}(t) = \sum_{y=0}^{n} \binom{n}{y} p^y (1 - p)^{n-y} t^y, \tag{1.34}$$

and this can be shown (see Problem 1.11) to be

$$\mathbb{p}(t) = (1 - p + pt)^n. \tag{1.35}$$

The original purpose for the pgf is to generate probabilities, as the name suggests. If the pgf can be found easily, as is often the case, then the coefficient of t^r in the pgf is the probability that the random variable takes the value r.

Another use of the pgf is to derive *moments* of a probability distribution. This is done as follows. The derivative of the pgf of a discrete random variable with respect to t, taken at the value $t = 1$, is the mean of that random variable. That is,

$$\mu = \left(\frac{d}{dt} \mathbb{p}(t) \right)_{t=1}. \tag{1.36}$$

The variance of a random variable with mean μ and whose probability distribution has pgf $\mathbb{p}(t)$ is found from

$$\sigma^2 = \left(\frac{d^2}{dt^2} \mathbb{p}(t) \right)_{t=1} + \mu - \mu^2. \tag{1.37}$$

The verification of these is left as an exercise (see Problem 1.12).

We end with a useful theorem about pgfs whose proof is discussed in Section B.14, and which applies to all random variables of interest to us.

Theorem. Suppose that two random variables have pgfs $\mathbb{p}_1(t)$ *and* $\mathbb{p}_2(t)$, *respectively, and that both pgfs converge in some open interval* I *containing* 1. *If* $\mathbb{p}_1(t) = \mathbb{p}_2(t)$ *for all* t *in* I, *then the two random variables have identical distributions.*

A consequence of this is that if the pgf of a random variable converges in an open interval containing 1, then the pgf completely determines the distribution of the random variable.

1.8   Continuous Random Variables

Some random variables, by their nature, are discrete, such as the number of successes in n Bernoulli trials. Other random variables, by contrast, are continuous. Measurements such as height, blood pressure, or the time taken until an event occurs, are all of this type. We denote a continuous random variable by X, and the observed value of the random variable by x. Continuous random variables usually take any value in some continuous range of values, for example $-\infty < X < \infty$ or $L \leq X < H$.

Probabilities for continuous random variables are not allocated to specific values, but rather are allocated to intervals of values. Each random variable X with range I has an associated *density function* $f_X(x)$, which is defined and positive for all x in the interval I, and the probability that the random variable takes a value in some given interval is obtained by integrating the density function over that interval. Specifically,

$$\text{Prob}(a < X < b) = \int_a^b f_X(x)\, dx. \tag{1.38}$$

This definition implies that the probability that a continuous random variable takes any specific nominated value is zero. It thus implies the further equalities

$$\begin{aligned}
\text{Prob}(a < X < b) &= \text{Prob}(a < X \leq b) \\
&= \text{Prob}(a \leq X < b) = \text{Prob}(a \leq X \leq b). \tag{1.39}
\end{aligned}$$

By elementary calculus, when $f_X(\cdot)$ is continuous on $[x, x+h]$, (1.38) gives

$$\lim_{h \to 0} \frac{\text{Prob}(x < X < x+h)}{h} = f_X(x). \tag{1.40}$$

We will use this identity several times in what follows, as well as the approximation that follows from it, namely that when $f_X(\cdot)$ is continuous on

$[x, x + h]$,
$$\text{Prob}(x < X < x + h) \approx f_X(x)h, \qquad (1.41)$$

when h is small.

As a particular case of equation (1.38), if the range of the continuous random variable X is either $L \leq X \leq H$, $L \leq X < H$, $L < X \leq H$, or $L < X < H$, then

$$\int_L^H f_X(x)\,dx = 1. \qquad (1.42)$$

These definitions of L and H are used throughout this and the next section.

A further important function associated with a continuous random variable X is its distribution function $F_X(x)$. This is the probability that the random variable takes a value x or less. The distribution function is given by

$$F_X(x) = \int_L^x f_X(u)\,du. \qquad (1.43)$$

This definition implies that $0 \leq F_X(x) \leq 1$, that $F_X(x)$ is nondecreasing in x, and the fundamental theorem of calculus shows that

$$f_X(x) = \frac{d}{dx} F_X(x). \qquad (1.44)$$

We reserve the notation $f_X(x)$ and $F_X(x)$ throughout this book to denote, respectively, the density function and the distribution function of the continuous random variable X.

1.9   The Mean, Variance, and Median of a Continuous Random Variable

1.9.1   Definitions

The mean μ and variance σ^2 of a continuous random variable having range (L, H) and density function $f_X(x)$ are defined respectively by

$$\mu = \int_L^H x f_X(x)\,dx \qquad (1.45)$$

and

$$\sigma^2 = \int_L^H (x - \mu)^2 f_X(x)\,dx. \qquad (1.46)$$

These definitions are the natural analogues of the corresponding definitions for a discrete random variable, and the remarks about the mean, and the variance, of a continuous random variable are very similar to those of a discrete random variable that were given on pages 14 and 16. In particular, remarks (i), (ii), (iii), (v), and (vii) for the mean of a discrete random variable remain unchanged for a continuous random variable (except for the replacement of "probability distribution function" with "density function" where necessary), and property (vi) requires only replacing a summation by an integration to give the analogous properties for a continuous random variable. That is, the mean value $E\left(g(X)\right)$ of the function $g(X)$ of the continuous random variable X is given by

$$E\left(g(X)\right) = \int_L^H g(x) f_X(x) dx, \qquad (1.47)$$

where $f_X(x)$ is the density function of X.

All of the variance properties given for a discrete random variable apply also for a continuous random variable.

For continuous random variables whose range is $(0, H)$ (for some positive H, possibly $+\infty$), an alternative formula for the mean (see Problem 1.8) is

$$\mu = \int_0^H \left(1 - F_X(x)\right) dx. \qquad (1.48)$$

For any continuous random variable arising in practice in bioinformatics there is a unique *median*, denoted here by M, having the property that

$$\text{Prob}(X < M) = \text{Prob}(X > M) = \frac{1}{2}. \qquad (1.49)$$

The mean and median coincide for random variables with symmetric density functions, both being at the point of symmetry, but the two are usually different for asymmetric density functions. Examples of mean, variance, and median calculations are given below.

1.9.2 Chebyshev's Inequality

The concepts of the mean and the variance of a random variable, discrete or continuous, allow us to prove an inequality that is of great use in probability theory and statistics.

Let X be a random variable, discrete or continuous, having mean μ and variance σ^2. Then Chebyshev's inequality states that for any positive constant d,

$$\text{Prob}(|X - \mu| \geq d) \leq \frac{\sigma^2}{d^2}. \qquad (1.50)$$

The proof of this statement is straightforward, and is given here for the continuous random variable case where the range of the random variable is $(-\infty, +\infty)$. The proof for any other range, and for a discrete random variable, is essentially identical. From the definition of σ^2,

$$\sigma^2 = \int_{-\infty}^{+\infty} (x-\mu)^2 f_X(x)\, dx$$

$$\geq \int_{-\infty}^{\mu-d} (x-\mu)^2 f_X(x)\, dx + \int_{\mu+d}^{+\infty} (x-\mu)^2 f_X(x)\, dx$$

$$\geq d^2 \int_{-\infty}^{\mu-d} f_X(x)\, dx + d^2 \int_{\mu+d}^{+\infty} f_X(x)\, dx$$

$$= d^2 \, \text{Prob}(|X-\mu| \geq d).$$

The result then follows immediately.

1.10 Five Important Continuous Distributions

1.10.1 The Uniform Distribution

A continuous random variable X has the *uniform* distribution if, for some constants a and b with $a < b$, its range is one of the intervals $I = [a,b]$, (a,b), $(a,b]$, or $[a,b)$ (interval notation is discussed in Section B.1) and its density function $f_X(x)$ is

$$f_X(x) = \frac{1}{b-a}, \quad \text{for } x \text{ in } I. \tag{1.51}$$

The density function of this uniform distribution is shown in Figure 1.9. The mean and variance of this distribution are respectively given by

$$\mu = \frac{a+b}{2}, \quad \sigma^2 = \frac{(b-a)^2}{12}. \tag{1.52}$$

The uniform distribution we consider most frequently is the particular case of (1.51) for which $a = 0$, $b = 1$. For this uniform distribution,

$$f_X(x) = 1, \quad F_X(x) = x, \quad 0 \leq x \leq 1. \tag{1.53}$$

From (1.52) the mean and variance of this distribution are, respectively, $1/2$ and $1/12$.

We use the term "uniform distribution" to describe any continuous uniform distribution and any discrete uniform distribution. When this is done the context makes clear which distribution is intended.

1.10.2 The Normal Distribution

The most important continuous random variable is one having the *normal*, or *Gaussian*, distribution. The (continuous) random variable X has a

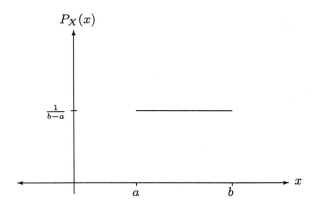

FIGURE 1.9. The density function of the uniform distribution $f_X(x) = \frac{1}{b-a}$.

normal distribution if it has range $(-\infty, \infty)$ and density function

$$f_X(x) = \frac{1}{\sqrt{2\pi}\beta}e^{-\frac{(x-\alpha)^2}{2\beta^2}},$$

where α and β are parameters of the distribution. The value of α is unrestricted, but it is required that $\beta > 0$.

It may be checked, either by using equation (1.45) or from symmetry that the mean of this distribution is α. Symmetry also implies that the median of this distribution is α. It is also possible, although less straightforward, to then use equation (1.46) to show that the variance of the distribution is β^2. Thus the distribution is usually written in the more convenient form

$$f_X(x) = \frac{1}{\sqrt{2\pi}\sigma}e^{-\frac{(x-\mu)^2}{2\sigma^2}}, \tag{1.54}$$

thus incorporating the mean μ and the variance σ^2 into the mathematical form of the normal distribution. A common notation is that a random variable having this distribution is said to be an $N(\mu, \sigma^2)$ random variable.

A particularly important normal distribution is the one for which $\mu = 0$ and $\sigma^2 = 1$, whose density function is shown in Figure 1.10. This is sometimes called the *standard normal* distribution, and this name arises for the following reason.

Suppose that a random variable X has the normal distribution (1.54), that is, with arbitrary mean μ and arbitrary variance σ^2. Then the "standardized" random variable Z, defined by $Z = (X - \mu)/\sigma$, has a normal

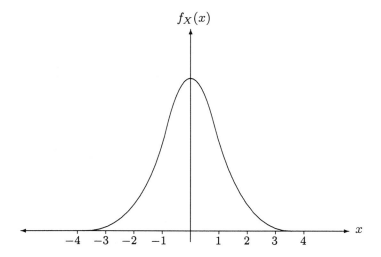

FIGURE 1.10. The density function for the standard normal distribution with $\mu = 0$, $\sigma = 1$.

distribution with mean 0, variance 1. The observed value z of Z is often called a *z-score*, so that if x is the observed value of X,

$$z\text{-score} = \frac{x - \mu}{\sigma}. \tag{1.55}$$

One of the uses of this standardization is the following. If X is a random variable having a normal distribution with mean 6 and variance 16, we cannot find $\mathrm{Prob}(5 < X < 8)$ by integrating the function in (1.54) in closed form. However, for the standardized variable Z, accurate approximations of such integrals are widely available in tables. These tables may be used, in conjunction with the standardization procedure, to find probabilities for a random variable having any normal distribution. Thus in the above example, we can rewrite $\mathrm{Prob}(5 < X < 8)$ in terms of the standard normal Z by

$$\mathrm{Prob}(5 < X < 8) = \mathrm{Prob}\left(\frac{5 - 6}{\sqrt{16}} < \frac{X - 6}{\sqrt{16}} < \frac{8 - 6}{\sqrt{16}}\right)$$
$$= \mathrm{Prob}(-0.25 < Z < 0.5), \tag{1.56}$$

and this probability is found from tables as 0.2902.

One application of the standardization process is that it shows that the probability that an observation from a normal random variable is within two standard deviations from the mean is .95. This *two standard deviation rule* is useful as a quick approximation for probabilities concerning random variables whose probability distribution is in some sense close to a normal distribution, although it can sometimes be misleading; see Problem 1.17.

In Section 14.9.3 we will need $E(|x - \mu|)$, where X is $N(\mu, \sigma)$. Equations (1.47) and (1.54) show that

$$E(|X - \mu|) = \int_{-\infty}^{+\infty} |x - \mu| \frac{1}{\sqrt{2\pi}\sigma} e^{-\frac{(x-\mu)^2}{2\sigma^2}} dx. \tag{1.57}$$

The symmetry of the integrand around $x = \mu$ shows that this integral is

$$2 \int_{\mu}^{+\infty} (x - \mu) \frac{1}{\sqrt{2\pi}\sigma} e^{-\frac{(x-\mu)^2}{2\sigma^2}} dx.$$

The change of variable $y = (x - \mu)^2/2\sigma^2$ shows that this is

$$\sigma \sqrt{\frac{2}{\pi}} \int_0^{+\infty} e^{-y} dy = \sigma \sqrt{\frac{2}{\pi}}. \tag{1.58}$$

1.10.3   The Normal Approximation to a Discrete Distribution

One of the many uses of the normal distribution is to provide approximations for probabilities for certain discrete random variables. The first approximation we consider is the normal approximation to the binomial. If the number of trials n in the binomial distribution (1.8) is very large, the normal distribution with mean np and variance $np(1 - p)$ provides a very good approximation. Figure 1.11 shows the approximation of the binomial with $p = \frac{1}{4}$ and $n = 20$ by the normal with $\mu = 5$ and $\sigma^2 = \frac{15}{4}$.

The normal approximation to the binomial is a consequence of the central limit theorem, which is discussed in Section 2.10.1. How large n must be depends on p and the desired degree of accuracy. For $p = \frac{1}{2}$ (the only case where the binomial probability distribution is symmetric) the approximation is good even for n as small as 20, as is shown in an example below. The closer p is to zero or one, the further the binomial distribution is from being symmetric, and the larger n must be for the normal distribution to fit the binomial distribution well.

If Y is binomial with parameters p and n, and X is normal with $\mu = np$ and $\sigma^2 = np(1-p)$, then for integers a and b, with $a < b$, the approximation is

$$\text{Prob}(a \le Y \le b) \approx \text{Prob}\left(a - \frac{1}{2} \le X \le b + \frac{1}{2}\right).$$

The reason for the term $\frac{1}{2}$ in this approximation can be seen from the graph in Figure 1.11. The rectangle whose area equals $\text{Prob}(Y = a)$ lies above the interval $(a - \frac{1}{2}, a + \frac{1}{2})$. Without the term $\frac{1}{2}$ in the approximation we would underestimate by $(\text{Prob}(Y = a) + \text{Prob}(Y = b))/2$. The term $\frac{1}{2}$ is known as the "continuity correction."

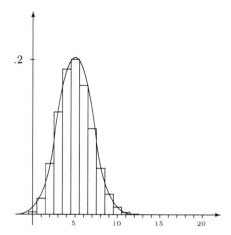

FIGURE 1.11. The normal approximation to the binomial random variable with $p = \frac{1}{4}$ and $n = 20$.

As an example, the probability that a binomial random variable with $p = \frac{1}{2}$ and $n = 20$ takes a value of 15 or more can be evaluated exactly from (1.8) to be 0.0207. The probability that a normal random variable with mean 10 and variance 5 (the appropriate mean and variance for this binomial distribution) exceeds the value 14.5 is 0.0222. This is a far more accurate approximation than that found without the continuity correction, which is 0.0127.

As a second example, the probability that a binomial random variable with $p = \frac{1}{4}$ and $n = 20$ takes a value of 9 or more is 0.0409. The normal approximation with a continuity correction is 0.0353, not as accurate as the approximation in the case $p = \frac{1}{2}$ for the same value of n. Indeed as p decreases, n must increase in order to achieve a good approximation.

A second application is to use the normal distribution to approximate the Poisson distribution. As can be seen from the example of a Poisson distribution in Figure 1.7, the Poisson has a normal-like shape when $\lambda = 5$. It follows from the central limit theorem, to be discussed in Section 2.10.1, that when λ is large, the Poisson is closely approximated by a normal with mean λ and variance λ. In practice, this approximation would not be very accurate when $\lambda < 20$, even with continuity correction.

1.10.4   The Exponential Distribution

A third probability distribution arising often in computational biology is the *exponential* distribution. A (continuous) random variable with this distribution has range $[0, +\infty)$ and density function

$$f_X(x) = \lambda e^{-\lambda x}, \quad x \geq 0. \tag{1.59}$$

Here the single positive parameter λ characterizes the distribution. The graph of this density function is shown in Figure 1.12.

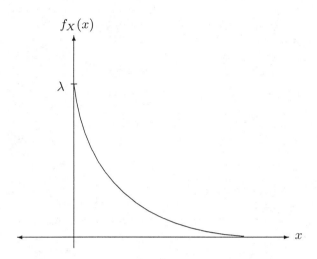

FIGURE 1.12. The density function of the exponential distribution, $f_X(x) = \lambda e^{-\lambda x}$.

The distribution function $F_X(x)$ is found by integration as

$$F_X(x) = 1 - e^{-\lambda x}, \quad x \geq 0. \tag{1.60}$$

We show in Section 1.13 that the exponential distribution has important properties in connection with spontaneous events.

Mean, Variance, and Median

The mean and variance of the exponential distribution are, respectively, $1/\lambda$ and $1/\lambda^2$ (see Problem 1.15). The median M, found by solving the equation

$$F_X(x) = \frac{1}{2},$$

is given by

$$M = \frac{\log 2}{\lambda}. \tag{1.61}$$

This value is just over two-thirds the mean, implying (as the graph of the density function shows immediately) that the exponential distribution is skewed to the right.

The Relation of the Exponential with the Geometric Distribution

The exponential distribution is the continuous analogue of the geometric distribution, as can be seen by comparing Figures 1.5 (page 11) and 1.12. Therefore, the exponential may be used as a continuous approximation to the geometric. This is discussed further in Section 2.11.2.

A further relationship between the exponential and geometric distributions is the following. Suppose that the random variable X has the exponential distribution (1.59) and that $Y = \lfloor X \rfloor$ is the integer part of X (i.e., the greatest integer less than or equal to X). Then $\mathrm{Prob}(Y = y) = \mathrm{Prob}(y \leq X < y + 1)$. By integration of (1.59), this probability is

$$\mathrm{Prob}(Y = y) = \left(1 - e^{-\lambda}\right) e^{-\lambda y}, \quad y = 0, 1, 2, \ldots. \tag{1.62}$$

With the identification $p = e^{-\lambda}$, this is in the form of the geometric probability distribution (1.10). It follows from equation (1.11) that

$$\mathrm{Prob}(Y \leq y - 1) = 1 - e^{-\lambda y}, \quad y = 1, 2, \ldots. \tag{1.63}$$

This relation of the exponential distribution with the geometric distribution is used often in computational biology and bioinformatics, so that we will sometimes refer to the geometric distribution as being given in the notation of equation (1.62) rather than in the notation of equation (1.10). In the notation of (1.62), the mean and variance of the geometric distribution are, respectively,

$$\mu = \frac{1}{e^{\lambda} - 1}, \quad \sigma^2 = \frac{e^{\lambda}}{(e^{\lambda} - 1)^2}. \tag{1.64}$$

The density function $f_D(d)$ of the *fractional part* D of X (defined by $D = X - \lfloor X \rfloor = X - Y$) can be shown to be

$$f_D(d) = \frac{\lambda e^{-\lambda d}}{1 - e^{-\lambda}}, \quad 0 \leq d < 1. \tag{1.65}$$

This density function does not depend on the value of X. The mean and variance of this distribution are found from (1.45) and (1.46) to be, respectively,

$$\mu = \frac{1}{\lambda} - \frac{1}{e^{\lambda} - 1}, \quad \sigma^2 = \frac{1}{\lambda^2} - \frac{1}{(e^{\lambda} - 1)} - \frac{1}{(e^{\lambda} - 1)^2}. \tag{1.66}$$

More Comments on Geometric-Like Random Variables

Random variables having a geometric-like distribution were defined by the asymptotic relation (1.12). It is useful to express this relation in the notation developed in this section, that is, with p replaced by $e^{-\lambda}$. With this notation, a discrete random variable Y has a geometric-like distribution if, as $y \to \infty$,

$$\mathrm{Prob}(Y \geq y) \sim C e^{-\lambda y}, \tag{1.67}$$

for some fixed positive constant $C < 1$. This equation, and geometric-like random variables, will be central to the theory of generalized random walks and of BLAST, developed respectively in Chapters 7 and 9.

1.10.5 The Gamma Distribution

The exponential distribution is a special case of the *gamma* distribution. The density function for the gamma distribution is

$$f_X(x) = \frac{\lambda^k x^{k-1} e^{-\lambda x}}{\Gamma(k)}, \quad x > 0. \tag{1.68}$$

Here λ and k are arbitrary positive parameters and $\Gamma(k)$ is the *gamma* function (see Section B.17). The value of k need not be an integer, but if it is, then $\Gamma(k) = (k-1)!$. The density function can take several different shapes, depending on the value of the parameter k (see Figure 1.13).

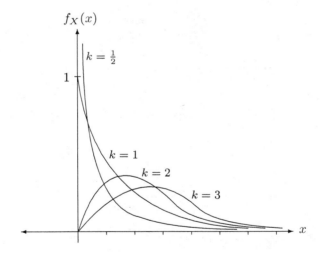

FIGURE 1.13. Examples of the density function of a gamma distribution with $\lambda = 1$ and $k = \frac{1}{2}, 1, 2, 3$.

The mean μ and variance σ^2 of the gamma distribution are given by

$$\mu = \frac{k}{\lambda}, \quad \sigma^2 = \frac{k}{\lambda^2}. \tag{1.69}$$

The exponential distribution is the special case of the gamma distribution where $k = 1$. Another important special case is where $\lambda = \frac{1}{2}$ and $k = \frac{1}{2}\nu$, with ν a positive integer, so that

$$f_X(x) = \frac{1}{2^{\nu/2}\,\Gamma(\frac{1}{2}\nu)}\, x^{\frac{1}{2}\nu-1}\, e^{-\frac{1}{2}x}, \quad x > 0. \tag{1.70}$$

This is called the *chi-square* distribution with ν degrees of freedom, and is important for statistical hypothesis testing. It can be shown that if Z is a standard normal random variable, then Z^2 is a random variable having the chi-square distribution with one degree of freedom. Further, the exponential distribution with $\lambda = \frac{1}{2}$ is the chi-square distribution with two degrees of freedom.

When k is a positive integer, the distribution function $F_X(x)$ of the gamma distribution may be found by repeated integration by parts, and is given by the sum of k Poisson distribution terms. This result is developed in Section 4.3. When k is not an integer, no simple expression exists for the distribution function of the gamma distribution.

1.10.6   The Beta Distribution

A continuous random variable X has a beta distribution with positive parameters α and β if its density function is

$$f_X(x) = \frac{\Gamma(\alpha+\beta)}{\Gamma(\alpha)\Gamma(\beta)}\, x^{\alpha-1}(1-x)^{\beta-1}, \quad 0 < x < 1. \tag{1.71}$$

Here $\Gamma(\cdot)$ is the gamma function, discussed above in connection with the gamma distribution. The mean μ and the variance σ^2 of this beta distribution are

$$\mu = \frac{\alpha}{\alpha+\beta}, \quad \sigma^2 = \frac{\alpha\beta}{(\alpha+\beta)^2(\alpha+\beta+1)}. \tag{1.72}$$

The uniform distribution (1.53) is the special case of this distribution for the case $\alpha = \beta = 1$.

1.11   The Moment-Generating Function

An important quantity for any random variable, discrete or continuous, or equivalently for its probability distribution, is its *moment-generating function*, or mgf for short. The moment-generating function is a function of a real variable θ and will be denoted by $\mathrm{m}(\theta)$. For a discrete random variable Y having probability distribution $P_Y(y)$, $\mathrm{m}(\theta)$ is defined by

$$\mathrm{m}(\theta) = E(e^{\theta Y}) = \sum_y e^{\theta y} P_Y(y), \tag{1.73}$$

the sum being taken over all possible values of Y. Clearly, $\mathrm{m}(0) = 1$, and for all random variables that we consider, $\mathrm{m}(\theta)$ exists (i.e., the sum or integral defining it converges) for θ in some interval $(-a, a)$ about 0, $a > 0$. In this definition θ is a "dummy" variable (as was the quantity t in the pgf). For discrete random variables, the substitution $t = e^\theta$ changes the probability-generating function to the moment-generating function.

Strictly speaking, we should denote this moment-generating function by $\mathrm{m}_Y(\theta)$, to indicate that it relates to the random variable Y. However, we use the simpler notation $\mathrm{m}(\theta)$ when the random variable in question is clear from the context.

The mgf of a continuous random variable X is the natural analogue of (1.73), namely

$$\mathrm{m}(\theta) = E(e^{\theta X}) = \int_L^H e^{\theta x} f_X(x) dx, \tag{1.74}$$

where L and H are the boundaries of the range of X, as in Section 1.8.

An analogue of the uniqueness property of pgfs discussed on page 21 holds for mgfs: If two mgfs agree in an open interval containing zero, the corresponding probability distributions are identical. In some cases an mgf is defined only for certain values of θ, while in other cases it exists for all θ. In the former case we restrict θ to those values for which the mgf exists.

Example 1. The mgf of the exponential distribution (1.59) is

$$\int_0^{+\infty} e^{\theta x} \lambda e^{-\lambda x} dx = \frac{\lambda}{\lambda - \theta}. \tag{1.75}$$

The integral on the left-hand side converges only when $\theta < \lambda$, so that θ is assumed to be restricted to these values.

Example 2. The mgf of the normal distribution (1.54) is

$$\int_{-\infty}^{+\infty} \frac{1}{\sqrt{2\pi}\sigma} e^{\theta x - \frac{(x-\mu)^2}{2\sigma^2}} dx.$$

Completing the square in the exponent, we find that this reduces to

$$e^{\mu\theta + \frac{1}{2}\sigma^2\theta^2}. \tag{1.76}$$

In this case θ can take any value in $(-\infty, \infty)$.

The moment-generating function is useful when it can be evaluated as a comparatively simple function, as in the above examples. This is not always possible; for the beta distribution (1.71), for example, the moment-generating function does not in general take a simple form.

The original purpose for the mgf was to generate moments, as the name suggests. Any moment of a probability distribution can be found by appropriate differentiation of the mgf $\mathrm{m}(\theta)$ with respect to θ, with the derivative being evaluated at $\theta = 0$. In particular, if a random variable has a distribution whose mgf is $m(\theta)$, the mean μ of that random variable is given by

$$\mu = \left(\frac{d\,m(\theta)}{d\theta} \right)_{\theta=0,} \qquad (1.77)$$

and the variance of that random variable is given by

$$\sigma^2 = \left(\frac{d^2\,m(\theta)}{d\theta^2} \right)_{\theta=0} - \mu^2. \qquad (1.78)$$

As an example, equation (1.75) shows that the mean of the exponential distribution (1.59) is

$$\mu = \left(\frac{\lambda}{(\lambda - \theta)^2} \right)_{\theta=0} = \frac{1}{\lambda},$$

and that the variance is

$$\sigma^2 = 2\left(\frac{\lambda}{(\lambda - \theta)^3} \right)_{\theta=0} - \mu^2 = \frac{1}{\lambda^2},$$

confirming the results of Problem 1.15, where these moments are found by integration.

It is often more useful to use the logarithm of the mgf than the mgf itself. If X is a random variable whose distribution has mgf $m(\theta)$, then the mean μ and the variance σ^2 of X are found from

$$\mu = \left(\frac{d\log m(\theta)}{d\theta} \right)_{\theta=0}, \qquad \sigma^2 = \left(\frac{d^2 \log m(\theta)}{d\theta^2} \right)_{\theta=0}. \qquad (1.79)$$

We conclude this section with two important properties of mgfs. First, let X be any continuous random variable with density function $f_X(x)$, $L < x < H$, and let $g(X)$ be any function of X. Then $g(X)$ is itself a random variable, and from (1.47) the moment-generating function of $g(X)$ can be written in terms of the density function for X as

$$m_{g(X)}(\theta) = \int_L^H e^{\theta g(x)} f_X(x)\, dx. \qquad (1.80)$$

A similar definition holds for a discrete random variable, namely

$$m_{g(Y)}(\theta) = \sum_y e^{\theta g(y)} P_Y(y), \qquad (1.81)$$

the summation being over all possible values of the random variable Y.

The second property of mgfs is important for BLAST theory, and we state it as a theorem.

Theorem 1.1. Let Y be a *discrete* random variable with mgf $m(\theta)$. *Suppose that Y can take at least one negative value (say $-a$) with positive probability $P_Y(-a)$ and at least one positive value (say b) with positive probability $P_Y(b)$, and that the mean of Y is nonzero. Then there exists a unique nonzero value θ^* of θ such that $m(\theta^*) = 1$.*

Proof. We prove the theorem for the case where the mgf is defined for all θ in $(-\infty, \infty)$. The general case is proved in a similar way.

Since all terms in the sum (1.73) defining $m(\theta)$ are positive,

$$m(\theta) > P_Y(-a)e^{-\theta a} \quad \text{and} \quad m(\theta) > P_Y(b)e^{\theta b}.$$

Thus $m(\theta) \to +\infty$ as $\theta \to -\infty$ and also as $\theta \to +\infty$. Further,

$$\frac{d^2 m(\theta)}{d\theta^2} = \sum_y y^2 P_Y(y)e^{\theta y}, \qquad (1.82)$$

and since this is positive for all θ, the curve of $m(\theta)$, as a function of θ, is convex. The definition (1.73) of an mgf shows that $m(0) = 1$. By (1.77), the mean of Y is the slope of $m(\theta)$ at $\theta = 0$, and this is nonzero by assumption. The above shows that if the mean of Y is negative, then the slope of the graph of $m(\theta)$ at $\theta = 0$ is negative, and together with the other properties of $m(\theta)$ determined above, its graph must be approximately as shown in the left-hand graph in Figure 1.14. Consequently, there is a unique positive value θ^* of θ such that $m(\theta^*) = 1$.

If the mean of Y is positive, then the graph of $m(\theta)$ is approximately as shown in the right-hand graph in Figure 1.14, and so in this case also there is a unique negative value θ^* of θ such that $m(\theta^*) = 1$.

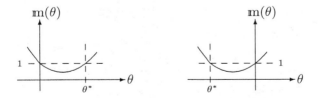

FIGURE 1.14.

The same conclusions also hold for continuous random variables, making the appropriate changes to the statement of the theorem.

1.12   Events

1.12.1   What Are Events?

So far, the development of probability theory has been in terms of random variables, that is, quantities that by definition take one or another numerical value. In many contexts, however, it is more natural, or more convenient, to consider probabilities relating to events rather than to random variables. For example, if we examine the nucleotides in two aligned DNA sequences, it is more natural to say that the event A_j occurs if the two sequences have the same nucleotide at position j, rather than to define a random variable X_j, taking the value $+1$ if the two sequences have the same nucleotide at this position and taking the value 0 if they do not. This section provides some elementary probability theory relating to events.

Our definition of an event is deliberately casual, since in bioinformatics a detailed sample-space-based definition is not necessary. We simply think of an event as something that either will or will not occur when some experiment is performed. For example, if we plan to roll a die, A_1 might be the event that the number turning up will be even and A_2 might be the event that the number turning up will a 4, 5, or 6. The experiment in this case is the actual rolling of the die, and none, one, or both of these two events will have occurred after the experiment is carried out.

A *certain event* is an event that must happen. In different contexts this might be described in apparently different, but in fact identical, ways. For example, in one context S_1 might be defined as the event that the number turning up on a die is less than 7, and in another context S_2 might be defined as the event that the number is divisible by 1. S_1 and S_2 are different ways of describing the same certain event.

1.12.2   Complements, Unions, and Intersections

Before considering the probability theory associated with events we consider some aspects of events themselves.

Associated with any event A is its complementary event "not A," usually denoted by $\bar{A}$. The event $\bar{A}$ is the event that A does not occur. For example, if A is the event "an even number turns up on the roll of a die," then $\bar{A}$ is the event "an odd number turns up."

If A_1 and A_2 are two events, the *union* of A_1 and A_2 is the event that either A_1 or A_2 (or both) occurs. In the die rolling example, if A_1 is the event that an even number turns up and A_2 is the event that the number turning up is 4, 5, or 6, the union of A_1 and A_2 is the event that the number turning up is either 2, 4, 5, or 6. The union of the events A_1 and A_2 is denoted by $A_1 \cup A_2$. For any event A, the event $A \cup \bar{A}$ is the certain event. More generally, in some cases at least one of A_1 or A_2 *must* occur, so that $A_1 \cup A_2$ is the certain event. In such a case we say that A_1 and A_2 are *exhaustive*.

The *intersection* of two events A_1 and A_2 is the event that both occur. In the die rolling example above this is the event that the number turning up is either 4 or 6. The intersection of the events A_1 and A_2 is denoted by $A_1 \cap A_2$ or, more frequently, simply by $A_1 A_2$.

In some cases the events A_1 and A_2 cannot occur together, that is, they are *mutually exclusive*. In this case the intersection of A_1 and A_2 is empty. We denote the empty event by $\emptyset$. For any event A, $A\bar{A} = \emptyset$.

These definitions extend in a natural way to any number of events. Thus the union of the events $A_1, A_2, \ldots, A_n$, denoted by $A_1 \cup A_2 \cup \cdots \cup A_n$, is the event that at least one of the events $A_1, A_2, \ldots, A_n$ occurs. The intersection of the events $A_1, A_2, \ldots, A_n$, denoted by $A_1 A_2 \cdots A_n$, is the event that all of the events $A_1, A_2, \ldots, A_n$ occur.

More complicated events such as $(A_1 A_2) \cup A_3$ may be defined, but the simple unions and intersections described above are sufficient for our purposes in this book.

1.12.3   Probabilities of Events

Probability refers to the assignment of a real number $P(A)$ to each event A. The numbers must satisfy the following "axioms of probability":

(1) $P(A) \geq 0$ for any event A.
(2) For any certain event S, $P(S) = 1$.
(3) For mutually exclusive events A_1 and A_2,

$$P(A_1 \cup A_2) = P(A_1) + P(A_2).$$

It follows from the axioms that for any event A, $0 \leq P(A) \leq 1$ and $P(\bar{A}) = 1 - P(A)$.

Often one wishes to find the probability that at least one of the events $A_1, A_2, \ldots, A_n$ occurs. If the events are mutually exclusive for all $i \neq j$, then it follows by mathematical induction from the above axioms that this probability is

$$P(A_1 \cup A_2 \cup \cdots \cup A_n) = \sum_i P(A_i). \qquad (1.83)$$

A more complicated formula arises when some of the events $A_1, A_2, \ldots, A_n$ are not mutually exclusive. The simplest case concerns two events, A_1 and A_2. The sum $P(A_1) + P(A_2)$ counts the probability of the intersection $A_1 A_2$ twice, whereas it should be counted only once (see Figure 1.15). This argument leads to

$$P(A_1 \cup A_2) = P(A_1) + P(A_2) - P(A_1 A_2). \qquad (1.84)$$

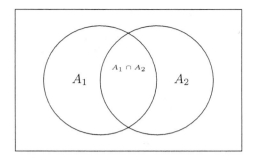

FIGURE 1.15. The probability of $A_1 \cup A_2$ counts $A_1 \cap A_2$ twice; therefore, $P(A_1 \cup A_2) = P(A_1) + P(A_2) - P(A_1 \cap A_2)$.

This formula can be generalized to find the probability of the union of an arbitrary number of events in terms of probabilities of intersections. This is discussed at length in Feller (1957), and we reproduce his results here without proof.

Define $S_1, S_2, \ldots, S_n$ by the formulae

$$S_1 = \sum_i P(A_i), \quad S_2 = \sum_{i<j} P(A_i A_j), \quad S_3 = \sum_{i<j<k} P(A_i A_j A_k), \ldots.$$

$$(1.85)$$

Then the probability P_1 of the union $A_1 \cup A_2 \cup \cdots \cup A_n$ is given by

$$P_1 = S_1 - S_2 + S_3 - \cdots + (-1)^{n-1} S_n. \qquad (1.86)$$

This implies that the probability that none of the events $A_1, A_2, \ldots, A_n$ occurs is

$$1 - S_1 + S_2 - S_3 - \cdots + (-1)^n S_n. \qquad (1.87)$$

Equation (1.86) may be generalized to show that the probability P_m that at least m of the events $A_1, A_2, \ldots, A_n$ occur is given by

$$S_m - \binom{m}{m-1} S_{m+1} + \binom{m+1}{m-1} S_{m+2} - \cdots + (-1)^{n-m} \binom{n-1}{m-1} S_n.$$

The probability $P_{[m]}$ that *exactly* m of the events $A_1, A_2, \ldots, A_n$ occur is given by

$$P_{[m]} = S_m - \binom{m+1}{m} S_{m+1} + \binom{m+2}{m} S_{m+2} - \cdots$$
$$+ (-1)^{n-m} \binom{n}{m} S_n. \qquad (1.88)$$

Example: Random subintervals of $(0,1)$. Suppose that n points are placed independently on the interval $(0,1)$, the position of each point having the uniform distribution (1.53). The probability that two of the points coincide is zero; therefore, they subdivide the interval $(0, 1)$ into $n + 1$ subintervals, with lengths $U_1, U_2, \ldots, U_{n+1}$, and $\sum_{j=1}^{n+1} U_j = 1$. For any given number u, $0 < u < 1$, let h be the largest integer for which $hu < 1$. It will be shown in Section 2.15.2 that if A_i is the event $\{U_i > u\}$, $i = 1, 2, \ldots, n + 1$ and if $i_1, \ldots, i_g$ are all distinct, then

$$P(A_{i_1} A_{i_2} \cdots A_{i_g}) = (1 - gu)^n \qquad (1.89)$$

for any g of the $n + 1$ events, provided that with h as defined above, $g \leq h$. When $g \geq h + 1$, the probability on the left-hand side of (1.89) is 0. Equation (1.86) then implies that the probability that at least one of the events $A_1, A_2, \ldots, A_{n+1}$ occurs is

$$\binom{n+1}{1}(1 - u)^n - \binom{n+1}{2}(1 - 2u)^n + \cdots + (-1)^{h+1}\binom{n+1}{h}(1 - hu)^n.$$
$$(1.90)$$

The application of these results to the analysis of a DNA sequence will be discussed in Section 5.5.

1.12.4 Conditional Probabilities

Suppose that a fair die is rolled once. If it is known that the number turning up is less than or equal to three, elementary arguments show that the probability that it is odd is $\frac{2}{3}$. We write this

$$P(\text{number is odd} \mid \text{number is less than or equal to 3}) = \frac{2}{3}. \qquad (1.91)$$

This is an example of a *conditional probability* calculation.

More precisely, suppose that A_1 and A_2 are two events and that $P(A_2) \neq 0$. Then the *conditional* probability that the event A_1 occurs, given that the event A_2 occurs, denoted by $P(A_1 \mid A_2)$, is given by the formula

$$P(A_1 \mid A_2) = \frac{P(A_1 A_2)}{P(A_2)}. \qquad (1.92)$$

This is a crucial formula. It is often presented as a definition in textbooks, but the reason for the formula can be seen from Figure 1.15, page 38. If it is given that the event A_2 occurs, the only part of the diagram of relevance is the right-hand circle, corresponding to the event A_2. Of the probability associated with this circle, the proportion $P(A_1 A_2)/P(A_2)$ corresponds to the event that A_1 occurs. This proportion is the right-hand side in (1.92).

Conditional probability calculations are often counterintuitive (some examples are given in the Problems), and it is essential that the definition (1.92) be used when calculating them. Clearly equation (1.92) makes sense only when $P(A_2) \neq 0$.

The calculation in (1.91) can be arrived at from (1.92) from the fact that if A_1 is the event that the number is odd and A_2 is the event that the number is less than or equal to three, then $A_1 A_2$ is the event that the number is 1 or 3. The probability of $A_1 A_2$ is then $\frac{1}{3}$ and that of A_2 is $\frac{1}{2}$. From this (1.91) follows.

In using (1.92), care must be exercised in assessing what the intersection event $A_1 A_2$ means. For example, suppose that a die (fair or unfair) is rolled twice. Let A_1 be the event that the number turning up on the first roll is y_1 and the number turning up on the second roll is y_2, and let A_2 be the event that the sum of the two numbers is y_3. Then the event $A_1 A_2$ is empty unless $y_1 + y_2 = y_3$, so that when $y_1 + y_2 \neq y_3$, $P(A_1 \mid A_2) = 0$. When $y_1 + y_2 = y_3$, the event A_1 implies the event A_2, so that $A_1 A_2 = A_1$. In this case $P(A_1 \mid A_2) = P(A_1)/P(A_2)$.

Note that $P(A_1 \mid A_2)$ does not necessarily equal $P(A_2 \mid A_1)$. If a fair die is rolled once and A_1 is the event that the number turning up is not a one, and A_2 be the event that the number is odd, then $P(A_1 \mid A_2) = \frac{2}{3}$, whereas $P(A_2 \mid A_1) = \frac{2}{5}$.

Equation (1.92) has various important consequences. It shows immediately, for example, that

$$P(A_1 A_2) = P(A_2)P(A_1 \mid A_2), \tag{1.93}$$

and in this form its truth is almost self-evident: The probability that both A_1 and A_2 occur is the probability that A_2 occurs multiplied by the probability that A_1 occurs *given* that A_2 occurs. Further, the left-hand side in equation (1.93) is symmetric in A_1 and A_2, so that the right-hand side may be replaced by $P(A_1)P(A_2 \mid A_1)$. This implies that

$$P(A_1)P(A_2 \mid A_1) = P(A_2)P(A_1 \mid A_2), \tag{1.94}$$

or

$$P(A_2 \mid A_1) = \frac{P(A_2)P(A_1 \mid A_2)}{P(A_1)}. \tag{1.95}$$

In this form the formula is used in Bayesian statistics; see equation (3.29).

A second implication of equation (1.92) is the following. Suppose that $B_1, B_2, \ldots, B_k$ is a set of mutually exclusive and exhaustive events, so that exactly one of these events must occur. Then for any event A,

$$P(A) = \sum_{j=1}^{k} P(AB_j).$$

Equation (1.92) then implies

$$P(A) = \sum_{j=1}^{k} P(B_j)P(A \mid B_j). \tag{1.96}$$

All the above formulae for events have direct analogues for probabilities of random variables. In fact, the latter probabilities may be derived immediately from the former. These are discussed at length in the next chapter.

1.12.5   Independence of Events

An event A_1 is *independent* of another event A_2 if

$$P(A_1) = P(A_1 \mid A_2). \tag{1.97}$$

The natural interpretation of this requirement is that knowledge of the occurrence of A_2 does not affect the probability of the occurrence of A_1. The definition of $P(A_1 \mid A_2)$ in (1.92) shows that A_1 is independent of A_2 if and only if

$$P(A_1 A_2) = P(A_1) \cdot P(A_2). \tag{1.98}$$

This equation provides a useful way to check for independence. Note that it follows from this that if A_1 is independent of A_2, then A_2 is independent of A_1, so that we simply say that A_1 and A_2 are independent.

As an example, if a fair die is to be rolled, the events A_1 (an even number turns up) and A_2 (4, 5, or 6 turns up) are not independent. Each of these events separately has probability $\frac{1}{2}$, and the probability of their intersection (the event that either a 4 or a 6 will turn up) is $\frac{1}{3}$. Thus equation (1.98) does not hold, and the events are not independent. On the other hand, if A_3 is the event that a 3, 4, 5, or 6 turns up, then A_1 and A_3 are independent. These statements all assume that the die is fair. If the die is unfair, they will usually no longer hold.

More generally, the events $A_1, A_2, \ldots, A_n$ are defined to be independent if and only if the requirements

$$P(A_i A_j) = P(A_i)P(A_j) \text{ for all distinct } i, j, \tag{1.99}$$

$$P(A_i A_j A_k) = P(A_i)P(A_j)P(A_k) \text{ for all distinct } i, j, k, \tag{1.100}$$

$$P(A_i A_j A_k A_l) = P(A_i)P(A_j)P(A_k)P(A_l) \text{ for all distinct } i, j, k, l,$$

$$\cdots$$

$$P(A_1 A_2 A_3 \cdots A_n) = P(A_1)P(A_2)P(A_3) \cdots P(A_n), \tag{1.101}$$

all hold.

As in the case of two random variables, this definition is equivalent to saying that the probability of any collection of the events is not affected by knowledge of any collection of the remaining events.

It is not sufficient for independence of the events $A_1, A_2, \ldots, A_n$ that the "pairwise" conditions (1.99) hold: All of the conditions (1.99)–(1.101) must be satisfied for the events to be independent. It is possible that the pairwise conditions (1.99) hold but that the "triple-wise" conditions (1.100) do not hold. The following example demonstrates this.

Example. Suppose that two fair dice are thrown, and that the events A_1, A_2, and A_3 are, respectively, "number on first die is odd," "number on second die is odd," "sum of the two numbers is odd." Clearly, both A_1 and A_2 have probability $\frac{1}{2}$ and it is not hard to show that A_3 also has probability $\frac{1}{2}$. Each pairwise intersection event has probability $\frac{1}{4}$; the event $A_1 A_3$, for example, that the first number is odd and the sum of the two numbers is odd, is the event "first number is odd, second number is even," and this has probability $\frac{1}{4}$. The probability of the triple-wise event $A_1 A_2 A_3$ is zero, since this event cannot occur. However, $P(A_1)P(A_2)P(A_3) = \frac{1}{8}$, so that condition (1.100) is not met, and the three events are not independent.

1.13   The Memoryless Property of the Geometric and the Exponential Distributions

Suppose that y_1 and y_2 are two positive integers and that a random variable Y having the geometric distribution (1.10) takes a value y_1 or more. What is the probability that it takes a value $y_1 + y_2$ or more? The intersection of the events $Y \geq y_1$ and $Y \geq y_1 + y_2$ is the event that $Y \geq y_1 + y_2$. In this case the conditional probability formula (1.92) becomes

$$\frac{\text{Prob}(Y \geq y_1 + y_2)}{\text{Prob}(Y \geq y_1)}, \qquad (1.102)$$

and from equation (1.11), this is $p^{y_1 + y_2}/p^{y_1} = p^{y_2}$. But this is just the unconditional probability that $Y \geq y_2$. Therefore,

$$\text{Prob}(Y \geq y_1 + y_2 \,|\, Y \geq y_1) = \text{Prob}(Y \geq y_2).$$

This is called the *memoryless* property of the geometric distribution.

The exponential distribution, like the geometric distribution, also has the memoryless property. This is shown as follows. If X is a random variable having the exponential distribution (1.46) and if $x_2 > 0$, then from (1.60),

$$\text{Prob}(X \geq x_2) = e^{-\lambda x_2}. \qquad (1.103)$$

From the conditional probability formula (1.92), if also $x_1 > 0$, then

$$\text{Prob}(X \geq x_1 + x_2 \,|\, X \geq x_1) = \frac{e^{-\lambda(x_1 + x_2)}}{e^{-\lambda x_1}} = e^{-\lambda x_2}. \qquad (1.104)$$

Thus

$$\text{Prob}(X \geq x_1 + x_2 \,|\, X \geq x_1) = \text{Prob}(X \geq x_2). \qquad (1.105)$$

An application of this property is as follows.

Cellular proteins and RNA molecules are regulated in many ways, involving the degradation and generation of molecules in a cell. Some molecules may age and eventually break down, while others are actively degraded. Still others may undergo spontaneous degradation. Molecules that age would be more likely to become nonfunctional when they are old than when they are young. Molecules that do not undergo any kind of aging process while still active, and which are not actively degraded by other means, would be as likely to degrade at any time, irrespective of their age. This latter phenomenon is, in effect, the memoryless property. The (random) lifetime of such a molecule is modeled by an exponential distribution.

A further property of the exponential distribution is relevant to this behavior. Suppose that h is small. The probability that a random variable having the exponential distribution (1.59) takes a value in the interval $(x, x + h)$, given that its value exceeds x, is, from (1.92) and (1.60),

$$\frac{\int_x^{x+h} \lambda e^{-\lambda u}\, du}{e^{-\lambda x}} = 1 - e^{-\lambda h} = \lambda h + o(h), \tag{1.106}$$

where the second equality follows from (B.21) and the $o(h)$ notation is discussed in Section B.8. In the case of cellular molecules, this implies that having lived to age at least x, the probability that a molecule degrades in the time interval $(x, x + h)$, where h is small, is approximately proportional to the length h of the interval. This example will be pursued further in Sections 2.11.1 and 4.1.

1.14   Entropy and Related Concepts

1.14.1   Entropy

Suppose that Y is a discrete random variable with probability distribution (1.3). The *entropy* $H(\mathbf{P})$ of this probability distribution is defined by

$$H(\mathbf{P}) = -\sum_y P_Y(y) \log P_Y(y), \tag{1.107}$$

the sum (as with all sums in this section) being taken over all values of y in the range of Y. Since this quantity depends only on the probabilities of the various values of Y, and not on the actual values themselves, it can be thought of, as the notation indicates, as being a function of the probability distribution $\mathbf{P}_Y = \{P_Y(y)\}$ rather than the random variable Y.

In some areas of computational biology the base of the logarithm in this definition is taken to be 2. The reason for this, in terms of "bits" of information, is discussed in Section B.10. Although we will use the base 2 in the definition (1.107) later in this book, for the moment we use natural logarithms (and the notation "log" for these).

The entropy of a probability distribution is a measure of the evenness of that distribution and thus, in a sense, of the unpredictability of any observed value of a random variable having that distribution. If there are s possible values for the random variable, the entropy is maximized when $P_Y(y) = s^{-1}$. In this case it takes the value $\log s$, and the value to be assumed by Y is in a sense maximally unpredictable. At the other extreme, if only one value of Y is possible, the entropy of the distribution is zero, and the value to be assumed by the random variable Y is then completely predictable.

Despite the fact that both the entropy and the variance of a probability distribution measure in some sense the uncertainty of the value of a random variable having that distribution, the entropy has an interpretation different from that of the variance. The entropy is defined only by the probabilities of the possible values of the random variable and not by the values themselves. On the other hand, the variance depends on these values. Thus if Y_1 takes the values 1 and 2 each with probability 0.5 and Y_2 takes the values 1 and 100 each with probability 0.5, the distributions of Y_1 and Y_2 have equal entropies but quite different variances.

1.14.2 Relative Entropy

Several procedures in bioinformatics use the *relative entropy* of two different probability distributions $P_{Y_0}(y)$ and $P_{Y_1}(y)$. It is convenient to denote the first of these probability distributions by $\boldsymbol{P}_0$ and the second by $\boldsymbol{P}_1$. We assume that these are different distributions but have the same range, that is, that $P_{Y_0}(y) > 0$ if and only if $P_{Y_1}(y) > 0$. There are two relative entropies associated with these two distributions, namely $H(\boldsymbol{P}_0\|\boldsymbol{P}_1)$ and $H(\boldsymbol{P}_1\|\boldsymbol{P}_0)$, defined respectively by

$$H(\boldsymbol{P}_0\|\boldsymbol{P}_1) = \sum_y P_{Y_0}(y) \log \frac{P_{Y_0}(y)}{P_{Y_1}(y)}, \qquad (1.108)$$

$$H(\boldsymbol{P}_1\|\boldsymbol{P}_0) = \sum_y P_{Y_1}(y) \log \frac{P_{Y_1}(y)}{P_{Y_0}(y)}, \qquad (1.109)$$

the summation in both cases being over all values in the (common) range of the two probability distributions. A proof (Durbin et al. (1998)) that both $H(\boldsymbol{P}_0\|\boldsymbol{P}_1)$ and $H(\boldsymbol{P}_1\|\boldsymbol{P}_0)$ are positive can be based on the fact that for positive x, $-\log x \geq 1 - x$, with equality holding only when $x = 1$. Then, for example,

$$H(\boldsymbol{P}_0 || \boldsymbol{P}_1) = \sum_y P_{Y_0}(y) \left(-\log \frac{P_{Y_1}(y)}{P_{Y_0}(y)} \right)$$

$$\geq \sum_y P_{Y_0}(y) \left(1 - \frac{P_{Y_1}(y)}{P_{Y_0}(y)} \right)$$

$$= \sum_y P_{Y_0}(y) - \sum_y P_{Y_1}(y) = 0, \qquad (1.110)$$

with equality holding only when $P_{Y_0}(y) = P_{Y_1}(y)$.

The relative entropy of two probability distributions measures in some sense the dissimilarity between them. However, since $H(\boldsymbol{P}_0 || \boldsymbol{P}_1)$ is not equal to $H(\boldsymbol{P}_1 || \boldsymbol{P}_0)$, the sometimes-used practice of calling one or the other of these quantities the distance between the two distributions is not appropriate, since $H(\boldsymbol{P}_0 || \boldsymbol{P}_1)$ and $H(\boldsymbol{P}_1 || \boldsymbol{P}_0)$ do not satisfy a basic requirement of a distance measure, namely that they be equal. On the other hand, the quantity $J(\boldsymbol{P}_0 || \boldsymbol{P}_1)$, defined by $J(\boldsymbol{P}_0 || \boldsymbol{P}_1) = H(\boldsymbol{P}_0 || \boldsymbol{P}_1) + H(\boldsymbol{P}_1 || \boldsymbol{P}_0)$, does satisfy this requirement. In the information theory literature, $J(\boldsymbol{P}_0 || \boldsymbol{P}_1)$ is called the *divergence* between the two distributions (see Kullback (1978)).

1.14.3 Scores and Support

We define the *support* $S_{0,1}(y)$ given by an observed value y of Y in favor of $\boldsymbol{P}_0$ over $\boldsymbol{P}_1$ by

$$S_{0,1}(y) = \log\left(P_{Y_0}(y) / P_{Y_1}(y) \right), \qquad (1.111)$$

and similarly we define $S_{1,0}(y)$, given by

$$S_{1,0}(y) = \log\left(P_{Y_1}(y) / P_{Y_0}(y) \right), \qquad (1.112)$$

as the support given by the observed value y of Y in favor of $\boldsymbol{P}_1$ over $\boldsymbol{P}_0$. The term "support" derives from the fact that if $S_{0,1}(y) > 0$, the observed value y of Y is more likely under $\boldsymbol{P}_0$ than it is under $\boldsymbol{P}_1$. If it happens, for some specific value of y of Y, that $P_{Y_0}(y) = P_{Y_1}(y)$, then both measures of support are zero, and this accords with the commonsense view that if Y is equally likely to take the value y under the two distributions, then observing the value y of Y affords no support for one distribution over the other. The definition of support also has the reasonable property that for any observed value y, $S_{0,1}(y) = -S_{1,0}(y)$.

If the true distribution of Y is $\boldsymbol{P}_0$, then $S_{0,1}(Y)$ is a random variable whose mean $I_{0,1}$ is found from equation (1.18) to be

$$I_{0,1} = \sum_y P_{Y_0}(y) S_{0,1}(y), \qquad (1.113)$$

which is identical to $H(\boldsymbol{P}_0||\boldsymbol{P}_1)$. Correspondingly, if the true distribution is $\boldsymbol{P}_1$, then the mean support for $\boldsymbol{P}_1$ over $\boldsymbol{P}_0$ is

$$I_{1,0} = \sum_y P_{Y_1}(y) S_{1,0}(y), \qquad (1.114)$$

and this is identical to $H(\boldsymbol{P}_1||\boldsymbol{P}_0)$. The notation $I_{0,1}$ and $I_{1,0}$ arise in the information theory context, and we have employed them here, although we will normally use the term "mean score" or "mean support" in discussing their applications in bioinformatics.

We introduce the concept of "support" here, since generalizations of the quantities $S_{0,1}(y)$ and $S_{1,0}(y)$ are used later in various hypothesis testing procedures. In particular, the BLAST testing procedure, to which we later pay much attention, uses generalized support statistics. A more general description of the concepts of support and information is given in Edwards (1992).

One single observation is not enough for us to assess which distribution we believe is more likely to be correct. In equation (2.75) we define the concept of the accumulated support afforded by a number of observations. In Section 8.5.1 we discuss the question of how much accumulated support is needed to reasonably decide between two values of a binomial parameter, and in Section 9.6 we will discuss accumulated support in the context of a BLAST procedure. We will later call $S_{1,0}(y)$ a "score" statistic, and in Section 9.2.4 we will consider the problem of the optimal choice of score statistics for the substitution matrices used in BLAST.

1.15   Transformations

Suppose that Y_1 is a discrete random variable and that $Y_2 = g(Y_1)$ is a one-to-one function of Y_1. Then if $y_2 = g(y_1)$,

$$\text{Prob}(Y_2 = y_2) = \text{Prob}(Y_1 = y_1).$$

This equation makes it easy to find the probability distribution of Y_2 from that of Y_1. In the case where the transformation is not one-to-one, a simple addition of probabilities yields the distribution of Y_2. For example, if Y_1 can take the values -1, 0, and $+1$, each with probability $\frac{1}{3}$, and if $Y_2 = Y_1^2$, then the possible values of Y_2 are 0 and 1, and Y_2 takes these values with respective probabilities $\frac{1}{3}$ and $\frac{2}{3}$.

The corresponding calculation for a continuous random variable is less straightforward. Suppose that X_1 is a continuous random variable and that $X_2 = g(X_1)$ is a one-to-one function of X_1. In most cases of interest this means that $g(x)$ is either a monotonically increasing, or a monotonically decreasing, function of x, and we assume that one of these two cases holds.

Let $f_1(x)$ be the density function of X_1 and let $f_2(x)$ be the density function of X_2, which we wish to find from that of X_1. Let $F_1(x)$ and $F_2(x)$ be the two respective distribution functions. (Here for typographic reasons we depart from the notation previously adopted for density functions and distribution functions.)

Suppose first that $g(x)$ is a monotonically increasing function of x and that $X_2 = g(X_1)$. Then

$$F_2(x) = \text{Prob}(X_2 < x) = \text{Prob}(X_1 < g^{-1}(x)) = F_1(g^{-1}(x)), \qquad (1.115)$$

where $g^{-1}(x)$ is the inverse function of $g(x)$. Differentiating both sides with respect to x and using the chain rule, we get

$$f_2(x) = \left(f_1(y) \div \frac{dg(y)}{dy} \right)_{y=g^{-1}(x)}. \qquad (1.116)$$

When X_2 is a monotonically decreasing function of X_1, similar arguments show that

$$f_2(x) = -\left(f_1(y) \div \frac{dg(y)}{dy} \right)_{y=g^{-1}(x)}. \qquad (1.117)$$

Both cases are covered by the equation

$$f_2(x) = \left| \left(f_1(y) \div \frac{dg(y)}{dy} \right) \right|_{y=g^{-1}(x)}. \qquad (1.118)$$

As an example, suppose that X_1 has density function $f_1(x) = 2x$, $0 < x < 1$, and that $X_2 = g(X_1) = X_1^3$. Then $dg(y)/dy = 3y^2$ and equation (1.118) shows that

$$f_2(x) = \left| \frac{2y}{3y^2} \right|_{y=\sqrt[3]{x}} = \frac{2}{3} x^{-1/3}, \quad 0 < x < 1. \qquad (1.119)$$

The form of this density function is quite different from that of X_1.

A particularly important function of X_1 is $X_2 = F(X_1)$, where $F(X_1)$ is the distribution function of X_1. Here application of (1.118) shows that

$$f_2(x_2) = 1, \quad 0 < x_2 < 1. \qquad (1.120)$$

That is, X_2 has the uniform distribution in $(0, 1)$, whatever the distribution of X_1 might be.

1.16   Empirical Methods

We often encounter random variables with distributions that cannot be calculated mathematically, or at least whose calculation appears difficult.

In other cases the mean and/or the variance can be calculated but not the entire probability distribution: An example is given in Section 2.4.2. Often we cannot even calculate these quantities. One consequence of currently available computational power, however, is that we can simulate an experiment many times and obtain "data" from each simulated replication. We can then approximate probability distributions and their means and variances using these simulation "data." The "plug-in" method discussed in Chapter 12 also uses this approach.

We illustrate the simulation approach with an example relating to the problem of gene mapping. Suppose we are given a DNA sequence consisting of L consecutive nucleotides. For convenience we will say that this sequence is of "length" L. A number of possibly overlapping segments of this sequence are observed, the first segment consisting of i_1 consecutive nucleotides, the second consisting of i_2 consecutive nucleotides, and so on. We will say that these segments are of respective lengths i_1, i_2, ... Figure 1.16 shows the four possible positions of a single segment of length $i_j = 3$ when $L = 6$. In general there are $L - i_j + 1$ possible positions for a segment of length i_j.

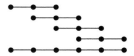

FIGURE 1.16. The 4 possible positions of a segment of length 3 in a sequence of length 6.

The direct-IBD mapping procedure of Cheung et al. (1998) produced data which can be represented as such a collection of segments. (The Cheung et al. segments were more complicated than those described here, but for the purposes of illustration, we have simplified the description of the data). Cheung et al. aimed at testing whether the various segments in the data showed a significant tendency to overlap. This was done by finding the maximum number M of segments that contain the same nucleotide. A significantly large value of M then supports the hypothesis that significant overlap of the segments does occur.

The elements of the theory used for testing hypotheses of this type will be described in Section 3.4. For our purposes now it is sufficient to say that the hypothesis testing procedure requires the calculation of the probability distribution of M when the segments are placed at random on the DNA sequence of length L. (By "random" we mean, for example, that the segment of length i_j occurs in each of the $L - i_j + 1$ possible positions with equal probability.) From this it is required to find the "random placement" probability that M is equal to or exceeds the observed value m of M.

When the number n of segments is large, it is quite difficult to find this probability by mathematical methods, particularly in the more complex case discussed by Cheung et al. Thus Cheung et al. used approximate and conservation calculations to find an upper bound for this probability.

The approximations used were sufficient for their present purposes, however as a general procedure the approximations are less efficient than is desirable. A different approach is to use either computation or simulation. If n and L are sufficiently small, it is possible to enumerate all possible configurations of the n segments by computer and from this obtain the exact distribution by counting. When n and L are large this approach is not feasible. An alternative and relatively simple approach is to replicate the experiment of randomly placing the n segments on a sequence of DNA of length L many times, and in this way obtain an empirical distribution of M, under the randomness assumption. The larger the number of replications, the closer the empirical distribution of M will be to the true one. There are simple computer programs for generating the uniform random variables necessary to do this simulation (see, for example, Press et al. (1992)).

For $n = 7$, $L = 100$, and $(i_1, \ldots, i_7) = (6, 5, 6, 5, 5, 8, 4)$ a simulation involving 10,000 replications gave the empirical distribution shown in Figure 1.17.

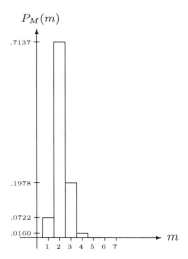

FIGURE 1.17. Empirical probabilities for M when $n = 7$ and $L = 100$. estimated probabilities are: $P_M(1) = .0722$, $P_M(2) = .7137$, $P_M(3) = .1978$, $P_M(4) = .0160$, $P_M(5) = .0003$. The values $M = 6$ and 7 were not observed in the simulation.

Suppose that, in a data set with $L = 100$ and 7 segments, with respective lengths i_1, ..., i_7 equal to 6,5,6,5,5,8,4, the observed value m of M is 4. From the empirical distribution, the estimated "randomness" probability that $M \geq 4$ is quite small, being .0163. We might then be unwilling to believe that the segments in the data were placed randomly, preferring the hypothesis that significant evidence of overlap occurs for some reason.

An empirical approach such as this was taken in Grant et al. (1999) to refine the approximate probability calculations in Cheung et al. (1998). This led to a probability estimate that was an order of magnitude more precise than the Cheung et al. value.

In the previous example, all that was needed was to generate random numbers from a uniform distribution. Often it is necessary, to perform the appropriate simulations, to generate numbers from other distributions. The result at the end of the previous section allows us to transform uniformly distributed random variables into random variables following any continuous distribution whose distribution function is available. Suppose that we wish to generate random values of a continuous random variable X_1 having distribution function $F_1(x)$. If we generate a uniformly distributed random variable X_2, and then define X_1 by $X_1 = F_1^{-1}(X_2)$, where F_1^{-1} is the inverse function of F_1, then X_1 is a random variable having the required density function. For example, (1.60) shows that the distribution function of a random variable having the exponential distribution (1.59) is $F_1(x) = 1 - e^{-\lambda x}$, $x \geq 0$. Thus if X_2 has a uniform distribution in $(0, 1)$, and if X_1 is defined by $X_1 = -\frac{1}{\lambda} \log(1 - X_2)$ (so that $X_2 = 1 - e^{-\lambda X_1} = F_1(X_1)$), then X_1 is a random variable having the exponential distribution (1.59).

A useful application of empirical methods in computational biology and bioinformatics concerns the distribution of test statistics used in statistical inference. The DNA segment example above provides an illustration of such an application. Further examples of this approach are given later in this book, in particular in connection with the complex theory of BLAST and of inference concerning phylogenetic trees. In some cases these examples demonstrate, using empirical methods, that various approximations commonly used for the distributions of test statistics are not as accurate as would be desired.

One might ask why, if empirical methods are so convenient, we should go to the trouble to calculate theoretical properties of distributions. One reason is that an exact theoretical distribution is preferred to an approximate empirical distribution is because the mathematical form of the distribution is often informative. Second, a drawback of empirical methods is that they are slow: In order to obtain reasonable estimates it might be necessary to perform hundreds of thousands of complicated simulations, which might take an unacceptably long time to run. Sometimes estimates have to be calculated in a matter of microseconds. When empirical methods are not adequate for this task, theoretical methods, even at the possible cost of some accuracy, become necessary.

Problems

1.1. Use equation (1.28) to prove that the mean of a random variable having the binomial distribution is np, and together with equation (1.23) show that the variance of this random variable is $np(1-p)$.

1.2. Use definition (1.16) and equation (B.18) to show that the mean of a random variable having the geometric distribution (1.10) is p/q, where $q = 1-p$. Also show that the mean of a random variable having the Poisson distribution (1.15) is λ.

1.3. Show that equations (1.21) and (1.23) follow from the variance definition (1.20).

1.4. Let Y be a random variable having a Poisson distribution with parameter λ. If λ is small (say less than 0.05), show that the probability that $Y \geq 1$ is approximately λ. *Hint:* Let A be the event $Y = 0$, so that the probability required is $P(\bar{A})$. Find $P(A)$ and then use the formula $P(\bar{A}) = 1 - P(A)$. Also use the approximation (B.21).

1.5. Suppose that the random variable Y has the Poisson distribution (1.15). Find the mean of the function $Y(Y-1)$. Then generalize your result to find the rth factorial moment of this distribution, that is,

$$E\big(Y(Y-1)\cdots(Y-r+1)\big),$$

where r is any positive integer less than or equal to n. *Hint:* Use an approach analogous to that leading to equation (1.28).

1.6. Use formula (1.23) and the results of Problem 1.5 for the case $r = 2$ to find the variance of a random variable having the Poisson distribution (1.15).

1.7. Suppose that a sequence of independent Bernoulli trials is conducted, with the same probability of success on each trial. The event that $r - 1$ or fewer successes occur in the first n trials implies the event that the number of trials needed to achieve r successes is $n + 1$ or more. Use this result to find a relation between the distribution functions of the binomial and the generalized geometric distributions.

1.8. Prove equation (1.48). *Hint:* Write the right-hand side as

$$\int_0^H 1 \cdot (1 - F_X(x))\, dx;$$

then use integration by parts.

1.9. Is Chebyshev's inequality (1.50) useful when $d = \sigma$? When X has a normal distribution, mean μ, variance σ^2, compare the exact probability that $|X - \mu| \geq d$ with the values found in Chebyshev's inequality for the cases $d = 2\sigma$, $d = 3\sigma$.

1.10. Use equation (B.19) and further manipulation to show that the variance of a random variable having the geometric distribution (1.10) is $p/q + (p/q)^2 = p/q^2$, where $q = 1 - p$.

1.11. Derive the expression (1.35) for the pgf of the binomial distribution. Also, show that the pgfs of the geometric and the Poisson distribution are, respectively, $(1 - p)/(1 - ps)$ and $e^{\lambda(s-1)}$.

1.12. Use definition (1.16), and equations (1.23) and (1.31) to prove equations (1.36) and (1.37).

1.13. Use the results of Problems 1.11 and 1.12 to rederive the results of Problems 1.1 and 1.2.

1.14. Use the pgfs found in Problem 1.11, together with (1.36) and (1.37), to find the variance of a random variable having (i) the Poisson distribution, (ii) the geometric distribution, and check that your results agree with those found respectively in Problem 1.6 and Problem 1.10.

1.15. Use equations (1.45) and (1.46) to prove that the mean of the exponential distribution (1.59) is $1/\lambda$ and that the variance is $1/\lambda^2$. (*Hint:* Use equations (B.41) and (B.42).) Confirm your calculation for the mean by using equations (1.60) and (1.48).

1.16. Show that the mean and variance of a uniform random variable are as given in (1.52).

1.17. Find the probability that a random variable having the exponential distribution (1.59) is within two standard deviations of its mean. Show further that although this probability is close to 0.95, the calculation is misleading so far as small values of the random variable are concerned.

1.18. Find the mean and variance given in (1.66) by using integration methods and (1.65).

1.19. Show that the mean and variance of a random variable having the chi-square distribution (1.70) with ν degrees of freedom are, respectively, ν and 2ν.

1.20. Find the mgfs of the binomial, the geometric, the Poisson, and the uniform distributions.

1.21. Use definitions (1.16) and (1.73) to prove (1.77), (1.78), and (1.79).

1.22. Suppose a discrete random variable Y takes the values -2 with probability 0.2, -1 with probability 0.3, and $+1$ with probability 0.5. Show that the mean of this random variable is negative. Confirm the result of Theorem 1.1, page 35, by showing that there is a unique positive solution θ^* for θ of the equation $m(\theta) = 1$, where $m(\theta)$ is the mgf of Y.

1.23. A fair die is to be rolled once. Define A_1 to be the event that an even number turns up and A_2 the event that a 4, 5, or 6 turns up. Define the event $A_1 \cup A_2$ and find its probability using a direct argument. Now check that the answer that you found agrees with that found by applying equation (1.84).

1.24. Two fair coins are tossed and you are told that at least one coin came up heads. What is the probability that both coins came up heads? Note: The two "instinctive" answers $\frac{1}{2}$ and $\frac{1}{4}$ are both incorrect: It is necessary to use the conditional probability formula (1.92).

1.25. Suppose that the random variable Y has the geometric-like distribution (1.12). Assuming for the moment that the probability given in (1.12) is exact rather than asymptotic, calculate $\text{Prob}(Y \geq y_1)$ for any $y_1 \geq 1$ and also calculate $\text{Prob}(Y \geq y_1 + 1)$. Then use equation (1.92) to find the conditional probability that $Y \geq y_1 + 1$, given that $Y \geq y_1$. Compare your result with that following (1.102).

Noting now that the probability in (1.12) is asymptotic rather than exact, what asymptotic statement can you make?

1.26. Suppose an unfair die has the property that the probability that the number i turns up is $\frac{i}{21}$. Are the events "an even number turns up" and "a number 4 or less turns up" independent? What is the corresponding result for a fair die?

1.27. Suppose that the various triple-wise intersection probabilities associated with three events A_1, A_2 and A_3 are generated from the formulae

$$P(A_1 A_2 A_3) = pqr + d,$$
$$P(A_1 A_2 \bar{A}_3) = pq(1 - r) - d,$$
$$P(A_1 \bar{A}_2 A_3) = p(1 - q)r - d,$$
$$P(A_1 \bar{A}_2 \bar{A}_3) = p(1 - q)(1 - r) + d,$$
$$P(\bar{A}_1 A_2 A_3) = (1 - p)qr - d,$$
$$P(\bar{A}_1 A_2 \bar{A}_3) = (1 - p)q(1 - r) + d,$$
$$P(\bar{A}_1 \bar{A}_2 A_3) = (1 - p)(1 - q)r + d,$$
$$P(\bar{A}_1 \bar{A}_2 \bar{A}_3) = (1 - p)(1 - q)(1 - r) - d,$$

where $0 < p, q, r < 1$ and d is chosen small enough so that all probabilities are positive. Show that the requirement (1.99) holds for such a probability scheme, but that the only case for which the events A_1, A_2, and A_3 are independent is when $d = 0$.

1.28. The classic example of the use of equations (1.86)–(1.88) arises through the procedure of taking a random permutation of the integers $(1, 2, \ldots, n)$. In any such random permutation, any given integer i will match its original position with probability n^{-1}. More generally, if A_i is the event that the integer i does indeed match its original position, then

$$P(A_i) = n^{-1},$$
$$P(A_i A_j) = \big(n(n - 1)\big)^{-1},$$
$$P(A_i A_j A_k) = \big(n(n - 1)(n - 2)\big)^{-1},$$

and so on. Use equation (1.87) to show that the probability that no number in the random permutation is in its original position is

$$1 - 1 + \frac{1}{2!} - \frac{1}{3!} + \cdots + (-1)^n \frac{1}{n!}. \tag{1.121}$$

Find the limit of this probability as $n \to +\infty$. Further, evaluate the expression (1.88) and comment on the limiting behavior of this expression as $n \to +\infty$.

1.29. Let $P_{Y_0}(y)$ and $P_{Y_1}(y)$ be two probability distributions defined on the same range R. It can be shown that $-\sum_{y \in R} P_{Y_0}(y) \log P_{Y_1}(y)$ is maximized (with respect to choice of $P_{Y_1}(y)$) by the choice $P_{Y_1}(y) = P_{Y_0}(y)$. Use this fact to show that the relative entropy (1.108) is nonnegative. *Hint:* Write $\log(P_{Y_0}(y)/P_{Y_1}(y))$ as $\log P_{Y_0}(y) - \log P_{Y_1}(y)$.

2
Probability Theory (ii): Many Random Variables

In almost every application of statistical methods we deal with the analysis of many observations. For example, if we wish to test whether a certain die is fair, we would roll it many times before making our assessment. Perhaps the most basic issue concerning many random variables is that of independence. The numbers appearing on the various rolls of a die are taken as independent random variables, as discussed below in Section 2.1.1. On the other hand, the random variables Y_i, $i = 1, \ldots, 6$, giving the respective numbers of times that i appears in a fixed number of rolls, comprise a *dependent* set of six random variables, since knowing the value of any five of them determines the sixth.

We will usually denote the number of independent random variables considered, for example the number of rolls of the die, by n. We will denote the number of dependent random variables considered, for example the Y_i's above, by k.

2.1 Many Discrete Random Variables

In this section we consider the case of n discrete random variables Y_1, $Y_2, \ldots, Y_n$. Associated to the *random vector* $\boldsymbol{Y} = (Y_1, Y_2, \ldots, Y_n)$ is the *joint probability distribution* $P_{\boldsymbol{Y}}(\boldsymbol{y})$ defined by

$$P_{\boldsymbol{Y}}(\boldsymbol{y}) = \mathrm{Prob}(Y_1 = y_1, Y_2 = y_2, \ldots, Y_n = y_n), \qquad (2.1)$$

where $\boldsymbol{y} = (y_1, y_2, \ldots, y_n)$ and y_i is a possible value of Y_i for each i.

2.1.1   The Independent Case

In this section we consider the case of n independent discrete random variables. Each of the n random variables $Y_1, Y_2, \ldots, Y_n$ will have a probability distribution $P_{Y_i}(y) = \text{Prob}(Y_i = y)$. The random variables $Y_1, Y_2, \ldots, Y_n$ are said to be *independent* if

$$P_Y(\boldsymbol{y}) = \prod_{i=1}^{n} P_{Y_i}(y_i) \tag{2.2}$$

for all possible combinations of values of $y_1, y_2, \ldots, y_n$. The conceptual interpretation of independence of several random variables is that the knowledge of the value of any subset of these random variables does not influence the probabilities of the values of any of the other variables. This interpretation is given below in mathematical terms in equation (2.38).

It often occurs that the independent random variables $Y_1, Y_2, \ldots, Y_n$ all have the same probability distribution. If this is the case, we say that they are *iid* (independently and identically distributed). As an example, in the case of rolling a die n times, we normally assume that the numbers $Y_1, Y_2, \ldots, Y_n$ appearing on rolls 1, 2, $\ldots$, n are iid.

The condition for independence of several random variables might appear to be less stringent than those given in (1.99)–(1.101) for the independence of events $A_1, A_2, \ldots, A_n$, but in fact, the conditions for independence of events and those for independence of random variables are identical. More specifically, if A_i is the event that $Y_i = y_i$, then the independence requirement (2.2) implies that all the conditions (1.99)–(1.101) are satisfied. Verification of this is left as an exercise (Problem 2.1).

Example 1. If $Y_1, Y_2, \ldots, Y_n$ are independent random variables each having the Poisson distribution (1.15), with parameters $\lambda_1, \ldots, \lambda_n$, respectively, then

$$P_Y(\boldsymbol{y}) = \frac{e^{-\sum \lambda_i} \prod \lambda_i^{y_i}}{\prod(y_i!)}, \tag{2.3}$$

the summation and both products being from 1 to n. If $Y_1, Y_2, \ldots, Y_n$ are iid, each having a Poisson distribution with parameter λ, then

$$P_Y(\boldsymbol{y}) = \frac{e^{-n\lambda} \lambda^{\sum y_i}}{\prod(y_i!)}. \tag{2.4}$$

Example 2. If the probability of success on each trial in a sequence of independent Bernoulli trials is p, and if $Y_i = 1$ when trial i results in success and $Y_i = 0$ when trial i results in failure, then $Y_1, Y_2, \ldots, Y_n$ are iid, and

$$P_Y(\boldsymbol{y}) = p^{\sum y_i}(1-p)^{n-\sum y_i} = p^y(1-p)^{n-y}, \tag{2.5}$$

where $y = \sum y_i$ is the total number of successes. The probability (2.5) differs from the binomial probability (1.8), since (1.8) includes the combinatorial term $\binom{n}{y}$. The binomial distribution gives the probability of y successes and $n - y$ failures regardless of the order of successes and failures, while (2.5) gives the probability of y successes and $n - y$ failures in a nominated order. As shown in Section B.4, the binomial term $\binom{n}{y}$ is the number of orderings of y successes and $n - y$ failures, and this observation explains the difference between the two probabilities.

The theory for independent random variables is far simpler than that for dependent random variables. We now turn to an example of the latter.

2.1.2   The Dependent Case and the Multinomial Distribution

In this section we consider the case of k dependent discrete random variables Y_1, $Y_2, \ldots$, Y_k. In this case the joint probability distribution of Y_1, $Y_2, \ldots$, Y_k does not factorize, as in (2.2), as the product of the separate probability distributions for each Y_i.

An important example of a joint discrete probability distribution where the individual random variables are dependent is the multinomial distribution. This is a direct generalization of the binomial distribution to the case where there are more than two possible outcomes on each trial.

Suppose that n independent trials are to take place, where n is a fixed number. We assume there are k possible outcomes on each trial, that successive trials are independent, and that there is some fixed probability p_i of outcome i on any trial ($i = 1, 2, \ldots, k$). An example is given by the case of n independent rolls of a die, where $k = 6$. Define Y_i as the number of times that outcome i occurs in the n trials ($i = 1, 2, \ldots, k$). Then the probability that $Y_i = y_i$, $i = 1, 2, \ldots, k$, is given by the *multinomial distribution formula*

$$P_Y(y) = \frac{n!}{\prod_i (y_i!)} \prod_i p_i^{y_i}, \qquad (2.6)$$

both products being over $i = 1, 2, \ldots, k$. This formula can be derived in a manner generalizing the derivation of the binomial distribution.

Here $Y_1, Y_2, \ldots, Y_k$ are dependent, since necessarily $\sum_j Y_j = n$. The most extreme case of dependence occurs for the binomial distribution ($k = 2$), since here the value of Y_2 is automatically determined as $n - Y_1$.

Each Y_i considered on its own has a binomial distribution with mean np_i and variance $np_i(1 - p_i)$.

2.1.3   Properties of Random Variables from pgfs

Let $Y_1, Y_2, \ldots, Y_n$ be independent discrete random variables, having probability distributions with respective pgfs $p_1(t), p_2(t), \ldots, p_n(t)$. Then the sum S_n, defined by

$$S_n = Y_1 + Y_2 + \cdots + Y_n, \tag{2.7}$$

is itself a random variable, and the pgf of its probability distribution is the product

$$\mathbb{p}_{S_n}(t) = \mathbb{p}_1(t) \cdot \mathbb{p}_2(t) \cdots \mathbb{p}_n(t). \tag{2.8}$$

The most direct proof of this statement uses the induction method described in Section B.18, and we outline it here. In the case $n = 2$, the probability that $S_2 = j$ is $\sum_i \text{Prob}(Y_1 = i) \cdot \text{Prob}(Y_2 = j - i)$. This sum is the coefficient of t^j in $\mathbb{p}_1(t) \cdot \mathbb{p}_2(t)$, so that (2.8) is true for the case $n = 2$. Suppose now that (2.8) is true for $n = m$ for some $m \geq 2$, so that

$$\mathbb{p}_{S_m}(t) = \mathbb{p}_1(t) \cdot \mathbb{p}_2(t) \cdots \mathbb{p}_m(t). \tag{2.9}$$

We write $Y_1 + \cdots + Y_{m+1}$ as $Y' + Y_{m+1}$, where $Y' = Y_1 + \cdots + Y_m$. By assumption, the pgf of Y' is as given in (2.9). The case $n = 2$ then shows that the pgf of $Y_1 + \cdots + Y_{m+1}$, that is of $Y' + Y_{m+1}$, is $(\mathbb{p}_1(t) \cdot \mathbb{p}_2(t) \cdots \mathbb{p}_m(t)) \cdot \mathbb{p}_{m+1}(t)$. This is of the same form as the right-hand side of (2.9), but with m replaced by $m + 1$. By induction, (2.8) is therefore true for all n.

An important particular case of (2.8) occurs when the random variables Y_i have the same probability distribution. If the common pgf of the random variables Y_i is $\mathbb{p}(t)$, then

$$\text{pgf of } S_n = \big(\mathbb{p}(t)\big)^n. \tag{2.10}$$

Suppose we recognize the pgf found in (2.8) as the pgf of a random variable Y that we know. The *uniqueness* theorem for pgfs given in Section 1.7 shows that $Y_1 + Y_2 + \cdots + Y_n$ has the same distribution as Y. Use of this theorem is often the most straightforward way of finding the distribution of the sum of several random variables.

Example 1. If $Y_1, Y_2, \ldots, Y_n$ are independent Poisson random variables, with respective parameters $\lambda_1, \lambda_2, \ldots, \lambda_n$, then equation (2.8), together with one of the calculations in Problem 1.11, shows that the pgf of $\sum_i Y_i$ is $e^{\sum_i \lambda_i (t-1)}$. The uniqueness property shows immediately that $\sum_i Y_i$ has a Poisson distribution with parameter $\sum_i \lambda_i$.

This result can be understood in an intuitive way by an example. Properties of the spontaneous degradation of cellular molecules were discussed in Section 1.13. It will be shown in Chapter 4 that the number of such molecules to degrade in any given time interval is a random variable having a Poisson distribution. Consider two nonoverlapping time intervals. We would expect that the distribution of the total number of molecules to degrade in the two time intervals combined is identical to the distribution of the number of molecules to degrade in a single time interval whose length is the sum of the two lengths of the two time intervals considered. The above result shows that this is indeed the case.

Example 2. The pgf (1.35) of the binomial distribution also follows from
equations (1.32) and (2.10). The number of successes in n trials is a sum,
namely the sum of the number of successes on trial 1 (either 0 or 1) the
number of successes on trial 2, (again 0 or 1), and so on. From (1.32), the
pgf of the number of successes on any one trial is $1 - p + pt$. Thus from
(2.10), the pgf of the total number of successes in n trials is $(1 - p + pt)^n$,
and this is (1.35).

By expanding (1.35) in powers of t and considering the coefficient of t^y
in this expansion, we arrive at (1.8) for the probability of y successes in n
trials. This calculation gives an alternative way of arriving at the binomial
distribution formula (1.8).

2.2 Many Continuous Random Variables

Suppose that $X_1, X_2, \ldots, X_n$ are continuous random variables. We define
their *joint range* as the set of all values $(x_1, x_2, \ldots, x_n)$ in n-dimensional
space that $(X_1, X_2, \ldots, X_n)$ can take. It is important to notice that if x_1 is
in the range of X_1 and x_2 is in the range of X_2, it is not necessarily the case
that (x_1, x_2) is in the range of (X_1, X_2). For example, suppose we observe
two cells until they both die. Let X_1 be the time it takes until the first of
the two cells dies, and let X_2 be the time it takes until both cells die. The
range of both X_1 and X_2 is $(0, \infty)$, but it is impossible that $X_1 > X_2$.

As in the case of a single continuous random variable, probabilities are
allocated to regions of the joint range of $X_1, X_2, \ldots, X_n$ and not to individual points. Associated with the random variables $X_1, X_2, \ldots, X_n$ is a
joint density function $f_{\boldsymbol{X}}(x_1, x_2, \ldots, x_n)$ such that for any region Q within
the joint range of $X_1, X_2, \ldots, X_n$ for which the integral in (2.11) exists,

$$\text{Prob}((X_1, X_2, \ldots, X_n) \text{ is in } Q)$$
$$= \int \cdots \int_Q f_{\boldsymbol{X}}(x_1, x_2, \ldots, x_n) dx_1 dx_2 \cdots dx_n. \quad (2.11)$$

A particular case of this equation is

$$\int \cdots \int_R f_{\boldsymbol{X}}(x_1, x_2, \ldots, x_n) dx_1 dx_2 \cdots dx_n = 1. \quad (2.12)$$

2.2.1 The Independent Case

Each one of the random variables $X_1, X_2, \ldots, X_n$, considered on its own,
will have its own density function, and we write the density function of X_i

as $f_{X_i}(x_i)$. $X_1, X_2, \ldots, X_n$ are said to be independent if their joint density function is the product of these respective density functions, that is, if

$$f_{\boldsymbol{X}}(x_1, x_2, \ldots, x_n) = \prod_{i=1}^{n} f_{X_i}(x_i) \qquad (2.13)$$

for all $(x_1, \ldots, x_n)$ in the joint range of $(X_1, \ldots, X_n)$. The conceptual meaning of independence of continuous random variables is the same as that for discrete random variables.

When the individual density functions $f_{X_1}(x_1), \ldots, f_{X_n}(x_n)$ of independent random variables are identical, we say (as in the case of discrete random variables) that the random variables are iid. For example, if X_1, $X_2, \ldots, X_n$ are iid, each having a normal distribution with mean μ and variance σ^2, in which case we say that $X_1, X_2, \ldots, X_n$ are NID(μ, σ^2), then their joint density function is

$$f_{\boldsymbol{X}}(x_1, x_2, \ldots, x_n) = \prod_{i=1}^{n} \frac{1}{\sqrt{2\pi}\sigma} e^{-\frac{(x_i-\mu)^2}{2\sigma^2}} = \left(\frac{1}{\sqrt{2\pi}\sigma}\right)^n e^{-\frac{\sum(x_i-\mu)^2}{2\sigma^2}},$$
$$-\infty < x_i < +\infty. \qquad (2.14)$$

The corresponding result for the joint distribution of n iid random variables having the exponential distribution (1.59) is

$$f_{\boldsymbol{X}}(x_1, x_2, \ldots, x_n) = \lambda^n e^{-\lambda \sum x_i}, \qquad 0 \leq x_i < +\infty. \qquad (2.15)$$

2.2.2   The Dependent Case

In the continuous case, as in the discrete case, we often have to deal with dependent random variables. When the random variables are dependent their joint density function does not factorize, as in (2.13), into the product of the respective density functions of the respective X_i. In many practical cases in biology it is assumed that the dependent random variables of interest have a multivariate normal distribution with arbitrary correlation structure between these random variables. We therefore defer further discussion of the dependent case until the topics of correlation and covariance have been discussed.

2.3   Moment-Generating Functions

Moment-generating functions (mgfs) have convolution properties parallel to those for probability-generating functions discussed in Section 2.1.3. These properties are identical for both discrete random variables and continuous random variables, and we develop them using continuous random variable notation.

Properties of mgfs. If $X_1, X_2, \ldots, X_n$ are independent random variables whose distributions have respective mgfs $\mathrm{m}_1(\theta), \ldots, \mathrm{m}_n(\theta)$, then their sum $S_n = X_1 + X_2 + \cdots + X_n$ is a random variable whose mgf is

$$\mathrm{m}_{S_n}(\theta) = \mathrm{m}_1(\theta) \cdot \mathrm{m}_2(\theta) \cdots \mathrm{m}_n(\theta). \tag{2.16}$$

In particular, if $X_1, X_2, \ldots, X_n$ are iid random variables whose common probability distribution has mgf $\mathrm{m}(\theta)$, then

$$\mathrm{m}_{S_n}(\theta) = (\mathrm{m}(\theta))^n. \tag{2.17}$$

Equations (1.77), (1.78), and (2.17) are useful for finding properties of sums of iid random variables. For example, suppose that $X_1, X_2, \ldots, X_n$ are iid random variables, each having mean μ and variance σ^2. From (2.17), the derivative of the mgf of their sum S_n is

$$n\,(\mathrm{m}(\theta))^{n-1}\,\frac{d\mathrm{m}(\theta)}{d\theta}. \tag{2.18}$$

Carrying out a further differentiation and then putting $\theta = 0$, we find from (1.77) and (1.78) that

$$\text{mean of } S_n = n\mu \tag{2.19}$$

and

$$\text{variance of } S_n = n\sigma^2. \tag{2.20}$$

When $X_1, X_2, \ldots, X_n$ are independent random variables with possibly different distributions, with respective means $\mu_1, \mu_2, \ldots, \mu_n$ and variances $\sigma_1^2, \sigma_2^2, \ldots, \sigma_n^2$, then equations (1.77), (1.78), and (2.16) show that

$$\text{mean of } S_n = \mu_1 + \mu_2 + \cdots + \mu_n, \tag{2.21}$$

and

$$\text{variance of } S_n = \sigma_1^2 + \sigma_2^2 + \cdots + \sigma_n^2. \tag{2.22}$$

More generally, if $S_n = a_1 X_1 + a_2 X_2 + \cdots + a_n X_n$, then

$$\text{mean of } S_n = a_1 \mu_1 + a_2 \mu_2 + \cdots + a_n \mu_n, \tag{2.23}$$

$$\text{variance of } S_n = a_1^2 \sigma_1^2 + a_2^2 \sigma_2^2 + \cdots + a_n^2 \sigma_n^2. \tag{2.24}$$

The former equation also holds when $X_1, X_2, \ldots, X_n$ are dependent, but other than in exceptional cases, not the latter. A proof of these claims is given on page 74.

Moment-generating functions for continuous random variables have a uniqueness property similar to that of discrete random variables. As in the discrete case, often the best way to find the distribution of some random variable is to find its mgf and recognize it as the mgf of a known distribution. This is often done in conjunction with equation (2.16) or equation (2.17). For example, the mgf of the gamma distribution (1.68) is

$$\int_0^{+\infty} \frac{\lambda^k \, x^{k-1} e^{-\lambda x} \, e^{\theta x}}{\Gamma(k)} \, dx = \left(\frac{\lambda}{\lambda - \theta}\right)^k. \tag{2.25}$$

The fact that this is the kth power of the mgf of the exponential distribution implies that the sum of k independent random variables, each having the exponential distribution (1.59), is a random variable having the gamma distribution (1.68).

In Chapter 7 we will consider "random walks" that consist of a sequence of random steps; each step goes up (a value of $+1$) with probability p and down (a value of -1) with probability $q = 1 - p$. Thus the value of a random step is similar to the number of successes in a Bernoulli trial, except that its possible values are -1 and $+1$ rather than 0 and 1. The mgf of a random step S is

$$m(\theta) = qe^{-\theta} + pe^{\theta}. \tag{2.26}$$

We can check (2.17) by considering the position, or displacement, of this random walk after two steps have been taken. The two-step random variable is the sum of two one-step random variables. The total displacement is either -2 (with probability q^2), 0 (with probability $2pq$), or $+2$ (with probability p^2). Thus the mgf of this "two-step" displacement is

$$q^2 e^{-2\theta} + 2pq + p^2 e^{2\theta},$$

and this is indeed the square of the expression given in (2.26). This result can be extended to the case of three steps (Problem 2.13) and any higher number of steps.

2.4   Indicator Random Variables

2.4.1   Definitions

Associated with an event A there is a particularly useful discrete random variable called the *indicator* random variable, for which we will normally use the special notation I, or I_A when it is necessary to indicate the dependence on A. This random variable is defined by

$$I_A = 1 \text{ if the event } A \text{ occurs,}$$
$$I_A = 0 \text{ if the event } A \text{ does not occur.}$$

It follows from this definition that I_A is a random variable having the Bernoulli distribution (1.6). The mean of I_A is simply the probability p

that the event A occurs. Let $A_1, A_2, \ldots, A_n$ be events and $I_1, I_2, \ldots, I_n$ their respective indicator functions. Then

$$\sum_{i=1}^{n} I_i$$

is the total number of the events that occur. The mean of a sum of random variables is the sum of the means, whether or not the random variables are independent, as noted after (2.24). Therefore, the mean of the number of the events $A_1, A_2, \ldots, A_n$ that occur is

$$E\left(\sum_{i=1}^{n} I_i \right) = \sum_{i=1}^{n} E(I_i) = p_1 + p_2 + \cdots + p_n. \qquad (2.27)$$

If $p_1 = p_2 = \cdots = p_n = p$, then $\sum I_i$ can be thought of as a generalization of a binomial random variable to the case of possibly dependent trials. In this case

$$E\left(\sum_{i=1}^{n} I_i \right) = np, \qquad (2.28)$$

the same as the binomial distribution mean found in Problem 1.1. In the derivation of the binomial distribution we required the trials to be independent, but the result just found shows that the mean np applies whether or not this is the case.

When the events $A_1, \ldots, A_n$ are independent, the variance of the number of the events $A_1, \ldots, A_n$ that occur is, from (2.22),

$$p_1(1 - p_1) + \cdots + p_n(1 - p_n). \qquad (2.29)$$

If $p_1 = p_2 = \cdots = p_n = p$, this variance is $np(1 - p)$, the binomial distribution variance found in Problem 1.1. Even when $p_1 = p_2 = \cdots = p_n = p$, the variance in the dependent case need not be, and usually is not, $np(1-p)$.

2.4.2   Example: Sequencing EST Libraries

In this example we apply (2.27) to answer a question regarding the generation of a database of ESTs (expressed sequence tags).

Genes are expressed by a two-step process. First they are transcribed from the DNA into molecules called messenger RNA (mRNA), and then the mRNAs are translated into proteins, which are sequences made up of the 20 amino acids. (For further details, see Appendix A.) Any gene is said to give rise to one or more[1] *species* of mRNA specific to that gene. An EST

---

[1]Some genes give rise to more than one mRNA, due to alternative splicing of introns and exons.

is a sequence on the order of a hundred or more contiguous bases of an mRNA. A typical cell has hundreds of thousands of mRNAs, representing thousands of different genes. Some genes are expressed at a high level in the cell, and thus there are many duplicate copies of the mRNA corresponding to such genes in the cell. Other genes are expressed at a low level and thus there are relatively few mRNAs from those genes in the cell. Any gene represented by L mRNAs will be said to be at abundance level L.

A database of ESTs is generated by repeated random sampling from the collection of mRNAs in the cell. Because of the way in which this is carried out, the sampling can be taken to be with replacement.

To illustrate (2.27), consider the following simplified example. Suppose there are 10,000 different species of mRNA in the cell, with the distribution given in Table 2.1. Suppose a database of S ESTs is to be created, so that a

copies per cell (abundance level L)	number of different mRNA species	number of mRNAs per abundance level
5	4000	20,000
50	3250	162,500
200	2500	500,000
1000	250	250,000
total 10,000		N=932,500

TABLE 2.1.

random sample of S ESTs is taken from the pool of all mRNAs in the cell. Define $J_L(S)$ to be the number of different species of abundance level L mRNA in the database. We are interested in the expected value $E(J_L(S))$, since from this expected value we can assess how large S should be in order that the mean percentage of any specified abundance class of mRNAs in the database reach some specified value. For example, one might be interested in determining how many ESTs must be generated in order to expect to have seen at least 50% of the different mRNA species that are expressed at the abundance level L.

The calculation of $E(J_L(S))$ using (1.16) is not easy, since obtaining the values $P_Y(y)$ in this case is difficult. However, the theory of indicator random variables leads to an immediate calculation for this mean. Let a be an mRNA species and define

$$I_a = \begin{cases} 1, & \text{if } a \text{ has been seen in the } S \text{ samples,} \\ 0, & \text{if } a \text{ has not been seen in the } S \text{ samples.} \end{cases}$$

Then the number of different species from abundance class L in the database of size S is $\sum_a I_a$, where the sum is taken over all mRNA species a in abundance class L. The I_a's are not independent, but they have the same mean for all a in abundance class L, and thus $\sum_a I_a$ is a generalization of a binomial random variable to the case of dependent trials, as discussed at the end of Section 2.4.1. Let p_L be the common mean of the I_a's in abundance class L. If there are n_L species of mRNA in abundance class L, then by (2.28), the required mean is $n_L p_L$. The probability p_L is most easily calculated as $1 - r_L$, where r_L is the probability that the specified species is not in the database. Then $r_L = (1 - L/N)^S$, where N is the total number of mRNAs in the cell. Thus the mean number of species of abundance level L seen in the sample is

$$n_L \left(1 - \left(1 - \frac{L}{N} \right)^S \right).$$

The total number of mRNAs in the data in the above table is 932,500. Thus, for example, the mean number of species of abundance level 50 expected in a sample of 10,000 is

$$3250 \left(1 - \left(1 - \frac{50}{932,500} \right)^{10,000} \right) = 1348.75.$$

This represents 41.5% of the different species in this abundance class. Table 2.2 gives percentage values for each abundance class for various values of S. It is interesting to note that even with $S =$ 50,000, the mean percentage of species seen in the lowest abundance class is only 23.52%.

abundance level	Size of EST database					
	1,000	5,000	10,000	50,000	250,000	1,000,000
5	0.53	2.65	5.22	23.52	73.83	99.53
50	5.22	23.52	41.5	93.15	100	100
200	19.31	65.78	88.29	100	100	100
1000	65.8	99.53	100	100	100	100

TABLE 2.2. Expected percent of mRNAs in each abundance class to be seen in an EST database of size S, given the distribution of Table 2.1.

As this example shows, it is difficult to obtain many of the lower abundance mRNAs by sampling in this way. If the goal is simply to find as many different species of mRNAs as possible, then one can sequence EST libraries from many different cell types, since the low-abundance mRNAs in one cell type might be expressed in moderate or high abundance in another cell type. However, some genes are always expressed at a low level, regardless of cell type. And often the goal is to know which genes are expressed in the specific cell type in question. In such cases there are procedures (called subtraction and normalization) to remove or lower the levels of the higher abundance classes. However, with these methods, any measure of relative abundance is lost.

2.5 Marginal Distributions

In this and the following sections we consider marginal and conditional distributions, expected values, covariance, and correlation. These concepts are of most interest for dependent random variables, so we use the notation k in these sections for the number of random variables of interest. All formulae derived, however, apply in the independent case also.

The *marginal* distribution of the subset $(Y_1, Y_2, \ldots, Y_i)$ of k discrete random variables $(Y_1, Y_2, \ldots, Y_k)$ is the distribution of $Y_1, Y_2, \ldots, Y_i$ ignoring $Y_{i+1}, Y_{i+2}, \ldots, Y_k$. If the joint distribution of $Y_1, Y_2, \ldots, Y_k$ is given, this marginal distribution is found by summation (in the discrete case) and integration (in the continuous case). Thus in the discrete case

$$\text{Prob}(Y_1 = y_1, \ldots, Y_i = y_i) = \sum_{y_{i+1}, \ldots, y_k} \text{Prob}(Y_1 = y_1, \ldots, Y_k = y_k), \quad (2.30)$$

the summation being over all possible values $y_{i+1}, \ldots, y_k$ such that $y_1, \ldots, y_k$ is in the joint range of $Y_1, \ldots, Y_k$. This formula is a generalization of the intuitively clear result applying when $i = 1$ and $k = 2$, for which the probability that $Y_1 = y_1$ is the sum, over all possible values of y_2, of the joint probability that $Y_1 = y_1$ and $Y_2 = y_2$.

Suppose that $Y_1, Y_2, \ldots, Y_k$ have the multinomial distribution (2.6). In a later application we will want the marginal distribution of $Y_1, Y_2, \ldots, Y_i$, for any $i < k$. Equation (2.30) can be used to give

$$\text{Prob}(Y_1 = y_1, \ldots, Y_i = y_i)$$
$$= \frac{n!}{y_1! \cdots y_i! m!} p_1^{y_1} \cdots p_i^{y_i} (1 - p_1 - \cdots - p_i)^m, \quad (2.31)$$

where $m = n - y_1 - \cdots - y_i$ (Problem 2.9).

The analogue of (2.30) in the continuous case is the following. Define $\boldsymbol{X}_i$ by $\boldsymbol{X}_i = (x_1, \ldots, x_i)$. If $X_1, \ldots, X_k$ have joint density function $f_{\boldsymbol{X}_k}(x_1, \ldots, x_k)$, the marginal density function $f_{\boldsymbol{X}_i}(x_1, \ldots, x_i)$ of $X_1, \ldots, X_i$ is given by

$$f_{\boldsymbol{X}_i}(x_1, \ldots, x_i) = \int \cdots \int_R f_{\boldsymbol{X}_k}(x_1, x_2, \ldots, x_k) dx_{i+1} \cdots dx_k, \quad (2.32)$$

the domain R of integration being over all possible values $x_{i+1}, x_{i+2}, \ldots, x_k$ such that $(x_1, \ldots, x_k)$ is in the joint range of $X_1, \ldots, X_k$. Care must be exercised in finding the correct domain of integration, since this can depend on the values of $X_1, X_2, \ldots, X_i$. The following is an example.

Example. Suppose that the joint density function of X_1 and X_2 is

$$f_{(X_1, X_2)}(x_1, x_2) = \frac{1}{2}, \tag{2.33}$$

the range of X_1 and X_2 being the square region bounded by the lines $X_2 = X_1$, $X_2 = -X_1$, $X_1 + X_2 = 2$ and $X_1 - X_2 = 2$ (see Figure 2.1). Although the expression $\frac{1}{2}$ for the joint density function of X_1 and X_2 is

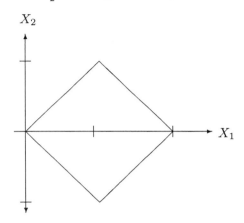

FIGURE 2.1. The region bounded by the lines $X_2 = X_1$, $X_2 = -X_1$, $X_1 + X_2 = 2$ and $X_1 - X_2 = 2$.

the product of a function of X_1 and a function of X_2, nevertheless X_1 and X_2 are not independent. This is shown as follows.

The marginal density function of X_1 is found by integrating out X_2 in the joint density function (2.33). When this is done, different terminals in the integration are needed for the two ranges $0 \leq X_1 \leq 1$ and $1 \leq X_1 \leq 2$. Specifically, when $0 \leq X_1 \leq 1$, the density function $f_{X_1}(x)$ of X_1 is given by

$$f_{X_1}(x_1) = \int_{-x_1}^{+x_1} \frac{1}{2} dx_2 = x_1, \tag{2.34}$$

and when $1 \leq X_1 \leq 2$, the density function of X_1 is

$$f_{X_1}(x_1) = \int_{x_1-2}^{2-x_1} \frac{1}{2} dx_2 = 2 - x_1. \tag{2.35}$$

Similarly, it is found that $f_{X_2}(x_2) = x_2$ $(0 \leq x_2 \leq 1)$, $f_{X_2}(x_2) = 2 - x_2$ $(1 \leq x_2 \leq 2)$. The product of these two marginal densities is not the same as the joint density (2.33), so that from the requirement (2.13), the two random variables are not independent.

For independent random variables, problems of the appropriate domain of integration do not arise: The range of any one random variable is in this case independent of the values of all other variables. In practice the procedure for independent random variables of going from the joint density function to the marginal density functions is often reversed: Given that the random variables are independent, joint density functions are obtained as the product of the individual density functions.

Whether the random variables of interest are independent or not, the marginal density function of any set of random variables is identical to their joint density function if all other random variables are ignored. Thus for continuous random variables we use the expressions "marginal density function of $X_1, \ldots, X_i$" and "joint density function of $X_1, \ldots, X_i$" interchangeably, with a similar identity for discrete random variables.

2.6   Conditional Distributions

The conditional probability formula for events (1.92) has a direct analogue for conditional distributions of random variables. Indeed, the conditional probability formula given in (2.36) below is another version of (1.92).

If Y_1 and Y_2 are two discrete random variables, then the conditional probability that $Y_1 = y_1$, given that $Y_2 = y_2$, is

$$\text{Prob}(Y_1 = y_1 \mid Y_2 = y_2) = \frac{\text{Prob}(Y_1 = y_1, Y_2 = y_2)}{\text{Prob}(Y_2 = y_2)}, \quad (2.36)$$

where it is assumed that $\text{Prob}(Y_2 = y_2) > 0$. Here the denominator on the right-hand side of the equation is the marginal probability that $Y_2 = y_2$, or, more simply, just the probability that $Y_2 = y_2$.

More generally, for discrete random variables $Y_1, Y_2, \ldots, Y_k$,

$$\text{Prob}(Y_{i+1} = y_{i+1}, \ldots, Y_k = y_k \mid Y_1 = y_1, \ldots, Y_i = y_i)$$
$$= \frac{\text{Prob}(Y_1 = y_1, \ldots, Y_k = y_k)}{\text{Prob}(Y_1 = y_1, \ldots, Y_i = y_i)}, \quad (2.37)$$

where the denominator of the right-hand side is the marginal distribution of $Y_1, \ldots, Y_i$, which is again assumed to be positive. One can, of course, condition on any subset of the Y_j's, not only $Y_1, \ldots, Y_i$.

When the random variables $Y_1, Y_2, \ldots, Y_k$ are independent, equation (2.2) shows that

$$\text{Prob}(Y_{i+1} = y_{i+1}, \ldots, Y_k = y_k \mid Y_1 = y_1, \ldots, Y_i = y_i)$$
$$= \frac{\prod_{j=1}^{k} \text{Prob}(Y_j = y_j)}{\prod_{j=1}^{i} \text{Prob}(Y_j = y_j)} = \text{Prob}(Y_{i+1} = y_{i+1}, \ldots, Y_k = y_k). \quad (2.38)$$

More generally, the conditional distribution of any subset of independent random variables, given the remaining random variables, is identical to the marginal distribution of that subset. Thus knowing the values of the remaining random variables does not change the probabilities of the values of the random variables in the subset of interest. This is perhaps the most intuitive interpretation of the word "independent."

The more interesting case of equation (2.37) arises when the random variables are dependent. As an example, suppose that the random variables $Y_1, Y_2, \ldots, Y_k$ have the multinomial distribution (2.6) and we wish to find the conditional probability of $(Y_{i+1}, Y_{i+2}, \ldots, Y_k)$, given $(Y_1 = y_1, Y_2 = y_2, \ldots, Y_i = y_i)$. Application of the probability formula (2.6), the marginal probability formula (2.31), and the conditional probability formula (2.37) shows that the probability $\text{Prob}(Y_{i+1} = y_{i+1}, \ldots, Y_k = y_k \mid Y_1 = y_1, \ldots, Y_i = y_i)$ is

$$\frac{m!}{\prod_{j=i+1}^{k}(y_j!)} \prod_{j=i+1}^{k} q_j^{y_j}, \tag{2.39}$$

where $m = n - y_1 - y_2 - \cdots - y_i = y_{i+1} + y_{i+2} + \cdots + y_k$ and $q_j = p_j/(p_{i+1} + p_{i+2} + \cdots + p_k)$. This is itself a multinomial distribution, having suitably amended parameter values.

A particularly important case arises for the conditional distribution of $Y_1, Y_2, \ldots, Y_n$ given their sum $S_n = \sum_j Y_j$. Even if the random variables $Y_1, Y_2, \ldots, Y_n$ are independent, the random variables $Y_1, Y_2, \ldots, Y_n$ and S_n are not, since S_n is determined by $Y_1, Y_2, \ldots, Y_n$. Thus finding the conditional distribution of $Y_1, Y_2, \ldots, Y_n$ given S_n is not necessarily straightforward.

The required conditional distribution is found as follows. The event that $Y_1 = y_1, \ldots, Y_n = y_n$ and that their sum S_n is equal to s, where $s = y_1 + \cdots + y_n$, is the same as the event that $Y_1 = y_1, \ldots, Y_n = y_n$. Thus

$$\text{Prob}(Y_1 = y_1, \ldots, Y_n = y_n \mid S_n = y_1 + \cdots + y_n = s)$$
$$= \frac{\text{Prob}(Y_1 = y_1, \ldots, Y_n = y_n)}{\text{Prob}(S_n = s)}. \tag{2.40}$$

As an example of the use of this formula, suppose that $Y_1, Y_2, \ldots, Y_n$ are independent Poisson random variables, having respective parameters $\lambda_1, \lambda_2, \ldots, \lambda_n$. Their sum has a Poisson distribution with parameter $\sum_i \lambda_i$ (Problem 2.3). In this case the right-hand side in equation (2.40) becomes

$$\frac{e^{-\sum \lambda_i} \prod \lambda_i^{y_i} / \prod(y_i!)}{e^{-\sum \lambda_i} (\sum \lambda_i)^{\sum y_i} / (\sum y_i)!}, \tag{2.41}$$

all sums and products being over $i = 1, 2, \ldots, n$. This reduces to

$$\frac{(\sum y_i)!}{\prod(y_i!)} \prod_i \left(\frac{\lambda_i}{\sum_j \lambda_j}\right)^{y_i}, \tag{2.42}$$

and with appropriate changes in notation, this becomes the multinomial probability distribution (2.6). In the particular case where $\lambda_1 = \lambda_2 = \cdots = \lambda_n = \lambda$, this conditional probability is independent of λ.

The formula corresponding to (2.37) in the continuous case is that the conditional density function of $X_1, \ldots, X_i$, given $X_{i+1}, \ldots, X_k$, is

$$\frac{f_{\boldsymbol{X}_i}(x_1, \ldots, x_k)}{f_{\boldsymbol{X}_k}(x_{i+1}, \ldots, x_k)}, \qquad (2.43)$$

where the denominator is assumed to be positive. As in the discrete case, this conditional density function is identical to the marginal density function of $X_1, \ldots, X_i$ for independent random variables.

The formula corresponding to (2.40) in the continuous case is that the conditional density function of $\boldsymbol{X} = X_1, \ldots, X_n$, given that their sum $S_n = X_1 + X_2 + \cdots + X_n$ takes the value s, is

$$f_{\boldsymbol{X} \mid S_n}(x_1, \ldots, x_n \mid s) = \frac{f_{\boldsymbol{X}}(x_1, \ldots, x_n)}{f_{S_n}(s)}, \qquad (2.44)$$

where the numerator on the right-hand side is the joint density function of $X_1, \ldots, X_n$ and the denominator is the density function of S_n.

As an example, if $X_1, \ldots, X_n$ are iid random variables having the exponential distribution (1.59), their joint density function (given in (2.15)) is

$$f_{\boldsymbol{X}}(x_1, x_2, \ldots, x_n) = \lambda^n e^{-\lambda \sum x_i}.$$

The sum S_n has a gamma distribution (see page 62), so its distribution is found from (1.68) by replacing k by n. That is, since n is an integer,

$$f_{S_n}(s) = \frac{\lambda^n s^{n-1} e^{-\lambda s}}{(n-1)!}.$$

The conditional density function of $X_1, \ldots, X_n$, given that $S_n = s$, is, from (2.44),

$$f_{\boldsymbol{X} \mid S_n}(x_1, \ldots, x_n \mid s) = \frac{(n-1)!}{s^{n-1}}, \qquad (2.45)$$

where the range of $X_1, \ldots, X_n$ is the set of $(x_1, x_2, \ldots, x_n)$ such that $x_i > 0$, $\sum x_i = s$. Note that this distribution is constant (as a function of $x_1, x_2, \ldots, x_n$), and further, is independent of λ.

Another important conditional density function arises when the range of a random variable is restricted in some way. Suppose, for example, that we wish to find the conditional density function of a continuous random variable X having density function $f_X(x)$, given that the value of X is in some interval (a, b) contained within the range (L, H) of X. Let $(x, x + h)$ be a small interval contained in (a, b). Then since the event $(x < X <$

$x+h)$ and $(a < X < b)$ is identical to the event $(x < X < x+h)$, equation
(1.92) gives

$$\text{Prob}(x < X < x + h \,|\, a < X < b) = \frac{\text{Prob}(x < X < x + h)}{\text{Prob}(a < X < b)}.$$

Dividing both sides by h and letting $h \to 0$, we find (by (1.40)) that the
conditional density function of X, given that $(a < X < b)$, is

$$\frac{f_X(x)}{\int_a^b f_X(u)\,du}. \tag{2.46}$$

As an example, the conditional density function of a continuous random
variable X having the exponential distribution (1.59), given that $0 < X <$
1, is

$$\frac{\lambda e^{-\lambda x}}{1 - e^{-\lambda}}, \quad 0 \le x < 1. \tag{2.47}$$

This is the same as the density function of the "fractional part" D of
X, given in (1.65). This is not a coincidence, and the result follows from,
and can be found directly by using, the "memoryless" property of the
exponential distribution.

As a more general example, the conditional distribution of an exponential
random variable X, given that $0 < X < L$, is

$$\frac{\lambda e^{-\lambda x}}{1 - e^{-\lambda L}}, \quad 0 \le x < L. \tag{2.48}$$

The mean of this distribution is

$$\int_0^L x \frac{\lambda e^{-\lambda x}}{1 - e^{-\lambda L}}\,dx = \frac{1}{\lambda} - \frac{L}{e^{\lambda L} - 1}. \tag{2.49}$$

We use this expression in Section 5.1 when considering shotgun sequencing.

A generalization of (2.46) to the case of two random variables X_1 and
X_2 is that the conditional joint density function of X_1 and X_2, given that
$(X_1, X_2) \in R$, for some region R of the joint range of X_1 and X_2, is

$$\frac{f_{X_1,X_2}(x_1, x_2)}{\int_R f_{X_1,X_2}(u, v)\,du\,dv}. \tag{2.50}$$

2.7 Expected Values

Suppose $\boldsymbol{X} = (X_1, X_2, \ldots, X_k)$ is a vector of continuous random variables,
dependent or independent, with joint range R and joint density function

$$f_{\boldsymbol{X}}(x_1, x_2, \ldots, x_k).$$

Let $g(X_1, X_2, \ldots, X_k)$ be some function of $X_1, X_2, \ldots, X_k$. Then the expected (or mean) value of $g(X_1, X_2, \ldots, X_k)$ is defined as

$$E(g(X_1, X_2, \ldots, X_k))$$
$$= \int \cdots \int_R g(x_1, x_2, \ldots, x_k) f_{\boldsymbol{X}}(x_1, x_2, \ldots, x_k) dx_1 dx_2 \cdots dx_k. \quad (2.51)$$

If $g(X_1, X_2, \ldots, X_k)$ is a function of $\boldsymbol{X}_i = (X_1, X_2, \ldots, X_i)$ only, written for convenience as $h(X_1, \ldots, X_i)$, its expected value can be found either from equation (2.51) or from the equation

$$E(g(X_1, X_2, \ldots, X_i))$$
$$= \int \cdots \int_{R'} h(x_1, x_2, \ldots, x_i) f_{\boldsymbol{X}_i}(x_1, x_2, \ldots, x_i) dx_1 dx_2 \cdots dx_i, \quad (2.52)$$

where $f_{\boldsymbol{X}_i}(x_1, x_2, \ldots, x_i)$ is the marginal density function of $X_1, X_2, \ldots,$ X_i and R' is the joint range of $(X_1, X_2, \ldots, X_i)$. On some occasions one of these equations is easier to use, and on other occasions the other is easier.

The parallel result for discrete random variables is that if $\boldsymbol{Y} = (Y_1, Y_2, \ldots, Y_k)$ is a vector of discrete random variables, dependent or independent, with joint range R and joint probability distribution $P_{\boldsymbol{Y}}(y_1, y_2, \ldots, y_k)$, then

$$E(g(Y_1, Y_2, \ldots, Y_k)) = \sum_R g(y_1, y_2, \ldots, y_k) P_{\boldsymbol{Y}}(y_1, y_2, \ldots, y_k). \quad (2.53)$$

If $g(Y_1, Y_2, \ldots, Y_k)$ is a function of $\boldsymbol{Y}_i = (Y_1, Y_2, \ldots, Y_i)$ only, say $h(Y_1, \ldots, Y_i)$, its expected value can be found either from equation (2.53) or from the equation

$$E(g(Y_1, Y_2, \ldots, Y_i)) = \sum_{R'} h(y_1, y_2, \ldots, y_i) P_{\boldsymbol{Y}_i}(y_1, y_2, \ldots, y_i), \quad (2.54)$$

where $P_{\boldsymbol{Y}_i}(y_1, y_2, \ldots, y_i)$ is the joint probability distribution of $Y_1, Y_2, \ldots,$ Y_i and R' is the joint range of $(Y_1, Y_2, \ldots, Y_i)$.

Independence of two or more random variables implies several simplifying properties concerning them. One useful property concerns expected values: If X_1 and X_2 are independent random variables, discrete or continuous, then

$$E(X_1 X_2) = E(X_1) E(X_2). \quad (2.55)$$

The proof of this claim follows from (2.13) and (2.51) in the continuous case and (2.2) and (2.54) in the discrete case. The converse statement is not true: In some cases equation (2.55) holds for dependent random variables.

In Sections 9.3.3 and 11.2.3 we will need the idea of a *conditional expectation*. For discrete random variables Y_1 and Y_2, the conditional expectation of Y_1 given Y_2 is the mean of the conditional distribution of Y_1 given Y_2. If the joint distribution of Y_1 and Y_2 is $P_Y(y_1, y_2)$ and the marginal distribution of Y_2 is $P_{Y_2}(y_2)$, this mean is denoted by $E(Y_1 \mid Y_2 = y_2)$, and is calculated (see equation (2.36)) from

$$E(Y_1 \mid Y_2 = y_2) = \frac{\sum_{y_1} y_1 P_Y(y_1, y_2)}{P_{Y_2}(y_2)}. \qquad (2.56)$$

The summation is taken over all possible values that Y_1 can take, given the value y_2 of Y_2; this might differ from one value of y_2 to another. That is, the conditional expectation is in general a function of y_2. A parallel definition holds for continuous random variables.

2.8   Covariance and Correlation

Two important concepts concerning the relationship between two dependent random variables is their covariance and their correlation. We introduce these concepts first for continuous random variables. Let X_1 and X_2 be continuous random variables with means μ_1 and μ_2 and variances σ_1^2 and σ_2^2, respectively. The *covariance* σ_{X_1, X_2} of X_1 and X_2 is defined by

$$\sigma_{X_1 X_2} = \iint_R (x_1 - \mu_1)(x_2 - \mu_2) f_X(x_1, x_2) dx_1 dx_2, \qquad (2.57)$$

the integration being taken over the joint range R of X_1 and X_2.

The covariance between two discrete random variables Y_1 and Y_2 is defined as

$$\sigma_{Y_1 Y_2} = \sum_{y_1, y_2} (y_1 - \mu_1)(y_2 - \mu_2) P_Y(y_1, y_2), \qquad (2.58)$$

where μ_i is the mean of Y_i $(i = 1, 2)$ and the summation is taken over all (y_1, y_2) in the joint range R of Y_1 and Y_2. The *correlation* $\rho_{1,2}$ between Y_1 and Y_2 is then defined by

$$\rho_{1,2} = \frac{\sigma_{Y_1 Y_2}}{\sigma_1 \sigma_2}. \qquad (2.59)$$

A parallel definition holds for the correlation between continuous random variables X_1 and X_2. In both the discrete and the continuous cases a correlation is a measure of the degree of linear association between the random variables involved. The numerical value of any correlation is always in the interval $[-1, +1]$ (see Problem 2.11).

The covariance and correlation between independent random variables are both zero, although the converse is not necessarily true (see Problem 2.6).

An important example in the discrete case is that of the covariance between Y_i and Y_j $(i \neq j)$ in the multinomial distribution (2.6). Application of (2.58) shows that

$$\sigma_{Y_i Y_j} = -n p_i p_j. \tag{2.60}$$

The variances and covariances of k random variables $X_1, X_2, \ldots, X_k$ can be gathered into a symmetric *variance–covariance matrix* Σ, whose ith diagonal element is the variance of X_i and whose (i, j) element $(= (j, i)$ element) is the covariance between X_i and X_j. The continuous random variables $X_1, X_2, \ldots, X_k$ have the *multivariate normal distribution* with mean vector $\boldsymbol{\mu} = (\mu_1, \ldots, \mu_k)$ and variance–covariance matrix Σ if their joint density function is

$$f_{\boldsymbol{X}}(\boldsymbol{x}) = \frac{1}{(2\pi)^{k/2} |\Sigma|^{1/2}} e^{-\frac{1}{2}(\boldsymbol{x}-\boldsymbol{\mu})' \Sigma^{-1} (\boldsymbol{x}-\boldsymbol{\mu})}, \quad -\infty < x_i < +\infty \text{ for all } i. \tag{2.61}$$

Here $|\Sigma|$ is the determinant of the matrix Σ and μ_i is the mean of X_i. Note that the exponent is the product of a row vector, a matrix, and a column vector, and is therefore a scalar.

One important use of covariances concerns the variance of a linear combination of random variables. Suppose $X_1, X_2, \ldots, X_k$ are random variables, discrete or continuous, with means $\mu_1, \mu_2, \ldots, \mu_k$ and variances $\sigma_1^2, \sigma_2^2, \ldots, \sigma_k^2$, respectively. Let σ_{ij} be the covariance between X_i and X_j. If $c_1, c_2, \ldots, c_k$ are given constants, then

$$\text{Var}(c_1 X_1 + \cdots + c_k X_k) = \sum_{i=1}^{k} c_i^2 \sigma_i^2 + 2 \sum_{i<j} c_i c_j \sigma_{ij}. \tag{2.62}$$

This result generalizes equation (2.22) in two ways, since it allows both general coefficients and dependence between the X_i's. Equation (2.62) reduces to (2.22) when $c_i = 1$ $(i = 1, 2, \ldots, k)$ and the random variables are independent. We outline the proof of it here. This outline is given for the continuous case; a parallel proof applies in the discrete case.

First, by equation (2.51), the mean of $c_1 X_1 + \cdots + c_k X_k$ is given by

$$E(c_1 X_1 + \cdots + c_k X_k)$$

$$= \int \cdots \int_R (c_1 x_1 + \cdots + c_k x_k) f_{\boldsymbol{X}}(x_1, x_2, \ldots, x_k) dx_1 dx_2 \cdots dx_k, \tag{2.63}$$

which is equal to

$$\sum_{i=1}^{k} c_i \int_R x_i f_{\boldsymbol{X}}(x_1, \ldots, x_k) dx_1 \cdots dx_k. \tag{2.64}$$

From equation (2.51),

$$\int_R x_i f_{\boldsymbol{X}}(x_1, \ldots, x_k) dx_1 \cdots dx_k = \mu_i, \tag{2.65}$$

so that

$$E(c_1 X_1 + \cdots + c_k X_k) = c_1 \mu_1 + \cdots + c_k \mu_k. \qquad (2.66)$$

This mean is independent of the correlation between any two of the random variables. The variance of $(c_1 X_1 + \cdots + c_k X_k)$ is thus given by

$$\text{Var}(c_1 X_1 + \cdots + c_k X_k)$$
$$= \int_D \cdots \int (c_1(x_1 - \mu_1) + \cdots + c_k(x_k - \mu_k))^2 \, f_X(x_1, \ldots, x_k) dx_1 \cdots dx_k.$$

$$(2.67)$$

Equation (2.62) follows by expanding the squared term on the right-hand side and then using the definition of variance and covariance, found from (2.51).

As an important special case, if the random variables $X_1, X_2, \ldots, X_k$ are independent, then

$$\text{Var}(c_1 X_1 + \cdots + c_k X_k) = \sum_{i=1}^{k} c_i^2 \sigma_i^2. \qquad (2.68)$$

Essentially identical calculations, replacing integrations by summations, and with the same conclusions, apply for discrete random variables.

2.9   Asymptotic Distributions

Let $X_1, X_2, X_3, \ldots$ be an infinite sequence of random variables, discrete or continuous, with respective distribution functions $F_1(x), F_2(x), F_3(x), \ldots$. There are various concepts of convergence for such a sequence. We consider here the concept of *convergence in distribution*. Suppose that X is a random variable whose distribution function is $F_X(x)$. Then if

$$\lim_{n \to \infty} F_n(x) = F_X(x)$$

for all points of continuity of $F_X(x)$, then we say that the sequence $\{X_n\}$ converges in distribution to X. All of the convergence results in this book, and in particular the central limit theorem described in Section 2.10.1, refer to convergence in distribution. If (as with the central limit theorem) the random variable X has a normal distribution, then we say that the sequence is *asymptotically normal*.

2.10   Derived Random Variables (i): Sums, Averages, and Minima

We have discussed above the joint distribution of n independent discrete random variables and the joint density function of n independent continuous random variables. These random variables can be used to define various *derived* random variables, for example their sum, their average, their minimum, and their maximum. The joint distribution of the n random variables defines, implicitly, the distribution of any such derived random variable. Many properties of derived random variables can be found almost immediately; for example, equations (2.21) and (2.22) show that the mean and the variance of the sum of n independent random variables can be written down immediately from the means and variances of the individual random variables.

Many statistical operations are carried out using derived random variables. To carry out these operations it is necessary to know the probability distribution of these random variables. In the discrete case, the calculations are usually straightforward in principle, although in practice they sometimes involve complicated steps involving combinatorial terms. The calculations in the continuous case involve transformation techniques and are less straightforward. Examples of transformation methods will be given in Section 2.15.

2.10.1   Sums and Averages

Suppose that $X_1, X_2, \ldots, X_n$ are iid random variables, discrete or continuous, each having a probability distribution with mean μ and variance σ^2. Perhaps the two most important random variables derived from $X_1, X_2, \ldots, X_n$ are their sum S_n, defined by

$$S_n = X_1 + X_2 + \cdots + X_n, \tag{2.69}$$

and their average $\bar{X}$, defined by

$$\bar{X} = \frac{X_1 + X_2 + \cdots + X_n}{n}. \tag{2.70}$$

It is important to emphasize that both of these quantities are *random variables*. In particular, the average $\bar{X}$ is a random variable, and must be distinguished carefully, as noted in Section 1.4, from a mean, which is a parameter.

The mean and variance of S_n are given in (2.19) and (2.20) as $n\mu$ and $n\sigma^2$ respectively. The corresponding formulae for the average are

$$\text{mean of } \bar{X} = \mu, \quad \text{variance of } \bar{X} = \frac{\sigma^2}{n}. \tag{2.71}$$

These formulae can be generalized in two ways. First, the same principle as is indicated in equations (2.71) applies to any well-behaved function of X. Thus if $g(X)$ is a function of X with finite mean and variance μ_g and σ_g^2 respectively, then the mean and variance of the average $\bar{g}$ of $g(X_1)$,

$g(X_2), \ldots, g(X_n)$ are

$$\text{mean of } \bar{g} = \mu_g, \quad \text{variance of } \bar{g} = \frac{\sigma_g^2}{n}. \tag{2.72}$$

The second generalization is the following. Suppose that $\boldsymbol{X}_1, \boldsymbol{X}_2, \ldots, \boldsymbol{X}_n$ are n iid vectors each of k random variables. The random variables within each vector can be discrete or continuous, independent or dependent. Write $\boldsymbol{X}_i = (X_{i1}, X_{i2}, \ldots, X_{ik})$ and suppose that $g(\boldsymbol{X}_i)$ is some function of X_{i1}, $X_{i2}, \ldots, X_{ik}$ with mean and variance μ_g and σ_g^2, respectively. From these vectors we can define the average $\bar{g}$ by

$$\bar{g} = \frac{g(\boldsymbol{X}_1) + g(\boldsymbol{X}_2) + \cdots + g(\boldsymbol{X}_n)}{n}. \tag{2.73}$$

Then the mean and variance of $\bar{g}$ are as given in (2.72).

It is important to emphasize that equations (2.71) and (2.72) are derived under the iid assumption. The statements about the mean remain the same in the case of dependent random variables. Generalizations of the variance statements to the dependent case can be found from (2.62).

Sums and averages have many important properties, of which we mention a few here.

The central limit theorem. Assume that $X_1, X_2, \ldots, X_n$ are iid, each with finite mean μ and finite variance σ^2. The simplest and most important version of the central limit theorem states that as $n \to \infty$, the standardized random variable $(S_n - n\mu)/(\sqrt{n}\sigma)$, which is identical to the standardized random variable $(\bar{X} - \mu)\sqrt{n}/\sigma$, converges in distribution to a random variable having the standardized normal distribution.

This result holds regardless of the (common) distribution of $X_1, X_2, \ldots$. The breadth of application of the central limit theorem is one reason for the importance of the normal distribution, since many random variables of interest to us are either sums or averages.

In the case where $X_1, X_2, \ldots, X_n$ are $\text{NID}(\mu, \sigma^2)$, both $\bar{X}$ and S_n have a normal distribution not only asymptotically but for all values of n, with respective means and variances given in (2.71), (2.19), and (2.20).

Particularly important cases of S_n and $\bar{X}$ arise when each X_i is a Bernoulli random variable with parameter p (see Section 1.3.1), and thus with mean p and variance $p(1-p)$. Here S_n is the total number of successes, and $\bar{X}$ is the *proportion* of successes, in n trials. The equations in (2.19) and (2.20) verify the mean np and the variance $np(1-p)$ for the binomial distribution, and the central limit theorem justifies the normal approximation to the binomial, discussed in Section 1.10.3. Equation (2.71) shows that

$$\text{mean of } \bar{X} = p, \quad \text{variance of } \bar{X} = \frac{p(1-p)}{n}, \tag{2.74}$$

and the central limit theorem shows that the distribution of $\bar{X}$ is approximately normal, with this mean and variance.

Accumulated support. In equations (1.111) and (1.112) we discussed the support that the observed value of one single random variable Y gave for one probability distribution over another. In practice we will normally use the observed values of many random variables before assessing which of two distributions the random variable of interest comes from. Suppose then that $Y_1, Y_2, \ldots, Y_n$ are iid random variables, and denote their observed values by $y_1, y_2, \ldots, y_n$. Then the total support given by these observations for distribution $\boldsymbol{P}_1$ over $\boldsymbol{P}_0$ is defined as the sum of the supports given by each individual observation; specifically,

$$\text{accumulated support for } \boldsymbol{P}_1 \text{ over } \boldsymbol{P}_0 = \sum_{j=1}^{n} S_{1,0}(y_j). \tag{2.75}$$

This accumulated support is the sum of the observed values of n iid random variables, and several of its properties follow from the properties of sums. In particular, the mean accumulated support given by n iid observations is n times the mean support given by one observation.

The reason for the additive definition of support will become clearer when we develop the concepts of support and accumulated support in the context of likelihood ratio tests (Chapter 3), substitution matrices (Chapter 5), sequential analysis (Chapter 8), and finally BLAST theory (Chapter 9).

Means, averages, and Chebyshev's inequality.

Chebyshev's inequality is often not useful in finding information about the probability behavior of $\bar{X}$, the central limit theorem being more useful for this purpose. For example, if $X_1, \ldots, X_{100}$ each have a distribution with mean μ and variance 25, Chebyshev's inequality states that $\text{Prob}(|\bar{X} - \mu)| \geq 1) \leq 0.25$, while the central limit theorem gives the generally more useful statement that $\text{Prob}(|\bar{X} - \mu)| \geq 1) \approx .0456$.

Random n. It is assumed above that the number n of random variables making up the sum S_n is fixed. In some cases, however, this number is also a random variable, which we denote by N. In such cases the derivation of the distribution of the sum, which we denote by S, is best found through pgfs. The analysis for the case of independent discrete random variables is as follows.

Suppose that $\text{Prob}(N = n) = P_n$ and put $\mathbb{p}(t) = \sum_n P_n t^n$, the pgf of N. We assume that the iid random variables $X_1, X_2, \ldots$ in the sum S each have the probability distribution of a random variable X, whose pgf is $\mathbb{q}(t)$.

Then from (1.96),

$$\text{Prob}(S = y) = \sum_n P_n \cdot \text{Prob}(S = y \,|\, N = n). \qquad (2.76)$$

Equation (2.10) shows the probability that $S = y$ given that $N = n$ is the coefficient of t^y in $[\mathbb{q}(t)]^n$. Then from (2.76),

$$\text{Prob}(S = y) = \text{ coefficient of } t^y \text{ in } \sum_n P_n \cdot (\mathbb{q}(t))^n$$

$$= \text{ coefficient of } t^y \text{ in } \mathbb{p}(\mathbb{q}(t)). \qquad (2.77)$$

Thus the pgf of S is $\mathbb{p}(\mathbb{q}(t))$, and from this we can find in principle (and often easily in practice) the complete probability distribution of S.

The mean and variance of S can be found using (1.36) and (1.37), together with the chain rule of differentiation. The mean of S is

$$E(S) = E(N)E(X), \qquad (2.78)$$

as might be expected, and the variance is found by using (1.37) replacing $\mathbb{p}(t)$ by $\mathbb{p}(\mathbb{q}(t))$ (see Problem 2.12).

As an example, suppose a fair die is rolled once, and if the number Y turns up we toss a fair coin Y times. Let S be the number of heads that then occur. Then S is the sum of a random number of Bernoulli random variables. Equation (2.78) shows the mean of S is $\frac{7}{2} \cdot \frac{1}{2} = \frac{7}{4}$. The pgf of Y is

$$\mathbb{p}(t) = \frac{1}{6}t + \frac{1}{6}t^2 + \cdots + \frac{1}{6}t^6,$$

and the pgf of the number of heads on each toss is

$$\mathbb{q}(t) = \frac{1}{2} + \frac{1}{2}t.$$

Thus the pgf of S is then

$$\mathbb{p}(\mathbb{q}(t)) = \frac{1}{6} \cdot \left(\frac{1}{2} + \frac{1}{2}t\right) + \frac{1}{6} \cdot \left(\frac{1}{2} + \frac{1}{2}t\right)^2 + \cdots + \frac{1}{6} \cdot \left(\frac{1}{2} + \frac{1}{2}t\right)^6$$

$$= \frac{21}{128} + \frac{5}{16}t + \frac{23}{128}t^2 + \frac{1}{6}t^3 + \frac{29}{384}t^4 + \frac{1}{48}t^5 + \frac{1}{384}t^6.$$

As an example of the use of this expression, the probability that $S = 1$ is the coefficient of t, namely $5/16$.

2.10.2   The Minimum of n Random Variables

Let X be a continuous random variable. Suppose that $X_1, X_2, \ldots, X_n$ are iid random variables each with the same distribution as X. We denote their minimum by $X_{\min}$. The minimum of $X_1, X_2, \ldots, X_n$ does not have the original distribution of the individual X_i's, and its density function is found as follows. To say that $X_{\min}$ is greater than or equal to some number x is equivalent to saying that all of the values $X_1, X_2, \ldots, X_n$ are greater than or equal to x. Then

$$\text{Prob}(X_{\min} \geq x) = (\text{Prob}(X \geq x))^n . \tag{2.79}$$

If we denote the density function and the distribution function of X by $f_X(x)$ and $F_X(x)$, respectively, and the distribution function of $X_{\min}$ by $F_{\min}(x)$, then equation (2.79) can be written

$$1 - F_{\min}(x) = (1 - F_X(x))^n . \tag{2.80}$$

The density function $f_{\min}(x)$ of $X_{\min}$ is found by differentiation (see equation (1.44)) to be

$$f_{\min}(x) = n\, f_X(x)\, (1 - F_X(x))^{n-1} . \tag{2.81}$$

As an example of the application of (2.79), suppose that $X_1, X_2, \ldots, X_n$ are independent random variables each having the exponential distribution (1.59). Then (1.103) and (2.79) show that

$$\text{Prob}(X_{\min} \geq x) = e^{-n\lambda x}, \quad x \geq 0, \tag{2.82}$$

so that

$$F_{\min}(x) = 1 - e^{-n\lambda x}, \quad x \geq 0.$$

Differentiation of both sides of this equation shows that the density function of $X_{\min}$ is

$$f_{\min}(x) = n\lambda e^{-n\lambda x}, \quad x \geq 0. \tag{2.83}$$

Thus $X_{\min}$ itself has an exponential distribution, but with parameter $n\lambda$ rather than λ. The result of Problem 1.15 then shows that

$$\text{mean of } X_{\min} = \frac{1}{n\lambda}, \quad \text{variance of } X_{\min} = \frac{1}{(n\lambda)^2}. \tag{2.84}$$

The possibility of modeling the lifetime of some proteins by an exponential distribution was discussed in Section 1.13, page 42. If there are n such molecules in a cell, (2.83) shows that the time until the first such molecule degrades also has an exponential distribution.

The uniform distribution. As a second example of the application of (2.79), and with a change in notation to one more convenient in a later application, suppose that $X_1, X_2, \ldots, X_k$ are independent random variables each having the uniform distribution with range $[0, L]$. Then from (2.81) the density function of $X_{\min}$ is

$$f_{\min}(x) = \frac{k(L - x)^{k-1}}{L^k}, \quad 0 \leq x \leq L. \tag{2.85}$$

The normal distribution. Since the distribution function for the normal distribution cannot be written in terms of elementary functions, a closed form for the distribution of $X_{\min}$ is not available. However, the mean and standard deviation of $X_{\min}$ can be approximated accurately by numerical methods, and these are available in tables for selected values of n.

Subintervals of $(0,1)$. If n points are placed randomly in the interval $(0,1)$, they will define subintervals of respective lengths $U_1, U_2, \ldots, U_{n+1}$. Since they must sum to 1, these lengths are not independent, so the procedure that led to equation (2.81) may not be used to find the density function of their minimum. To find this density function it is necessary instead to use the theory developed in the example of Section 1.12.3, page 38. The case $g = n + 1$ of equation (1.89) shows that for any u in $(0, (n+1)^{-1})$, the probability that the lengths of all $n + 1$ intervals exceed u is

$$(1 - (n+1)u)^n. \tag{2.86}$$

This is also the probability that the minimum length $U_{\min}$ exceeds u, so that if $F_{\min}(u)$ and $f_{\min}(u)$ are respectively the distribution function and the density function of $U_{\min}$, then

$$F_{\min}(u) = 1 - (1 - (n+1)u)^n, \qquad 0 < u < (n+1)^{-1}, \tag{2.87}$$

$$f_{\min}(u) = n(n+1)(1 - (n+1)u)^{n-1}, \quad 0 < u < (n+1)^{-1}. \tag{2.88}$$

These results will be used in the theory of r-scans, discussed in Section 5.5.

2.11 Derived Random Variables (ii): The Maximum of n Random Variables

2.11.1 Distributional Properties: Continuous Random Variables

Several important tests in bioinformatics are carried out by examining the observed value of the *maximum* of several random variables. For this reason we consider this maximum in far greater detail than the corresponding minimum.

Suppose that $X_1, X_2, \ldots, X_n$ are iid random variables and that the maximum of these random variables is denoted by $X_{\max}$. The density function of $X_{\max}$ is, as with $X_{\min}$, different from that of the individual X_i's, and is found by an argument analogous to that used in finding the density function of $X_{\min}$.

To say that $X_{\max}$ is less than or equal to some number x is equivalent to saying that all of the values $X_1, X_2, \ldots, X_n$ are less than or equal to x.

This implies that, if X is a random variable having the same distribution as each X_i, then

$$\text{Prob}(X_{\max} \leq x) = \big(\text{Prob}(X \leq x)\big)^n. \tag{2.89}$$

In terms of distribution functions, if X has density function $f_X(x)$ and distribution function $F_X(x)$, and $F_{\max}(x)$ is the distribution function of $X_{\max}$, then

$$F_{\max}(x) = \big(F_X(x)\big)^n. \tag{2.90}$$

Equivalently,

$$\text{Prob}(X_{\max} \geq x) = 1 - \big(F_X(x)\big)^n. \tag{2.91}$$

The density function $f_{\max}(x)$ of $X_{\max}$ is therefore

$$f_{\max}(x) = n f_X(x) \big(F_X(x)\big)^{n-1}. \tag{2.92}$$

The uniform distribution. As an example of the application of (2.92), consider the maximum of n independent random variables, each having the uniform distribution with range $[0, L]$. The density function of $X_{\max}$ is

$$f_{\max}(x) = \frac{n \, x^{n-1}}{L^n}, \quad 0 \leq x \leq L. \tag{2.93}$$

Elementary integration shows that the mean and variance of this distribution are, respectively,

$$\frac{nL}{(n+1)}, \quad \frac{nL^2}{(n+1)^2(n+2)}. \tag{2.94}$$

The exponential distribution. The density function of the maximum of n independent random variables, each having the exponential distribution (1.59), is, from (1.59), (1.60), and (2.92),

$$f_{\max}(x) = n\lambda e^{-\lambda x}(1 - e^{-\lambda x})^{n-1}, \quad 0 \leq x < +\infty. \tag{2.95}$$

The distribution function $F_{\max}(x)$ of $X_{\max}$ is

$$F_{\max}(x) = (1 - e^{-\lambda x})^n. \tag{2.96}$$

It is not easy, for the exponential distribution, to find the mean and variance of $X_{\max}$ using (1.45), (1.46), and (2.95).[2] However, a simple indirect way of finding these, relying on the memoryless property of the exponential distribution and the mean and the variance of the *minimum* of n exponential random variables, is illustrated by the following example.

---

[2] At least compared to the straightforward calculation for $X_{\min}$. This shows that the theory for maximum and minimum must be handled separately.

As in Section 1.13 we assume that the lifetimes of certain cellular proteins until degradation have an exponential distribution. Assuming this, and starting at time 0, we follow the fate of a cohort of n proteins. The mean time until one or another protein degrades is the mean of the minimum of n exponential random variables, and from (2.84) this mean is $1/(n\lambda)$. By similar arguments the variance of this time is $1/(n\lambda)^2$.

The memoryless property of the exponential distribution implies that the mean value of the further time until the next protein degrades is independent of the time taken until the first one degraded. However, there are now only $n - 1$ proteins involved, so that the mean of the time until the next degradation is, by the same argument as that just used, $1/((n - 1)\lambda)$, and the variance of this time is $1/\big((n - 1)\lambda\big)^2$. Continuation of this argument shows that the time until the final protein degrades has mean

$$\frac{1}{\lambda} + \frac{1}{2\lambda} + \cdots + \frac{1}{n\lambda} \tag{2.97}$$

and variance

$$\frac{1}{\lambda^2} + \frac{1}{(2\lambda)^2} + \cdots + \frac{1}{(n\lambda)^2}. \tag{2.98}$$

These are, respectively, the mean and variance of X_{max}.

It is curious that as $n \to \infty$ the expression for the mean of $X_{\max}$ diverges to $+\infty$, whereas the expression for the variance converges (see equations (B.7)–(B.9)). For large n, (B.7) and (B.9) show that the mean and variance of $X_{\max}$ can be approximated by

$$\text{mean of } X_{\max} \approx (\gamma + \log n)/\lambda, \quad \text{variance of } X_{\max} \approx \pi^2/6\lambda^2, \tag{2.99}$$

where γ is Euler's constant $0.577216\ldots$. Thus the mean of $X_{\max}$ grows (very slowly) at the approximate rate $(\log n)/\lambda$, while the variance becomes essentially constant, remaining within 1% of $\pi^2/6\lambda^2$ for all $n \geq 100$. The fact that the variance of $X_{\max}$ does not diverge to $+\infty$ as $n \to \infty$ is quite important in bioinformatics, and we will return to this point later.

Subintervals of $(0, 1)$. In Section 2.10.2 we discussed properties of the minimum of the $n + 1$ subinterval lengths defined by n points placed randomly in the interval $(0, 1)$. Here we discuss the density function of the maximum of these lengths, noting that the dependence between the lengths precludes us from using equation (2.92).

The probability of the event that at least one subinterval has length exceeding u is given in equation (1.90). This is also the probability that the length $U_{\max}$ of the maximum subinterval length exceeds u, and may therefore by written as $1 - F_{\max}(u)$. The density function $f_{\max}(u)$ of $U_{\max}$ is then found, by differentiation of the expression (1.90), to be

$$f_{\max}(u) = n \sum_{j=1}^{h} (-1)^{j+1} j \binom{n+1}{j} (1 - ju)^{n-1}, \tag{2.100}$$

where h is the largest integer for which $1 - hu > 0$. This density function takes different forms for different values of u. When $\frac{1}{2} < u < 1$ only the first term appears in the density function; when $\frac{1}{3} < u < \frac{1}{2}$ only the first two terms appear; and so on.

2.11.2 Distributional Properties: Discrete Random Variables

We now consider properties of maxima of iid discrete integer-valued random variables. Our approach is based on equation (2.89), which applies also in the discrete case with the change of notation from X to Y. Equation (2.89) implies that if $Y_{\max}$ is the maximum of n discrete iid random variables $Y_1, Y_2, \ldots, Y_n$ with common distribution function $F_Y(y)$, then

$$\text{Prob}(Y_{\max} \leq y) = \left(F_Y(y)\right)^n, \tag{2.101}$$

$$\text{Prob}(Y_{\max} \geq y) = 1 - \left(F_Y(y - 1)\right)^n, \tag{2.102}$$

so that

$$\text{Prob}(Y_{\max} = y) = \left(F_Y(y)\right)^n - \left(F_Y(y - 1)\right)^n. \tag{2.103}$$

It follows from (2.101) that when n is large, very sudden changes in the distribution function of $Y_{\max}$ can occur from one value of y to the next, implying in turn that the distribution of $Y_{\max}$ is tightly concentrated around its mean.

From equation (2.103), the mean $\mu_{\max}$ and the variance and $\sigma^2_{\max}$ of $Y_{\max}$ are given, respectively, by

$$\mu_{\max} = \sum_y y\left((F_Y(y))^n - (F_Y(y - 1))^n\right) \tag{2.104}$$

and

$$\sigma^2_{\max} = \sum_y y^2\left((F_Y(y))^n - (F_Y(y - 1))^n\right) - \mu^2_{\max}. \tag{2.105}$$

Since several statistical procedures in bioinformatics use the maximum of n discrete random variables, an evaluation of the probabilities (2.103), (2.104), and (2.105) is often required. Since n is often large and $F_Y(y - 1)$ and $F_Y(y)$ are raised to the nth power, small errors in estimating $F_Y(y-1)$ and $F_Y(y)$ will compound into large errors in (2.103), (2.104), and (2.105), so very precise estimates of $F_Y(y-1)$ and $F_Y(y)$ are needed. We will return to this issue in Section 2.12.3.

The Geometric Distribution

The maximum of a number of iid geometric-like random variables is used in BLAST theory. (Geometric-like random variables are defined on page 11.) As an introduction to the relevant theory, we now discuss properties of the maximum of n independent geometrically distributed random variables in some detail.

Suppose that $Y_1, Y_2, \ldots, Y_n$ are independent random variables each having the geometric distribution (1.10). Then equations (1.11) and (2.101) show that when y is an integer,

$$\text{Prob}(Y_{\max} \leq y) = (1 - p^{y+1})^n, \tag{2.106}$$

and from this it follows immediately that

$$\text{Prob}(Y_{\max} \geq y) = 1 - (1 - p^y)^n \tag{2.107}$$

and

$$\text{Prob}(Y_{\max} = y) = (1 - p^{y+1})^n - (1 - p^y)^n. \tag{2.108}$$

Thus the mean $\mu_{\max}$ of Y_{max} is given by

$$\mu_{\max} = \sum_{y=0}^{\infty} y \left((1 - p^{y+1})^n - (1 - p^y)^n \right), \tag{2.109}$$

and the variance $\sigma_{\max}^2$ of $Y_{\max}$ is given by

$$\sigma_{\max}^2 = \sum_{y=0}^{\infty} y^2 \left((1 - p^{y+1})^n - (1 - p^y)^n \right) - \mu_{\max}^2. \tag{2.110}$$

The relation between the geometric distribution and the exponential distribution often makes it convenient to write the geometric distribution in the reparametrized form (1.62), that is, with p replaced by $e^{-\lambda}$. In this notation,

$$\text{Prob}(Y_{\max} \leq y) = (1 - e^{-\lambda(y+1)})^n, \tag{2.111}$$

$$\text{Prob}(Y_{\max} \geq y) = 1 - (1 - e^{-\lambda y})^n, \tag{2.112}$$

$$\text{Prob}(Y_{\max} = y) = (1 - e^{-\lambda(y+1)})^n - (1 - e^{-\lambda y})^n, \tag{2.113}$$

$$\mu_{\max} = \sum_{y=0}^{\infty} y \left((1 - e^{-\lambda(y+1)})^n - (1 - e^{-\lambda y})^n \right), \tag{2.114}$$

$$\sigma_{\max}^2 = \sum_{y=0}^{\infty} y^2 \left((1 - e^{-\lambda(y+1)})^n - (1 - e^{-\lambda y})^n \right) - \mu_{\max}^2. \tag{2.115}$$

We use the two sets of notation interchangeably, since for some purposes one notation is preferred and for other purposes the other is preferred.

Until recently, the calculation of $\mu_{\max}$ and $\sigma_{\max}^2$ as given in equations (2.109) and (2.110), and similar calculations for other discrete random variables, often required significant computing effort for large values of

n, and an approach using continuous distribution approximations to discrete distributions was used to find approximate properties of quantities like $Y_{\max}$. Current computing power implies that direct computation, using only properties of the discrete random variable itself, is now often possible. Nevertheless, continuous approximations are still sometimes used and are valuable for theoretical considerations, and we begin our discussion of them by considering the approximation of the distribution of the maximum of geometric random variables by using the maximum of exponential random variables.

This approximation is based on the fact (discussed on page 30) that the integer part of a random variable having the exponential distribution (1.59) has the geometric distribution (1.62). The corresponding result for maxima is that the integer part of the maximum of n exponential random variables with density function (1.59) has the distribution of the maximum of n random variables having the geometric distribution (1.62). This is seen as follows.

If $X_{\max}$ is the largest of n random variables having the exponential distribution (1.59), and if $Y_{\max} = \lfloor X_{\max} \rfloor$ is the integer part of $X_{\max}$, then

$$\text{Prob}(Y_{\max} = y) = \text{Prob}(y \leq X_{\max} < y + 1). \tag{2.116}$$

From equation (2.96) this is

$$(1 - e^{-\lambda(y+1)})^n - (1 - e^{-\lambda y})^n, \tag{2.117}$$

which is identical to the required expression (2.113).

When $n > 1$ the density function of D, the fractional part $X_{\max} - \lfloor X_{\max} \rfloor$ of $X_{\max}$, is not identical to (1.65) and does not have mean and variance given by (1.66). Instead, even for comparatively small n, the density function of D can be shown to be very close to the uniform distribution (1.53), and this approximation becomes increasingly accurate as n increases. Thus to a close approximation D has a mean of $\frac{1}{2}$ and variance $\frac{1}{12}$ when n is large. Various properties of $Y_{\max}$ can then be obtained by writing $Y_{\max} = \lfloor X_{\max} \rfloor$ as $X_{\max} - D$. For example, equations (2.66) and (2.97) imply that to a close approximation,

$$E(Y_{\max}) = E(X_{\max}) - E(D) \approx \frac{1}{\lambda} + \frac{1}{2\lambda} + \cdots + \frac{1}{n\lambda} - \frac{1}{2}. \tag{2.118}$$

Thus from equation (B.7),

$$E(Y_{\max}) \approx \frac{\gamma + \log n}{\lambda} - \frac{1}{2}. \tag{2.119}$$

Further, it can be shown that the covariance between D and $X_{\max}$ approaches zero as $n \to +\infty$. Thus equations (2.68) and (2.98), together with the variance given immediately below (1.53), imply that to a close approximation,

$$\text{Var}(Y_{\max}) \approx \text{Var}(X_{\max}) + \text{Var}(D)$$
$$\approx \frac{1}{\lambda^2} + \frac{1}{(2\lambda)^2} + \cdots + \frac{1}{(n\lambda)^2} + \frac{1}{12}, \qquad (2.120)$$

so that for large n, from equation (B.9),

$$\text{Var}(Y_{\max}) \approx \frac{\pi^2}{6\lambda^2} + \frac{1}{12}. \qquad (2.121)$$

This expression shows that the variance of $Y_{\max}$ shares the property of the variance of the maximum $X_{\max}$ of n exponential random variables in that it does not diverge to $+\infty$ as $n \to \infty$.

The accuracy of these approximations can be assessed from the values given in Table 2.3, which displays the respective approximations (2.118) and (2.119) for the mean and the respective approximations (2.120) and (2.121) for the variance of $Y_{\max}$.

These values show the accuracy of these various approximations increasing as n increases. In Section 6.3 we use the approximation (2.119) for the mean and (2.121) for the variance when $n = 75{,}000$, and for this value of n these approximations are extremely accurate.

2.11.3   An Asymptotic Formula for the Distribution of $X_{\max}$

We begin by considering a continuous random variable W whose density function is

$$f_W(w) = e^{-w-e^{-w}}, \quad -\infty < w < +\infty, \qquad (2.122)$$

and whose distribution function, found by integration, is

$$F_W(w) = e^{-e^{-w}}, \quad -\infty < w < +\infty. \qquad (2.123)$$

This random variable has mean γ (Euler's constant) and variance $\pi^2/6$ (see Problem 2.15).

Suppose that $X_{\max}$ is the maximum of n iid continuous random variables and has mean $\mu_{\max}$ and standard deviation $\sigma_{\max}$. Then the derived random variable W', defined by

$$W' = \frac{\pi(X_{\max} - \mu_{\max})}{\sigma_{\max}\sqrt{6}} + \gamma, \qquad (2.124)$$

has the same mean and variance as the distribution (2.122). The identity between the two goes further than this: For continuous random variables having finite moments of all orders, and whose domain is of the form

$n = 5$

	Mean			Variance		
	exact	approximations		exact	approximations	
p	(2.109)	(2.118)	(2.119)	(2.110)	(2.120)	(2.121)
1/4	1.140	1.147	1.077	0.867	0.845	0.939
1/3	1.576	1.578	1.490	1.307	1.296	1.446
e^{-1}	1.782	1.783	1.687	1.554	1.547	1.728
1/2	2.794	2.794	2.655	3.133	3.130	3.507

$n = 10$

	Mean			Variance		
	exact	approximations		exact	approximations	
p	(2.109)	(2.118)	(2.119)	(2.110)	(2.120)	(2.121)
1/4	1.616	1.612	1.577	0.881	0.890	0.939
1/3	2.167	2.166	2.121	1.363	1.367	1.446
e^{-1}	2.429	2.429	2.380	1.631	1.633	1.728
1/2	3.726	3.726	3.655	3.309	3.309	3.507

$n = 20$

	Mean			Variance		
	exact	approximations		exact	approximations	
p	(2.109)	(2.118)	(2.119)	(2.110)	(2.120)	(2.121)
1/4	2.093	2.095	2.077	0.919	0.914	0.939
1/3	2.775	2.775	2.752	1.408	1.406	1.446
e^{-1}	3.098	3.098	3.073	1.680	1.680	1.728
1/2	4.690	4.690	4.655	3.406	3.406	3.507

TABLE 2.3. Exact values and two approximations for the mean and variance of the maximum (Y_{max}) of $n = 5$, 10, and 20 geometric random variables, for selected values of $p = e^{-\lambda}$. Relevant equation numbers are noted.

$(A, +\infty)$ for some finite value A (the domain usually of interest to us), the asymptotic ($n \to \infty$) distribution of W' is the same as (2.122).

The case of interest to us is when, for large n, $\mu_{max} \approx a \log n$ and $\sigma_{max} \approx b$, where a and b are constants. In this case, when X_{max} and x are both close to μ_{max},

$$\text{Prob}(X_{max} \leq x) \approx e^{-e^{-\left(\pi(x-\mu_{max})/(\sigma_{max}\sqrt{6})+\gamma\right)}}, \tag{2.125}$$

or equivalently

$$\text{Prob}(X_{max} \geq x) \approx 1 - e^{-e^{-\left(\pi(x-\mu_{max})/(\sigma_{max}\sqrt{6})+\gamma\right)}}. \tag{2.126}$$

This result can be checked for the case of the maximum X_{max} of n iid exponential random variables. For large n, the mean μ_{max} and the standard deviation σ_{max} of X_{max} are approximately, from (2.99), $(\gamma + \log n)/\lambda$ and

$\pi/\lambda\sqrt{6}$, respectively. Substitution of these values into (2.125) leads to

$$\text{Prob}(X_{\max} \leq x) \approx e^{-ne^{-\lambda x}}. \tag{2.127}$$

When $x = (\log n + u)/\lambda$ this becomes

$$\text{Prob}(X_{\max} \leq x) \approx e^{-e^{-u}}, \tag{2.128}$$

while the exact probability $(1 - e^{-\lambda x})^n$ given in (2.96) is

$$\left(1 - \frac{e^{-u}}{n}\right)^n. \tag{2.129}$$

The discussion following (B.3) indicates that when $n \geq 50$ the approximation (2.128) is within 1% of this exact value, and that the approximation rapidly improves as n increases.

The approximation (2.125) is sometimes used to obtain distributional properties of the maximum of n NID$(0,1)$ random variables. However, in this case the convergence of the asymptotic distribution of $X_{\max}$ to the values in (2.125) is very slow, and thus (2.125) might not give accurate approximations if n is moderate.

2.11.4 Bounds for the Distributions of the Maxima of Geometric and Geometric-Like Random Variables

The previous section shows that there exists an asymptotic distribution for the maximum of n iid continuous random variables. For many *discrete* random variables of importance in bioinformatics, on the other hand, analogous asymptotic distributions for the maximum of n iid random variables do not exist. In particular, this is true of the maximum of n iid random variables having the geometric or geometric-like distributions (1.10) and (1.12). It is therefore important to find approximations, or bounds, for the distribution of the maximum of these random variables.

We use a bounding process employing asymptotic results for continuous random variables. If $X_{\max}$ is the maximum of n iid continuous random variables, and if $Y_{\max} = \lfloor X_{\max} \rfloor$ is the integer part of $X_{\max}$, then $Y_{\max}$ is a discrete random variable and

$$X_{\max} - 1 < Y_{\max} \leq X_{\max}.$$

Thus for any positive integer y,

$$\text{Prob}(X_{\max} \leq y) \leq \text{Prob}(Y_{\max} \leq y) \leq \text{Prob}(X_{\max} \leq y + 1). \tag{2.130}$$

Since an asymptotic form for probabilities for the continuous random variable $X_{\max}$ is available in (2.125), the inequalities in (2.130) can be used to find asymptotic bounds for $Y_{\max}$. We now give two examples of this bounding procedure.

Let $X_{\max}$ be the maximum of n iid random variables each having the exponential distribution (1.59), and put $Y_{\max} = \lfloor X_{\max} \rfloor$. Then the argument surrounding equations (2.116) and (2.117) shows that $Y_{\max}$ has the same distribution as the maximum of n iid random variables, each having the geometric distribution (1.62). Application of the approximation (2.129) and the bounds in (2.130) shows that to a close approximation,

$$e^{-ne^{-\lambda y}} \leq \mathrm{Prob}(Y_{\max} \leq y) \leq e^{-ne^{-\lambda(y+1)}}. \tag{2.131}$$

From this it follows that to the same level of approximation,

$$1 - e^{-ne^{-\lambda y}} \leq \mathrm{Prob}(Y_{\max} \geq y) \leq 1 - e^{-ne^{-\lambda(y-1)}}, \tag{2.132}$$

for any positive integer y. In the case of a geometric distribution these bounds are not necessary, since for this case exact formulae for $\mathrm{Prob}(Y_{\max} \leq y)$ and $\mathrm{Prob}(Y_{\max} \geq y)$ are available in equations (2.111) and (2.112). Using an argument similar to that at the end of the previous section, when n is large and y is of the form $y = (\log n + u)/\lambda$, where u is a constant, the upper bound in (2.131) is essentially the exact probability given in (2.111) and the lower bound in (2.132) is essentially the exact value of $\mathrm{Prob}(Y_{\max} \geq y)$. Thus for large n any y close to $(\log n)/\lambda$,

$$\mathrm{Prob}(Y_{\max} \geq y) \approx 1 - e^{-ne^{-\lambda y}}. \tag{2.133}$$

The upper bound in (2.131) and thus the lower bound in (2.132) hold for small y, even for $y = 0$.

The above discussion suggests how a parallel calculation for a random variable having a geometric-like distribution is carried out. If $Y_{\max}$ is the maximum of n iid random variables, each having the geometric-like distribution given in (1.67), then a calculation analogous to that leading to (2.131) gives

$$e^{-nCe^{-\lambda y}} \leq \mathrm{Prob}(Y_{\max} \leq y) \leq e^{-nCe^{-\lambda(y+1)}}. \tag{2.134}$$

Thus

$$1 - e^{-nCe^{-\lambda y}} \leq \mathrm{Prob}(Y_{\max} \geq y) \leq 1 - e^{-nCe^{-\lambda(y-1)}}. \tag{2.135}$$

Whereas, as noted above, the upper bound in (2.131) and the lower bound in (2.132) hold even for small y, the corresponding result does not necessarily apply for the geometric-like distribution. The bounds (2.134) and (2.135) are ultimately based on the *asymptotic* expression (1.67) applying for large values of y (as well as n). They might not hold for small values of y, even when n is large.

2.12   Introduction to P-Values for $Y_{\max}$

2.12.1   Introduction

In Section 3.4.2 we will introduce and define the concept of a *P-value*, a quantity widely used in statistical hypothesis testing. For our present purposes we consider this concept in the case of $Y_{\max}$, the maximum of n iid discrete random variables with unknown probability distribution. The P-value for an observed value y of $Y_{\max}$ is $\text{Prob}(Y_{\max} \geq y)$, the probability being calculated under some hypothesized distribution of the random variables. The smaller the P-value, the less likely it is that the hypothesized distribution is the correct one. All P-value calculations in this chapter are calculated assuming this hypothesized distribution.

When $Y_{\max}$ is the maximum of n iid random variables each hypothesized to have the geometric distribution (1.62), equation (2.133) shows that, to a close approximation,

$$P\text{-value} \approx 1 - e^{-n\,e^{-\lambda y}}. \tag{2.136}$$

This expression can be written in the form

$$P\text{-value} \approx 1 - e^{-e^{-\lambda(y - \log n/\lambda)}}. \tag{2.137}$$

If we denote by $\mu_{\max}$ and $\sigma^2_{\max}$ the approximate mean and variance of the maximum $Y_{\max}$ of n iid random variables having the geometric distribution, given respectively in the expressions (2.119) and (2.121), we get, to a close approximation,

$$\frac{\log n}{\lambda} \approx \mu^* - \frac{\gamma}{\lambda}, \quad \lambda \approx \frac{\pi}{\sigma^* \sqrt{6}},$$

where

$$\mu^* = \mu_{\max} + \frac{1}{2}, \quad (\sigma^*)^2 = \sigma^2_{\max} - \frac{1}{12}. \tag{2.138}$$

The approximation (2.137) can then be written in the more cumbersome, but more easily generalized, form

$$P\text{-value} \approx 1 - e^{-e^{-(\pi(y - \mu^*)/\sigma^* \sqrt{6} + \gamma)}}. \tag{2.139}$$

2.12.2   P-Values for Geometric-Like Random Variables

If $Y_{\max}$ is hypothesized to be the maximum of n iid random variables each having some null hypothesis geometric-like distribution, bounds for the P-value for an observed value y of $Y_{\max}$, namely $\text{Prob}(Y_{\max} \geq y)$, are provided for large values of y by (2.135) as

$$1 - e^{-nCe^{-\lambda y}} \leq P\text{-value} \leq 1 - e^{-nCe^{-\lambda(y-1)}}. \tag{2.140}$$

If a conservative P-value is wanted, that is to say a number known to be equal to or greater than the true P-value, the inequalities (2.140) show immediately that for large values of y a conservative P-value is

$$1 - e^{-nCe^{-\lambda(y-1)}}. \tag{2.141}$$

In many applied cases $nCe^{-\lambda(y-1)}$ is quite small. When this is so, the approximation (B.21) can be used to show that for large values of y an approximate conservative P-value is

$$nCe^{-\lambda(y-1)}. \tag{2.142}$$

2.12.3   P-Values for the Generalized Geometric Distribution

The similarity between the approximation (2.139) on the one hand, and the approximation (2.126) on the other, is clear. Since (2.126) applies for the maximum of a wide range of continuous random variables, it is tempting to use the approximation (2.139) to find P-values for the maximum $Y_{\max}$ of n iid random variables having any discrete distribution, where $\mu_{\max}$ and $\sigma_{\max}^2$ denote the mean and variance of $Y_{\max}$, and μ^* and $(\sigma^*)^2$ are defined as in equation (2.138). This approximation is made in several areas of bioinformatics; a specific example will be discussed in Section 6.3 in connection with the comparison of two DNA sequences. Thus a check on the accuracy of this approximation is desirable.

Equation (2.139) was derived from the geometric distribution. It has, however, been used as an approximation for other distributions, including the generalized geometric, substituting the mean and variance of the distribution in question. We now assess the accuracy of this procedure in the case where this distribution is the generalized geometric.

In Table 2.4 we display the mean $\mu_{\max}$ and the variance $\sigma_{\max}^2$ of the maximum of 75,000 independent random variables, each having the generalized geometric distribution (1.14). These are calculated from equations (2.104) and (2.105) and the theory in Section C. The calculations assume that $p = \frac{1}{4}$ and considers the cases $k = 0, 1, \ldots, 5$.

k	0	1	2	3	4	5
$\mu_{\max}$	8.013	10.559	12.812	14.919	16.933	18.883
$\sigma_{\max}^2$	0.939	1.055	1.152	1.244	1.331	1.413

TABLE 2.4. The mean $\mu_{\max}$ and the variance $\sigma_{\max}^2$ of the maximum of 75,000 iid generalized geometric random variables, for various values of k. $p = \frac{1}{4}$.

With the means and variances from Table 2.4 in hand, it is possible to use (2.139) to obtain approximate P-values, which can then be compared with *exact* P-values calculated from equation (2.102) and the theory of Section C. Both sets of values are displayed in Table 2.5.

y	$k=0$ Exact	Approx	$k=1$ Exact	Approx	$k=2$ Exact	Approx
7	0.990	0.990				
8	0.682	0.682				
9	0.249	0.249	1.000	1.000		
10	0.069	0.069	0.891	0.892		
11	0.018	0.018	0.456	0.455	1.000	1.000
12	0.004	0.004	0.152	0.152	0.940	0.934
13	0.001	0.001	0.044	0.044	0.563	0.558
14			0.012	0.012	0.214	0.218
15			0.003	0.003	0.067	0.071
16					0.006	0.007

y	$k=3$ Exact	Approx	$k=4$ Exact	Approx	$k=5$ Exact	Approx
14	0.950	0.952				
15	0.604	0.603				
16	0.247	0.245	0.943	0.937		
17	0.082	0.082	0.604	0.597		
18	0.025	0.026	0.256	0.258	0.923	0.918
19	0.008	0.008	0.089	0.094	0.597	0.572
20	0.002	0.002	0.029	0.032	0.249	0.251
21			0.009	0.010	0.089	0.093
22			0.003	0.003	0.030	0.033

TABLE 2.5. Exact and approximate P-values for the maximum of 75,000 iid generalized geometric random variables, for various values of k and y. Exact values are from equation (2.102) and the theory of Section C. Approximations are obtained using equations (2.138), (2.139), and the values in Table 2.4, $p = \frac{1}{4}$

The case $k = 0$ corresponds to the geometric distribution, for which the approximation using (2.139) is very accurate. The values in Table 2.5 show that the approximation using (2.139) is also good for all values of k up to 5, at least for the given range of values of y, which covers cases of practical interest. However, as k increases, the values in Table 2.5 show that the accuracy of the approximation near the mean of $Y_{\max}$ decreases.

Apart from assessing the accuracy of the approximation (2.139), the calculations in Table 2.5 illustrate a point made in Section 2.11.2, that very sharp changes in the distribution function of the maximum of n iid random variables occur near the mean when n is large. This implies that unless very accurate approximations are used for the mean and variance in (2.139), serious errors in P-value approximations can arise when the approach via (2.139) is used. We return to this matter in Section 6.3, where we apply the maximum of generalized geometric distributions to testing the significance of a DNA sequence alignment.

2.13   Large Observations and Rare Events

The analysis in the previous sections of the behavior of $Y_{\max}$ will be employed later when the observed value of such a maximum is used to test statistical hypotheses associated with the BLAST procedure. This testing procedure would be strengthened if we focus not only on the maximum, or largest, observation, but on several large observations. This can be done by fixing some large number A such that the probability that the observed value of any random variables of interest should exceed A is small, and then to use the *number of observations* whose values exceed A in the testing procedure. This number has an approximate Poisson distribution, and in Chapter 9 we illustrate how this can be used in a concrete statistical testing procedure in connection with BLAST.

2.14   Order Statistics

2.14.1   Definition

The random variables $X_{\min}$ and $X_{\max}$ are examples of *order statistics* of continuous random variables. In this section we discuss these order statistics in more detail.

Suppose that X is a continuous random variable and that $X_1, \ldots, X_n$ are iid random variables each with the same distribution as X. Let $X_{(1)}$ be the smallest of the X_i's, $X_{(2)}$ the second smallest, and so on up to $X_{(n)}$, the largest. We call these *order statistics*. The order statistic $X_{(1)}$ is identical to $X_{\min}$, and $X_{(n)}$ is identical to $X_{\max}$. Because the probability that two continuous random variables both take the same value is zero, these order statistics are distinct.

The density function of $X_{(i)}$ is found as follows. We let h be small and ignore events whose probability is $o(h)$. Then the event that $x_{(i)} < X_{(i)} < x_{(i)} + h$, is the event that $i - 1$ of the random variables are less than $x_{(i)}$, that one of the random variables is between $x_{(i)}$ and $x_{(i)} + h$, and that the remaining random variables exceed $x_{(i)} + h$. This is a multinomial event with n trials and $k = 3$ outcomes on each trial. Use of the approximation $\text{Prob}(y < X < y + h) \approx f_X(y)h$ when h is small, (2.6) shows that the probability of the event $x_{(i)} < X_{(i)} < x_{(i)} + h$ is

$$\frac{n!}{(i-1)!(n-i)!} \left(F_X\big(x_{(i)}\big) \right)^{i-1} f_X\big(x_{(i)}\big) \, h \left(1 - F_X\big(x_{(i)} + h\big) \right)^{n-i}. \quad (2.143)$$

Thus by (1.40)

$$f_{X_{(i)}}\big(x_{(i)}\big) = \frac{n!}{(i-1)!(n-i)!} \left(F_X\big(x_{(i)}\big) \right)^{i-1} f_X\big(x_{(i)}\big) \left(1 - F_X\big(x_{(i)}\big) \right)^{n-i}. $$
$$(2.144)$$

The expressions (2.81) and (2.92) are particular cases of this density function.

In this formula the expressions $f_X(\cdot)$ and $F_X(\cdot)$ on the right-hand side refer, as the suffix indicates, to the original density function and distribution function of the random variables.

Suppose next that $i < j$ and that h_1 and h_2 are small. Then arguing as above, and ignoring terms of order $o(h_1)$ and $o(h_2)$, the probability of the joint event that $x_{(i)} < X_{(i)} < x_{(i)} + h_1$ and that $x_{(j)} < X_{(j)} < x_{(j)} + h_2$ is the probability of the event that $i - 1$ of the random variables are less than $x_{(i)}$, that one random variable is between $x_{(i)}$ and $x_{(i)} + h_1$, that $j - i - 1$ of the random variables lie between $x_{(i)} + h_1$ and $x_{(j)}$, that one random variable is between $x_{(j)}$ and $x_{(j)} + h_2$, and that $n - j$ random variables exceed $x_{(j)} + h_2$. From this argument the joint density function $f_{X_{(i)}, X_{(j)}}(x_{(i)}, x_{(j)})$ of $X_{(i)}$ and $X_{(j)}$ is

$$\frac{n!}{(i-1)!(j-i-1)!(n-j)!} \left(F_X(x_{(i)})\right)^{i-1} f_X(x_{(i)})$$

$$\times \left(F_X(x_{(j)}) - F_X(x_{(i)})\right)^{j-i-1} f_X(x_{(j)}) \left(1 - F_X(x_{(j)})\right)^{n-j}. \quad (2.145)$$

An important particular case is that where $i = 1, j = n$. Here (2.145) reduces to the simpler form

$$f_{X_{(1)}, X_{(n)}}(x_{(1)}, x_{(n)})$$

$$= n(n-1) f_X(x_{(1)}) \left(F_X(x_{(n)}) - F_X(x_{(1)})\right)^{n-2} f_X(x_{(n)}). \quad (2.146)$$

As above, $f_X(\cdot)$ and $F_X(\cdot)$ refer to the original density function and distribution function of the random variables.

One can continue in this way, finding the joint density function of any number of order statistics. Eventually, the joint density function

$$f_{X_{(1)}, X_{(2)}, \dots, X_{(n)}}(x_{(1)}, x_{(2)}, \dots, x_{(n)})$$

of all n order statistics is found as

$$f_{X_{(1)}, X_{(2)}, \dots, X_{(n)}}(x_{(1)}, x_{(2)}, \dots, x_{(n)}) = n! \prod_{i=1}^{n} f_X(x_{(i)}). \quad (2.147)$$

The meaning of this equation is seen more clearly when it is used to find the conditional distribution of the original observations $X_1, X_2, \dots, X_n$, given the order statistics $X_{(1)}, X_{(2)}, \dots, X_{(n)}$. Using the conditional probability formula (2.37), this is, from (2.147), $1/n!$. This simply states that if the order statistics are given, all of the $n!$ allocations of the actual observations to the order statistics' values are equally likely, each having probability $1/n!$. This observation is the basis of many *nonparametric* and *computationally intensive* statistical procedures, discussed in more detail in Chapters 3 and 12.

Order statistics for discrete random variables have much more complicated distributions than those for continuous random variables because of the possibility that two or more of the random variables take the same value. Thus we do not consider them here.

2.14.2   Example: The Uniform Distribution

As an example of order statistics, suppose that $X_1, X_2, \ldots, X_n$ are iid, each having the uniform distribution with range $(0, L)$. Then (2.144) shows that the density function of $X_{(i)}$ is

$$f_{X_{(i)}}(x_{(i)}) = \frac{n!}{(i-1)!\,(n-i)!} x_{(i)}^{i-1} (L - x_{(i)})^{n-i} L^{-n}. \qquad (2.148)$$

In the case $L = 1$, this is a beta distribution with parameters $\alpha = i$, $\beta = n - i + 1$. From equation (1.72), the mean and variance of $X_{(i)}$ are

$$\text{mean of } X_{(i)} = \frac{i}{n+1}, \quad \text{variance of } X_{(i)} = \frac{i(n-i+1)}{(n+1)^2(n+2)}. \qquad (2.149)$$

For general L,

$$\text{mean of } X_{(i)} = \frac{Li}{n+1}, \quad \text{variance of } X_{(i)} = \frac{L^2 i(n-i+1)}{(n+1)^2(n+2)}. \qquad (2.150)$$

Equation (2.145) shows that the joint density function of $X_{(i)}$ and $X_{(j)}$ is

$$f_{X_{(i)}, X_{(j)}}(x_{(i)}, x_{(j)})$$
$$= \frac{n!}{(i-1)!\,(j-i-1)!\,(n-j)!} x_{(i)}^{i-1} (x_{(j)} - x_{(i)})^{j-i-1} (L - x_{(j)})^{n-j} L^{-n}. \qquad (2.151)$$

In the important particular case $i = 1$, $j = n$, this is

$$f_{X_{(1)}, X_{(n)}}(x_{(1)}, x_{(n)}) = n(n-1)(x_{(n)} - x_{(1)})^{n-2} L^{-n}. \qquad (2.152)$$

Finally, equation (2.147) shows that the joint density function of all n order statistics is given by

$$f_{X_{(1)}, X_{(2)}, \ldots, X_{(n)}}(x_{(1)}, x_{(2)}, \ldots, x_{(n)}) = n! L^{-n}. \qquad (2.153)$$

2.14.3   The Sample Median

The sample median $\hat{M}$ is an important random variable, and in this section we define it and briefly consider some of its properties.

Consider n iid continuous random variables $X_1, X_2, \ldots, X_n$, or equivalently their order statistics $X_{(1)}, X_{(2)}, \ldots, X_{(n)}$. When n is odd, so that we can write $n = 2m + 1$, the sample median $\hat{M}$ is defined as $X_{(m+1)}$: Half of the observed sample values are less than $\hat{m} = x_{(m+1)}$, and half exceed this value. When n is even, so that we can write $n = 2m$, the convention is to define the sample median $\hat{M}$ as $(X_{(m)} + X_{(m+1)})/2$, so that the observed value $\hat{m}$ of the sample median is the average $(x_{(m)} + x_{(m+1)})/2$ of the two central observations.

The sample median $\hat{M}$ is a random variable, and when n is odd, its probability distribution is identical to that of $X_{(m+1)}$, found from equation (2.143) with $n = 2m + 1$, $i = m + 1$. From this distribution the mean, the variance, and other properties of the sample median can in principle be found. In practice this might involve difficult problems of integration. These problems can be even greater when n is even. An example of a direct way of finding the mean and variance of $X_{(m+1)}$ when the iid random variables $X_1, X_2, \ldots, X_n$ have an exponential distribution is given in Problem 2.20.

When the random variables are discrete it becomes possible that the observed values of several random variables are equal. Because of this the theory for the distribution of the sample median for discrete random variables is more complex than it is for continuous random variables, and therefore we do not consider the discrete case here.

2.15   Transformations

2.15.1   Theory

Suppose that $X_1, X_2, \ldots, X_n$ are continuous random variables and let $U_1 = U_1(X_1, X_2, \ldots, X_n)$, $U_2 = U_2(X_1, X_2, \ldots, X_n), \ldots, U_n = U_n(X_1, X_2, \ldots, X_n)$ be functions of $X_1, X_2, \ldots, X_n$. Suppose that there is a one-to-one relation between $(X_1, X_2, \ldots, X_n)$ and $(U_1, U_2, \ldots, U_n)$, that the transformation from $(X_1, X_2, \ldots, X_n)$ to $(U_1, U_2, \ldots, U_n)$ is such that all partial derivatives in the two Jacobian matrices defined below exist, and that the two Jacobians defined below are always nonzero.

If the joint density function of $X_1, X_2, \ldots, X_n$ is $f_{\boldsymbol{X}}(x_1, x_2, \ldots, x_n)$, arguments extending those given in Section 1.15 to the n-dimensional case show that the joint density function of $U_1, U_2, \ldots, U_n$ is given by

$$f_{\boldsymbol{U}}(u_1, u_2, \ldots, u_n) = f_{\boldsymbol{X}}(x_1, x_2, \ldots, x_n)|J^{-1}|, \qquad (2.154)$$

where J is the Jacobian

$$
J = \begin{vmatrix}
\frac{\partial U_1}{\partial X_1} & \frac{\partial U_1}{\partial X_2} & \cdots & \frac{\partial U_1}{\partial X_n} \\
\frac{\partial U_2}{\partial X_1} & \frac{\partial U_2}{\partial X_2} & \cdots & \frac{\partial U_2}{\partial X_n} \\
\vdots & \vdots & \ddots & \vdots \\
\frac{\partial U_n}{\partial X_1} & \frac{\partial U_n}{\partial X_2} & \cdots & \frac{\partial U_n}{\partial X_n}
\end{vmatrix}
$$

and the right-hand side in (2.154) is computed as a function of $U_1, U_2, \ldots, U_n$. An equivalent formula is

$$
f_U(u_1, u_2, \ldots, u_n) = f_X(x_1, x_2, \ldots, x_n)|J^*|, \qquad (2.155)
$$

where J^* is the Jacobian

$$
J^* = \begin{vmatrix}
\frac{\partial X_1}{\partial U_1} & \frac{\partial X_1}{\partial U_2} & \cdots & \frac{\partial X_1}{\partial U_n} \\
\frac{\partial X_2}{\partial U_1} & \frac{\partial X_2}{\partial U_2} & \cdots & \frac{\partial X_2}{\partial U_n} \\
\vdots & \vdots & \ddots & \vdots \\
\frac{\partial X_n}{\partial U_1} & \frac{\partial X_n}{\partial U_2} & \cdots & \frac{\partial X_n}{\partial U_n}
\end{vmatrix}
$$

with the right-hand side in (2.155) again being expressed as a function of $U_1, U_2, \ldots, U_n$. Sometimes (2.154) is the easier formula to use, sometimes (2.155).

The transformation from the joint density function of $X_1, \ldots, X_n$ to that of $U_1, \ldots, U_n$ is often used to find the density function of a *single* random variable. Suppose that we wish to find the density function of U_1 only. We first choose $n-1$ "dummy" variables $U_2, \ldots, U_n$, of no direct interest to us. We then use transformation techniques to find the joint density function of $U_1, \ldots, U_n$. Having found this, we integrate out $U_2, \ldots, U_n$ to find the desired (marginal) density function of U_1, as described in equation (2.32). With a sufficiently careful choice of the dummy variables, this seemingly roundabout procedure is often the most efficient way of finding this density function. As noted in the discussion below equation (2.32), care must be exercised in finding the correct domain of integration for $U_2, \ldots, U_n$, since this can depend on the value of U_1. An example of the use of a dummy variable in evaluating a density function by transformation methods, of direct relevance to BLAST, is given in Appendix D.

2.15.2   Example: The Uniform Distribution (Continued)

In this section we continue the analysis of properties of the uniform distribution, begun in Section 2.14.2, focusing on two applications that are used later in bioinformatics contexts. In both cases we consider n iid random variables $X_1, X_2, \ldots, X_n$, each having the uniform distribution on $(0, 1)$, and in both cases interest centers on the associated order statistics $X_{(1)}$, $X_{(2)}, \ldots, X_{(n)}$.

The density function of any chosen order statistic is given by (2.148) with $L = 1$. In the particular case $j = 1$, the density function of the smallest order statistic is

$$f_{X_{(1)}}(x_{(1)}) = n(1 - x_{(1)})^{n-1}. \tag{2.156}$$

This follows also from equation (2.81). If some probability α is given, the value $K(n, \alpha)$ such that $\text{Prob}(X_{(1)} \leq K(n, \alpha)) = \alpha$ is found by integration as

$$K(n, \alpha) = 1 - \sqrt[n]{1 - \alpha}. \tag{2.157}$$

This expression will also appear in Section 3.8, and we return to it in Section 12.4, where it forms the basis of an approach to hypothesis testing in microarray analysis.

We next find properties of the subintervals of $(0, 1)$ defined by the order statistics $X_{(j)}$, $j = 1, 2, \ldots, n$. These order statistics divide the interval $(0, 1)$ into $n + 1$ subintervals of lengths $U_1, U_2, \ldots, U_{n+1}$, where

$$U_1 = X_{(1)}, \quad U_2 = X_{(2)} - X_{(1)}, \quad \ldots, \quad U_{n+1} = 1 - X_{(n)}. \tag{2.158}$$

These lengths were discussed in the example of Section 1.12.3, page 38, and our first aim here is to prove equation (1.89), the key result of that section.

The joint density function of $U_1, U_2, \ldots, U_n$ may be found from the joint density function $n!$ of $X_{(1,)} X_{(2)}, \ldots, X_{(n)}$ by transformation techniques. The Jacobian of this transformation can be shown to be 1, so that the joint density function of $U_1, U_2, \ldots, U_n$ is

$$f_{U_n}(u_1, u_2, \ldots, u_n) = n!, \quad u_j > 0, \quad \sum_{j=1}^{n} u_j \leq 1. \tag{2.159}$$

Since $U_{n+1} = 1 - (U_1 + U_2 + \cdots + U_n)$, the value of U_{n+1} is determined by the values of $U_1, U_2, \ldots, U_n$. Thus the joint density of $U_1, U_2, \ldots, U_{n+1}$ is

$$f_{U_{n+1}}(u_1, u_2, \ldots, u_{n+1}) = n!, \quad u_j > 0, \quad \sum_{j=1}^{n+1} u_j = 1. \tag{2.160}$$

The form of this density function shows that the joint density function of any subset of g of the lengths $U_1, U_2, \ldots, U_{n+1}$ is independent of the subset chosen. It is therefore sufficient for our purposes to discuss the joint density function $f_{U_g}(u_1, u_2, \ldots, u_g)$ of $U_1, U_2, \ldots, U_g$. This is found as the marginal probability

$$f_{U_g}(u_1, \ldots, u_g) = \int_0^{w_{g+1}} \int_0^{w_{g+2}} \cdots \int_0^{w_n} n! \, du_n \cdots du_{g+1}, \tag{2.161}$$

where $w_j = 1 - u_1 - u_2 - \cdots - u_{j-1}$. This integration yields

$$f_{U_g}(u_1, \ldots, u_g) = \frac{n!}{(n-g)!} (1 - u_1 - \cdots - u_g)^{n-g}. \tag{2.162}$$

Assuming that $1 - ug > 0$, the probability that each U_j, $j = 1, 2, \ldots, g$, exceeds u is then found by integration as

$$\int_u^1 \int_u^{w_2} \cdots \int_u^{w_g} \frac{n!}{(n-g)!} (1 - u_1 - \cdots - u_g)^{n-g} \, du_g \cdots du_1. \qquad (2.163)$$

The value of this integral is $(1 - ug)^n$, thus justifying the claim made in equation (1.89).

The joint density function (2.159) can be derived in a different way. Suppose that $X_1, X_2, \ldots, X_{n+1}$ are iid random variables having the exponential distribution (1.59), and put $S = \sum_{j=1}^{n+1} X_j$. Define U_j by $U_j = X_j/S$, $j = 1, 2, \ldots, n$, and $U = S$. We now make a transformation from the joint density function of $X_1, X_2, \ldots, X_{n+1}$ to the joint density function of $U_1, U_2, \ldots, U_n, U$. From the definitions of $U_1, U_2, \ldots, U_n$ we can write $X_j = U_j U$, $j = 1, 2, \ldots, n$. The first n diagonal elements in the $(n+1) \times (n+1)$ matrix defining the Jacobian in (2.155) all have the value U and the final diagonal element has value 1. All entries below the main diagonal are zero. This implies that the value of the Jacobian is U^n. From this the joint density function of $U_1, U_2, \ldots, U_n, U$ is

$$\lambda^{n+1} e^{-\lambda U} U^n, \quad U \geq 0, \ U_j \geq 0, \ \sum_{j=1}^n U_j \leq 1. \qquad (2.164)$$

Note that the form of this joint density function is independent of $U_1, U_2, \ldots, U_n$. The marginal density function of $U_1, U_2, \ldots, U_n$ is found from (2.32) as

$$\lambda^{n+1} \int_0^{+\infty} e^{-\lambda U} U^n \, dU = n!, \quad U_j \geq 0, \ \sum_{j=1}^n U_j \leq 1. \qquad (2.165)$$

This is identical to (2.159). The marginal density function of U is similarly found to be

$$f_U(u) = \lambda^{n+1} e^{-\lambda U} U^n / n!, \quad U > 0, \qquad (2.166)$$

confirming a result found in Section 2.3 using moment-generating functions.

This approach to finding the joint density function (2.159) has several useful properties, of which we mention one. A comparison of (2.164), (2.165), and (2.166) shows that U is independent of $U_1, U_2, \ldots, U_{n+1}$ and hence of the order statistics $U_{(1)}, U_{(2)}, \ldots, U_{(n+1)}$. Then from equation (2.55), $E(X_{(j)}) = E(UU_{(j)}) = E(U)E(U_{(j)})$. Since $X_{(j)}$ is the jth largest of $n+1$ iid random variables each having the exponential distribution (1.59), the mean of $X_{(j)}$ is, from equation (2.97),

$$\frac{1}{\lambda} \left(\frac{1}{n+1} + \frac{1}{n} + \cdots + \frac{1}{n-j+2} \right).$$

Since the mean of U is $(n+1)/\lambda$, the mean of $U_{(j)}$ is

$$E\left(U_{(j)}\right) = \frac{1}{n+1}\left(\frac{1}{n+1} + \frac{1}{n} + \cdots + \frac{1}{n-j+2}\right). \tag{2.167}$$

This is perhaps the easiest way to establish the mean value of $U_{(j)}$. A particular case of equation (2.167) is that

$$E\left(U_{(n+1)}\right) = \frac{1}{n+1}\left(\frac{1}{n+1} + \frac{1}{n} + \cdots + \frac{1}{2} + \frac{1}{1}\right). \tag{2.168}$$

This expression may in principle be found from the density function (2.100), but the above derivation is more efficient. Both the density function (2.100) and the mean (2.168) will be discussed further in Section 5.5, where the theory of r-scans is described.

Problems

2.1 Let $\boldsymbol{Y} = (Y_1, Y_2, \ldots, Y_n)$ be a random vector such that

$$P_Y(\boldsymbol{y}) = \prod_{i=1}^{n} P_{Y_i}(y_i) \tag{2.169}$$

for all possible combinations of values of $\boldsymbol{y} = (y_1, y_2, \ldots, y_n)$. If A_i is the event that $Y_i = y_i$, then show (2.169) implies that all the conditions (1.99)–(1.101) are satisfied.

2.2 In the discussion of the random permutation example following equation (1.88) it was shown that if A_i is the event that the number i occurs in its correct position, then p_i, the probability of the event A_i, is n^{-1}. Use this result in conjunction with equation (2.27) to find the mean of the number of events $A_1, A_2, \ldots, A_n$ to occur.

2.3. Suppose that $Y_1, Y_2, \ldots, Y_n$ are independent random variables, each having a Poisson distribution, the parameter of the distribution of Y_j being λ_j. Find the pgf of the distribution of their sum S_n, and thus show that the probability distribution of S_n is Poisson with parameter $\sum \lambda_j$.

2.4. Use (2.16) and (1.79) to prove (2.21) and (2.22), and use (2.17) and (1.79) to prove (2.19) and (2.20).

2.5. Suppose we are given a DNA sequence consisting of 10 consecutive nucleotides. Three segments of this sequence are to be chosen at random, one consisting of 3 consecutive nucleotides, a second consisting of 4 consecutive nucleotides, and the third consisting of 5 consecutive nucleotides. By "random" we mean the segment of 3 nucleotides can be in any of the 8 possible positions with equal probability, and similarly the segment of 4 nucleotides can be in any of the 7 possible positions with equal probability, and the segment of 5 in any of the 6 possible positions with equal probability. Let Y be the number of positions (out of 10) that are in all three segments. Then Y has observable values 0, 1, 2, 3. What is the expected value of Y, $E(Y)$? *Hint:* Use indicator random variables.

2.6. Show that although the random variables in the example in Section 2.5 are dependent, the covariance between them is 0.

2.7. Use (2.46) to generalize the result given in (2.47) by showing that if X is a random variable having the exponential probability distribution (1.59), and if it is given that $a \leq X < a+1$, then $X - a$ has the distribution (2.47).

2.8. Use (2.17), together with the formula for the mgf of a normal random variable, mean μ, variance σ^2 (found in Problem 1.20) to find the mgf of the distribution of the sum of n independent random variables, each having this normal distribution. What conclusion do you draw about the distribution of this sum?

2.9. Prove equations (2.31) and (2.39).

2.10. Use equation (2.48) to derive (2.49).

2.11. Suppose X_1 and X_2 are possibly dependent random variables, each with mean 0, variance 1 (so that $EX_i^2 = 1$ for $i = 1, 2$). Since $(X_1 - X_2)^2$ is never negative, $E\left((X_1 - X_2)^2\right) \geq 0$. By expanding the term $(X_1 - X_2)^2$, show that $E(X_1 X_2) \leq 1$. Apply a similar argument to $(X_1 + X_2)^2$ to show that $E(X_1 X_2) \geq -1$. It follows from these two facts that $|E(X_1 X_2)| \leq 1$.

Now suppose X_1 and X_2 are arbitrary random variables with respective means and standard deviations μ_i and σ_i, $i = 1, 2$. Apply the above conclusion to $X_i' = (X_i - \mu_i)/\sigma_i$ (for $i = 1, 2$) to show that $|\rho| \leq 1$, where ρ is the correlation between X_1 and X_2.

2.12. Use equation (2.77) to establish equation (2.78), and carry out one further differentiation to show that the variance of S is

$$E(N) \operatorname{Var}(X) + E(X)^2 \operatorname{Var}(N).$$

2.13. For the random walk taking a step up with probability p and a step down with probability $q = 1 - p$, find the possible values of the total displacement of the random walk after three steps, together with their probabilities. Thus find the mgf of this displacement, and check that it is the cube of the expression given in (2.26).

2.14. Prove equation (2.96) by appropriate integration.

2.15. Show that the mgf of the "extreme value" density (2.122) is $\Gamma(1-\theta)$, and thus find the mean and variance of this distribution. *Hint:* In finding the mgf, make the change of variable $u = e^{-w}$, and be careful about using the appropriate terminals in the ensuing integration. To find the mean and variance, use equations (1.77) and (1.79) and the properties of the gamma function given in Section B.17.

2.16. Let $X_1, X_2, \ldots, X_n$ be iid random variables coming from a continuous probability distribution with median θ. Find the probability distribution of the number of these random variables that are less than θ.

2.17. Let $X_1, X_2, \ldots, X_n$ be independent random variables, each having the exponential distribution (1.59). Use (2.144) to find the density function of $X_{(2)}$. From this, find the mean and variance of $X_{(2)}$ and relate these to the calculations leading to equations (2.97) and (2.98).

2.18 The following problem is relevant to linkage analysis. Suppose that a parent of genetic type Mm has three children. Then the parent transmits the M gene to each child with probability $\frac{1}{2}$, and the genes that are transmitted to each of the three children are independent. Let $I_1 = 1$ if children 1 and 2 had the same gene transmitted (that is, both received M or both received m), and $I_1 = 0$ otherwise. Similarly, let $I_2 = 1$ if children 1 and 3 had the same gene transmitted, $I_2 = 0$ otherwise, and let $I_3 = 1$ if children 2 and 3 had the same gene transmitted, $I_3 = 0$ otherwise.

   (i) Show that while these three random variables are pairwise independent, they are not independent.

   (ii) Show that despite the conclusion of (i), the variance of $I_1 + I_2 + I_3$ is the sum of the variances of I_1, I_2, and I_3.

   (iii) Explain your result in terms of the various covariances between I_1, I_2, and I_3 and the pairwise independence of I_1, I_2, and I_3.

2.19. Prove the results given in (2.150).

2.20. Let $X_1, X_2, \ldots, X_n$ be independent random variables, each having the exponential distribution (1.59). For $n = 2m+1$ an odd integer, use the argument that led to equations (2.97) and (2.98) to show that the mean and variance of the sample median $X_{(m+1)}$ are, respectively,

$$E(X_{(m+1)}) = \frac{1}{(2m+1)\lambda} + \frac{1}{(2m)\lambda} + \cdots + \frac{1}{(m+1)\lambda},$$

$$\text{Var}(X_{(m+1)}) = \frac{1}{(2m+1)^2\lambda^2} + \frac{1}{(2m)^2\lambda^2} + \cdots + \frac{1}{(m+1)^2\lambda^2}.$$

2.21. *Continuation.* Use the approximation formula (B.7) to show that as $n \to +\infty$, the mean of $X_{(m+1)}$ approaches the exponential distribution median given in equation (1.61).

2.22. *Continuation.* Use the asymptotic results

$$1 + \frac{1}{2} + \cdots + \frac{1}{n} \approx \log n + \gamma + \frac{1}{2n}$$

and

$$\log\left(\frac{2m+1}{m}\right) \approx \log 2 + \frac{1}{2m}$$

to show that the $E(X_{(m+1)})$ exceeds the median (1.61) by an amount $\frac{1}{2n\lambda} + O(n^{-2})$.

2.23. *Continuation.* Use the approximation

$$\frac{1}{a^2} + \frac{1}{(a+1)^2} + \cdots + \frac{1}{b^2} \approx \frac{1}{a - \frac{1}{2}} - \frac{1}{b + \frac{1}{2}}$$

to show that the variance of the sample median $X_{(m+1)}$ is approximately $\frac{1}{n\lambda^2}$ when n is large and odd.

2.24. *Continuation.* When $n = 2m$ is even, the sample median is defined as $(X_{(m)} + X_{(m+1)})/2$. Use the asymptotic results given in Problem 2.22 to show that when n is even, the mean of this sample median also exceeds the median (1.61) by an amount $\frac{1}{2n\lambda} + O(n^{-2})$.

3
Statistics (i): An Introduction to Statistical Inference

3.1 Introduction

Statistics is the method by which we analyze data in whose generation chance has played some part. In practice, it consists of two main areas: estimation and hypothesis testing. Both of these are used extensively in bioinformatics. In this chapter we give a brief introduction to estimation and hypothesis testing ideas: A more complete discussion of both is given in Chapter 8.

3.2 Classical and Bayesian Methods

There are two main approaches to both estimation and hypothesis testing, each deriving from a broad view of the way in which we should conduct statistical inference. These are the "classical," or "frequentist," approach on the one hand, and the "Bayesian" approach on the other. There is controversy among some statisticians as to the theoretical underpinnings of each of these approaches. There is also a more pragmatic approach among some practitioners who are willing to use whatever approach leads to a useful answer. We do not attempt to summarize the views of both camps in detail, especially since there is no unique Bayesian or classical position, and give only a brief sketch of their respective approaches.

There are two main arguments supporting the Bayesian approach to statistical inference.

(i) The Bayesian approach asks the right question in a hypothesis testing procedure, namely, "What is the probability that this hypothesis is true, given the data?" rather than the classical approach, which asks a question like, "Assuming that this hypothesis is true, what is the probability of the observed data?"

(ii) Prior knowledge and reasonable prior concepts can be built into a Bayesian analysis. For example, suppose that a fair-looking coin is tossed three times and gives three heads. A Bayesian might well claim that the classical estimate of the probability of heads, in this case 1, is unreasonable, and does not take into account the information that the coin seems reasonably symmetric.

Behind this approach there is the broad feeling that a probability is a measure of belief in a proposition, rather than the frequentist interpretation that a probability of an event is in some sense the long-term frequency with which it occurs.

The main arguments for using a classical approach to statistical inference can perhaps best be stated by giving the classical approach counterargument to the Bayesian positions outlined above.

The question, "What is the probability that this hypothesis is true?" usually cannot be answered without making assumptions about prior probabilities of hypotheses that might not be justified. Bayesians often choose prior probabilities for mathematical convenience rather than from any objective scientific basis. Further, in research, the prior distribution that a Bayesian needs for his/her inferences about new phenomena cannot be known with certainty. These problems are not overcome by using so-called uninformative priors.

Bayesians have replies to these views, and the debate continues. We make no comment on these positions here, other than to note that while the focus in this book is on classical hypothesis testing methods, in particular that associated with BLAST, we describe in Section 6.6 a procedure that uses Bayesian concepts and that might well not work if classical methods were employed.

3.3   Classical Estimation Methods

3.3.1   Unbiased Estimation

Much of the theory concerning estimation of parameters is essentially the same for both discrete and continuous random variables, so in this section we use the notation X for both.

Let X be a random variable having a probability distribution $P_X(x)$ (for discrete random variables) or density function $f_X(x)$ (for continuous random variables). Suppose that this distribution depends on some unknown parameters. How may we estimate these parameters from observed values of X?

The observed value x of X on its own will usually not be sufficient to allow for a good estimate. We must consider repeating the experiment that generated this value an (ideally large) number of times to give n observations $x_1, x_2, \ldots, x_n$. We think of these as the observed values of n iid random variables $X_1, X_2, \ldots, X_n$, each X_i having a probability distribution $P_{X_i}(x)$ identical to $P_X(x)$ (for discrete random variables) or density function $f_{X_i}(x)$ identical to $f_X(x)$ (for continuous random variables).

An *estimator* of a parameter θ is some function of the random variables $X_1, X_2, \ldots, X_n$ and thus may be written $\hat{\theta}(X_1, X_2, \ldots, X_n)$, a notation that emphasizes that this estimator is itself a random variable. For convenience we generally use the shorthand notation $\hat{\theta}$. This estimator is said to be an *unbiased* estimator of θ if its mean value $E(\hat{\theta})$ is equal to θ. The quantity $\hat{\theta}(x_1, x_2, \ldots, x_n)$, calculated from the observed values $x_1, x_2, \ldots, x_n$ of $X_1, X_2, \ldots, X_n$, is called the *estimate* of θ.

In this section we consider four examples of estimation, and defer a more detailed discussion to Chapter 8. Specifically, we consider estimation of the mean μ and the variance σ^2 of any probability distribution, estimation of the probability p of the Bernoulli distribution (1.6), and estimation of the parameters $\{p_i\}$ in the multinomial distribution (2.6).

It is natural to estimate the mean μ of a probability distribution by the observed sample average $\bar{x}$. Since the mean value of the random variable $\bar{X}$ is μ (from equation (2.71)), $\hat{\mu} = \bar{X}$ is an unbiased estimator of μ. Since the variance of $\bar{X}$ decreases as the sample size n increases (see (2.71)), the observed average $\bar{x}$ is more and more likely to be close to μ as n increases.

This can be quantified using Chebyshev's inequality (1.50), or, to a close approximation, by the central limit theorem. If the variance σ^2 of X is known and the sample size n is large, the central limit theorem of Section 2.10.1, page 77, and the approximate two standard deviation rule of Section 1.10.2, page 26, can be used to show that

$$\text{Prob}\left(\mu - \frac{2\sigma}{\sqrt{n}} < \bar{X} < \mu + \frac{2\sigma}{\sqrt{n}}\right) \approx 0.95. \tag{3.1}$$

This can be written in the equivalent form

$$\text{Prob}\left(\bar{X} - \frac{2\sigma}{\sqrt{n}} < \mu < \bar{X} + \frac{2\sigma}{\sqrt{n}}\right) \approx 0.95, \tag{3.2}$$

which provides an approximate 95% *confidence interval* for μ, in the sense that the probability that the random interval

$$\left(\bar{X} - \frac{2\sigma}{\sqrt{n}}, \bar{X} + \frac{2\sigma}{\sqrt{n}}\right) \tag{3.3}$$

contains μ is approximately 95%. Given the observed values $x_1, x_2, \ldots, x_n$ of $X_1, X_2, \ldots, X_n$, the observed value of this interval is

$$\left(\bar{x} - \frac{2\sigma}{\sqrt{n}}, \bar{x} + \frac{2\sigma}{\sqrt{n}}\right). \tag{3.4}$$

Often in practice the variance σ^2 is unknown, so that (3.4) is not immediately applicable. However, an unbiased estimator of the variance σ^2 of any distribution is provided by the estimator S^2, defined by

$$S^2 = \frac{\sum_{i=1}^{n}(X_i - \bar{X})^2}{n-1}. \tag{3.5}$$

Correspondingly, the estimate s^2 of σ^2 is

$$s^2 = \frac{\sum_{i=1}^{n}(x_i - \bar{x})^2}{n-1}. \tag{3.6}$$

From the above,

$$\frac{S^2}{n} = \frac{\sum_{i=1}^{n}(X_i - \bar{X})^2}{n(n-1)}$$

is an unbiased estimator of the variance σ^2/n of $\bar{X}$, and is estimated from the data by

$$\frac{\sum_{i=1}^{n}(x_i - \bar{x})^2}{n(n-1)}. \tag{3.7}$$

An approximate 95% confidence interval for μ is then

$$\left(\bar{X} - \frac{2S}{\sqrt{n}}, \bar{X} + \frac{2S}{\sqrt{n}}\right), \tag{3.8}$$

and given the data, the observed value of this interval is

$$\left(\bar{x} - \frac{2s}{\sqrt{n}}, \bar{x} + \frac{2s}{\sqrt{n}}\right). \tag{3.9}$$

Such an estimated confidence interval is useful, since it provides a measure of the accuracy of the estimate $\bar{x}$.

The estimation of a binomial parameter p is carried out by using the theory of the binomial distribution directly and not by using the above general theory. Using the above theory will lead to slightly different results. Let Y have the binomial distribution parameter p and index n. Equation (2.74) shows that the mean value of Y/n is p and the variance of Y/n is $p(1-p)/n$. Thus

$$\hat{p} = Y/n \tag{3.10}$$

is an unbiased estimator of p. If y successes were obtained when the trials were carried out, the estimate of p is y/n and the estimate of the variance of $\hat{p}$ is

$$\frac{y(n-y)}{n^3}. \tag{3.11}$$

An approximate 95% confidence interval for p calculated from this is

$$\left(\frac{y}{n} - 2\sqrt{\frac{y(n-y)}{n^3}}, \frac{y}{n} + 2\sqrt{\frac{y(n-y)}{n^3}} \right). \tag{3.12}$$

The estimation of the multinomial parameter p_i is carried out in a similar way. Let Y_i be the number of times that outcome i occurs in n iid multinomial trials. Then the mean value of Y_i/n is p_i, and the variance of Y_i/n is $p_i(1-p_i)/n$. Thus $\hat{p}_i = Y_i/n$ is an unbiased estimator of p_i. If, when the trials are carried out, outcome i occurs y_i times, then the estimate of p_i is y_i/n and the estimate of the variance of $\hat{p}_i$ is

$$\frac{y_i(n-y_i)}{n^3}. \tag{3.13}$$

3.3.2 Biased Estimators

The four estimators $\bar{X}$, S^2, $\hat{p}$, and $\hat{p}_i$ discussed above are unbiased estimators of the parameters μ, σ^2, p, and p_i, respectively. In some cases *biased estimators* of a parameter arise: $\hat{\theta}$ is a biased estimator of θ if the mean value $E(\hat{\theta})$ of $\hat{\theta}$ differs from θ, and its *bias* is defined as $E(\hat{\theta}) - \theta$. A biased estimator of a parameter might be preferred to an unbiased estimator if its mean square error (defined below) is smaller than that of the unbiased estimator. An example of this is given in Problem 8.6. Another reason why biased estimators are relevant is that in some cases no unbiased estimator of a parameter exists. More generally, there might be no unbiased estimator of some function of a parameter. For example, although there is an unbiased estimator of the parameter p in a binomial distribution, there is no unbiased estimator of p^{-1} (see Problem 3.3).

When $\hat{\theta}$ is a biased estimator of θ its accuracy is usually assessed by its mean square error (MSE) rather than its variance. The MSE is defined by

$$\mathrm{MSE}(\hat{\theta}) = E\left((\hat{\theta} - \theta)^2 \right). \tag{3.14}$$

Thus the MSE of an unbiased estimator is its variance, and more generally it can be shown that

$$\text{MSE}(\hat{\theta}) = \text{Var}(\hat{\theta}) + \left(E(\hat{\theta}) - \theta\right)^2, \tag{3.15}$$

where $\text{Var}(\hat{\theta})$ is the variance of $\hat{\theta}$. It often happens, when $\hat{\theta}$ is a function of n random variables, that the bias of $\hat{\theta}$ is proportional to n^{-1}. In this case it follows from equation (3.15) that the MSE and the variance of $\hat{\theta}$ differ by a term proportional to n^{-2}, so that when n is large the two are close.

3.4   Classical Hypothesis Testing

3.4.1   General Principles

Classical statistical hypothesis testing involves the test of a *null hypothesis* against an *alternative hypothesis*. The procedure consists of five steps, the first four of which are completed before the data to be used for the test are gathered, and relate to probabilistic calculations that set up the statistical inference process.

We illustrate these steps by using the two DNA sequences given in (1.1). We call this the "sequence-matching" example and will refer to it several times.

The first step is to declare the null hypothesis H_0 and the alternative hypothesis H_1. In the sequence-matching example the null hypothesis might be one of uniformity and independence. That is, the null hypothesis might specify that each of the four nucleotides is generated with probability 0.25 at any site in any sequence, independently of all other nucleotides in both sequences, and in particular of the nucleotide at the same site in the other sequence.

The alternative hypothesis might specify that the probability of a match at any site is some value larger than 0.25, for example 0.35. So, if the (unknown) probability of a match at any site is p, the null hypothesis then claims that $p = 0.25$ and the alternative hypothesis then claims that $p = 0.35$.

The choice of null and alternative hypotheses should be made before the data are seen. To decide on a hypothesis as a result of the data is to introduce a bias into the procedure, invalidating any conclusion that might be drawn from it.

A hypothesis can be *simple* or *composite*. A simple hypothesis specifies the numerical values of all unknown parameters in the probability distribution of interest. In the above example, both null and alternative hypotheses are simple. A composite alternative does not specify all numerical values of all the unknown parameters. In the sequence-matching example, the alternative hypothesis "p exceeds 0.25" is composite. It is also *one-sided* $(p > 0.25)$ as opposed to *two-sided* $(p \neq 0.25)$.

The nature of the alternative hypothesis is determined by the context of the test, in particular whether it is one-sided up (that is the unknown parameter θ exceeds some specified value θ_0), one-sided down ($\theta < \theta_0$), or two-sided ($\theta \neq \theta_0$). In many cases in bioinformatics the natural alternative is both composite and one-sided. The sequence-matching case is an example: Unless there is some reason to choose a specific alternative such as $p = 0.35$, it seems more reasonable to choose the composite alternative $p > 0.25$.

Since the decision as to whether H_0 or H_1 is accepted will be made on the basis of data derived from some random process, it is possible that an incorrect decision will be made, that is, to reject H_0 when it is true (a *Type I error*), or to accept H_0 when it is false (a *Type II error*). When testing a simple null hypothesis against a simple alternative it is not possible, with a fixed predetermined sample size, to ensure that the probabilities of making a Type I error and a Type II error are both arbitrarily small. This difficulty is resolved in practice by observing that there is often an asymmetry in the implications of making the two types of error. In the sequence-matching case, for example, there might be more concern about making the false positive claim of a similarity between the two sequences when there is no such similarity, and less concern about making the false negative conclusion that there is no similarity when there is. For this reason a procedure frequently adopted is to fix the numerical value α of the Type I error at some acceptably low level (usually 1% or 5%), and not to attempt to control the numerical value of the Type II error. Step 2 of the hypothesis testing procedure consists in choosing the numerical value for the Type I error.

In practice, and in particular when discrete random variables are involved, it may be impossible to arrive at a procedure having exactly the Type I error chosen. This is illustrated in Step 4 below. In all later references to the choice of a Type I error, this caveat is to be taken as understood.

For the test of a simple null hypothesis against a simple alternative, again with a fixed sample size, the choice of the Type I error implicitly determines the numerical value β of the Type II error, or equivalently of the *power* of the test, defined as the probability $1 - \beta$ of rejecting the null hypothesis when the alternative is true.

When the alternative hypothesis is composite, Step 2 again consists in choosing the numerical value α of the Type I error. In this case there is no unique Type II error and thus no unique value for the power of the test. If the alternative hypothesis leaves one parameter unspecified, the power of the test depends on the value of this parameter, and can be described by a *power curve*, giving the probability that the null hypothesis is rejected as a function of that parameter.

Step 3 in the procedure consists in determining a *test statistic*. This is the quantity calculated from the data whose numerical value leads to acceptance or rejection of the null hypothesis. In the sequence-matching example one possible test statistic is the total number Y of matches. There is a substantial body of statistical theory associated with the optimal choice of test statistic; this is discussed in detail in Chapter 8. A poor choice of test statistic will lead to an inefficient testing procedure.

Step 4 consists in determining those observed values of the test statistic that lead to rejection of H_0. This choice is made so as to ensure that the test has the numerical value for the Type I error chosen in Step 2. We illustrate this step with the sequence-matching example. In this example, the total number Y of matches is chosen as the test statistic. Then in both the case of a simple alternative hypothesis such as "$p = 0.35$" and in the case of the composite alternative hypothesis "$p > 0.25$," the null hypothesis $p = 0.25$ is rejected in favor of the alternative when the observed value y of Y is sufficiently large, that is, if y is greater than or equal to some *significance point* K. If for example the Type I error is chosen as 5%, K is found from the requirement

$$\text{Prob(null hypothesis is rejected when it is true)}$$
$$= \text{Prob}(Y \geq K \,|\, p = 0.25) = 0.05. \tag{3.16}$$

The caveat discussed in Step 3 should be kept in mind here: It is often impossible to find a value of K such that (3.16) is satisfied exactly, and the choice of K made in practice is found by a conservative rounding procedure. For example, when $\alpha = .05$, $p = .25$ and $n = 100$, $\text{Prob}(Y \geq 32) = .069$ and $\text{Prob}(Y \geq 33) = .044$. In this case, we use the conservative value 33 for K.

For very long sequences, a normal approximation to the binomial might be employed. For example, if the sequences both have length 1,000,000, and the Type I error is chosen as 5%, K is determined by the requirement

$$\text{Prob}\left(X \geq K - \frac{1}{2}\right) = 0.05, \tag{3.17}$$

where X is a random variable having a normal distribution with mean $1{,}000{,}000(0.25) = 250{,}000$ and variance $1{,}000{,}000(0.25)(0.75) = 187{,}500$ and the continuity correction $\frac{1}{2}$ has been employed. The resulting value of K is 250,712.81; in practice the conservative value 250,713 would be used.

In the above example the null hypothesis is rejected if Y is sufficiently large. If the alternative hypothesis had specified a value of p that is less than 0.25, then the null hypothesis would be rejected for sufficiently small Y.

The final step (Step 5) in the testing procedure is to obtain the data, and to determine whether the observed value of the test statistic is equal to or more extreme than the significance point calculated in Step 4, and to reject the null hypothesis if it is.

3.4.2   P-Values

A testing procedure equivalent to that just described involves the calculation of a so-called *P-value*, or *achieved significance level*. Here Step 4 in the above sequence, the calculation of the significance point K, is not carried out. Instead, once the data are obtained, we calculate the null hypothesis probability of obtaining the observed value of the test statistic or one more extreme in the direction indicated by the alternative hypothesis. This probability is called the *P*-value. If the *P*-value is *less than* the chosen Type I error, the null hypothesis is rejected. This procedure always leads to a conclusion identical to that based on the significance point approach.

For example, the null hypothesis ($p = 0.25$) probability of observing 11 matches or more in a sequence comparison of length 26 (as in (1.1)) is found from the binomial distribution to be about 0.04. This is then the *P*-value associated with the observed number 11. If in the length 1000 sequence-matching case there are 278 matches, the *P*-value is, to a close approximation,

$$\text{Prob}(X \geq 277.5),$$

where X has a normal distribution with mean 250 and variance 187.5. This gives a *P*-value of 0.022. If the Type I error had been chosen to be 1%, the null hypothesis would then not be rejected, agreeing with the conclusion obtained using the significance point 283.

Before the experiment is conducted, the eventual *P*-value is a random variable. If the test statistic is continuous and the null hypothesis is true, the probability distribution of this *P*-value is the continuous uniform distribution (1.53). This follows from the fact that a *P*-value is the probability that the test statistic is more extreme than its observed value. In the case where the alternative hypothesis corresponds to small values of the test statistic, the *P*-value is $F_X(x)$, where x is the observed value of the test statistic and $F_X(x)$ is the distribution of the test statistic when the null hypothesis is true. The discussion following equation (1.120) then shows that the *P*-value has the uniform distribution (1.53) as claimed. A similar argument holds when the alternative hypothesis corresponds to large values of the test statistic or to both large and small values.

3.4.3   Hypothesis Testing Examples

Example 1. A classic test in statistics concerns the unknown mean μ of a normal distribution with known variance σ^2. One case of this is the test of the null hypothesis $\mu = \mu_0$ against the one-sided alternative hypothesis $\mu > \mu_0$. If this test is carried out using the observed values of random variables $X_1, X_2, \ldots, X_n$ having the normal distribution in question, the statistical theory of Chapter 8 leads to the use of $\bar{X}$ as an optimal test statistic and the rejection of the null hypothesis if the observed value $\bar{x}$ of $\bar{X}$ is "too much larger" than μ_0. More precisely, if the Type I error is 5%, the null hypothesis is rejected if

$$\bar{x} \geq \mu_0 + 1.645\sigma/\sqrt{n}. \tag{3.18}$$

This calculation uses the known variance σ^2/n of $\bar{X}$ given in (2.71) and the standardization procedure described in Section 1.10.2. A more realistic situation arises when σ^2 is unknown, in which case a *one-sample t-test* is used. Instead of describing this test we will describe below in Example 2 the more frequently occurring *two-sample t-test*.

If the alternative hypothesis had been $\mu < \mu_0$, the test statistic would again be $\bar{X}$, but the null hypothesis would now be rejected if the observed value $\bar{x}$ of $\bar{X}$ were less than $\mu_0 - 1.645\sigma/\sqrt{n}$. If the alternative hypothesis had been two-sided, that is "$\mu \neq \mu_0$," the null hypothesis would be rejected if $\bar{x} < \mu_0 - 1.96\sigma/\sqrt{n}$ or if $\bar{x} > \mu_0 + 1.96\sigma/\sqrt{n}$. This shows that the nature of the alternative hypothesis determines the values of the test statistic that lead to rejection of the null hypothesis. It will be shown in Section 3.4.5 that in some cases it can also determine the choice of the test statistic itself.

Example 2. A protein coding gene is a segment of the DNA that codes for a particular protein (or proteins). In any given cell type at any given time, this protein may or may not be needed. Each cell will generate the proteins it needs, which will usually be some small subset of all possible proteins. If a protein is generated in a cell, we say that the gene coding for this protein is *expressed* in that cell type. Furthermore, any given protein can be expressed at many different levels. One cell type might need more copies of a particular protein than another cell type. When this happens we say that the gene is *differentially expressed* between the two cell types. There are several techniques for measuring the level of gene expression in a cell type. All of these methods are subject to both biological and experimental variability. Therefore, one cannot simply measure the level of expression once in each cell type to compare for differential expression. Instead, one must repeat each experiment several times and perform a statistical test of the hypothesis that they are expressed at the same or different levels.

Suppose that the expression levels in two cell types are to be compared. In statistical terms, this comparison can be phrased as the test of the equality of two unknown means. Our aim is to measure the expression levels of m cells of one type and compare these with the expression levels of n cells of another type. Suppose that before the experiment the measurements $X_{11}, X_{12}, \ldots, X_{1m}$ from the first cell type are thought of as m NID(μ_1, σ^2) random variables, and the measurements $X_{21}, X_{22}, \ldots, X_{2n}$ from the second cell type are thought of as n NID(μ_2, σ^2) random variables. The null hypothesis states that $\mu_1 = \mu_2$ ($= \mu$, unspecified). We assume for the moment that the alternative hypothesis is $\mu_1 \neq \mu_2$. The value of the common variance σ^2 is unknown and is unspecified under both hypotheses.

The theory in Chapter 8 shows that under the assumptions made, the appropriate test statistic is t, defined by

$$t = \frac{(\bar{X}_1 - \bar{X}_2)\sqrt{mn}}{S\sqrt{m+n}},$$ (3.19)

with S defined from

$$S^2 = \frac{\sum_{i=1}^{m}(X_{1i} - \bar{X}_1)^2 + \sum_{i=1}^{n}(X_{2i} - \bar{X}_2)^2}{m+n-2}.$$

The null hypothesis probability distribution of t is well known (as the t distribution with $m+n-2$ degrees of freedom), and significance points of this distribution are widely available. This enables a convenient assessment of the significance of the observed value of t.

In reality, these expression levels cannot always be expected to have normal distributions, nor should the variances of the two types be expected to be equal. These two assumptions were made in the above t-test procedure, and the significance points of the t distribution are calculated assuming that both assumptions hold. Thus in practice it might not be appropriate to use the t-test to test for differential expression. This problem also arises in other practical situations, and leads to the introduction of alternative testing procedures that do not rely on the normality and equal variances assumptions. The equal variance assumption is discussed further in Section 8.4, and the normal distribution assumption will be discussed in Section 3.5.

Example 3. In this example we consider a test of the null hypothesis that prescribes specific values for the probabilities $\{p_i\}$ in the multinomial distribution (2.6). The alternative hypothesis considered here is composite and leaves these probabilities unspecified. This can be used to test for prescribed probabilities for the four nucleotides in a DNA sequence.

Let Y_i be the number of observations in category i. A test statistic often used for this testing procedure is X^2, defined by

$$X^2 = \sum_{i=1}^{k} \frac{(Y_i - np_i)^2}{np_i}.$$ (3.20)

Sufficiently large values of the observed value

$$\sum_{i=1}^{k} \frac{(y_i - np_i)^2}{np_i}$$ (3.21)

of X^2 lead to rejection of the null hypothesis. The quantity (3.21) may be thought of as a measure of the discrepancy between the observed values $\{y_i\}$ and the respective null hypothesis means $\{np_i\}$.

When the null hypothesis is true and n is large, X^2 has approximately the chi-square distribution (1.70) with $\nu = k - 1$ degrees of freedom. Thus the statistic X^2 is frequently referred to as the "chi-square statistic," and this explains the notation X^2. Tables of the significance points of the chi-square distribution are widely available for all values of ν likely to arise in practice.

The choice of the test statistic (3.20) is not arrived at from the optimality statistical theory of Section 8.4.2. Instead, that theory and the discussion following (8.53) leads to the test statistic

$$2\sum_i Y_i \log \frac{Y_i}{np_i}. \tag{3.22}$$

When the null hypothesis is true and the sample size is large, the numerical values of the statistics (3.20) and (3.22) are usually quite close, and this can be thought of as a justification for the use of (3.20).

Example 4. Association tests. The chi-square procedure described in Example 3 has several generalizations, one of which we describe here. Data often

		column					Total
		1	2	3	$\cdots$	c	
row	1	Y_{11}	Y_{12}	Y_{13}	$\cdots$	Y_{1c}	$y_{1\cdot}$
	2	Y_{21}	Y_{22}	Y_{23}	$\cdots$	Y_{2c}	$y_{2\cdot}$
	$\vdots$	$\vdots$	$\vdots$	$\vdots$	$\ddots$	$\vdots$	$\vdots$
	r	Y_{r1}	Y_{r2}	Y_{r3}	$\cdots$	Y_{rc}	$y_{r\cdot}$
Total		$y_{\cdot 1}$	$y_{\cdot 2}$	$y_{\cdot 3}$	$\cdots$	$y_{\cdot c}$	y

TABLE 3.1. Two-way table data.

arise in the form of a two-way table such as Table 3.1. As an example, if there are two rows (corresponding to males and females) and two columns (corresponding to left- and right-handed), the data in the four positions in the table give, for some sample of individuals, the numbers of left-handed males, of right-handed males, of left-handed females, and of right-handed females. The null hypothesis is that there is no association between row categorization and column categorization; in the above example this is that there is no association between gender and handedness. In other words, the null hypothesis claims that the probabilities that a male is (respectively) left- or right-handed are the same as corresponding probabilities for females.

The row and column totals are not of direct interest and can often be chosen in advance. They are thus written in lowercase in Table 3.1. The testing procedure described below is valid only if the y observations leading to the counts $\{Y_{jk}\}$ in the table are all independent of each other. Thus if two of the observations in the handedness example come from identical twins, then the independence requirement is not met. However, all individuals are ultimately related, so that their genetic material is also not entirely independent. Thus association tests using DNA sequences must be used with care. The extent to which this observation significantly invalidates use of the chi-square procedure clearly depends on the test and the data used, and in practice the procedure is often carried out without much attention being paid to this point.

Given the row and column totals, it can be shown that when the null hypothesis is true, Y_{jk} is a random variable with mean value E_{jk}, where $E_{jk} = y_{j.} y_{.k}/y$. Thus when the null hypothesis is true, the observed value of Y_{jk} should be close to E_{jk}. This argument leads to the frequently used chi-square test statistic

$$\sum_{jk} \frac{(Y_{jk} - E_{jk})^2}{E_{jk}}, \tag{3.23}$$

which can be regarded as a measure of the difference of the Y_{jk} and E_{jk} values. When the null hypothesis is true and the independence requirement discussed above holds, the statistic (3.23) has an asymptotic chi-square distribution with $\nu = (r-1)(c-1)$ degrees of freedom.

Some examples of the application of this test in a genetic context are given in Sections 5.2, 5.3.4, and 6.1, and a theoretical discussion is given in Example 4 of Section 8.4.2.

The theory of Section 8.4.2 leads to the use of the test statistic

$$2\sum_{jk} Y_{jk} \log \frac{Y_{jk}}{E_{jk}} \tag{3.24}$$

as the appropriate test statistic, rather than the statistic (3.23). The numerical values of the two statistics are usually close when the null hypothesis is true and the sample size is large.

3.4.4   Likelihood Ratios, Information, and Support

In this section we briefly discuss aspects of the choice of test statistic in a hypothesis testing procedure. It will be shown in Chapter 8 that in testing a simple null hypothesis against a simple alternative hypothesis, a reasonable optimality argument leads to the use of the ratio of the probability of the data under the alternative hypothesis to the probability of data under the null hypothesis as the test statistic. For obvious reasons, this is called the *likelihood ratio*.

Consider the sequence-matching example and define the indicator variable Y_i by $Y_i = 1$ if the pair in position i match and $Y_i = 0$ if they do not, for $i = 1, 2, \ldots, n$. If the null hypothesis probability of a match is 0.25 and the alternative hypothesis probability is 0.35, the likelihood ratio is

$$\frac{(0.35)^{\sum Y_i}(0.65)^{(n-\sum Y_i)}}{(0.25)^{\sum Y_i}(0.75)^{(n-\sum Y_i)}}. \tag{3.25}$$

This simplifies to

$$\left(\frac{7}{5}\right)^Y \left(\frac{13}{15}\right)^{n-Y}, \tag{3.26}$$

where $Y = \sum Y_i$ is the (random) total number of matches. The logarithm of this expression is

$$Y\log\left(\frac{7}{5}\right) + (n-Y)\log\left(\frac{13}{15}\right). \tag{3.27}$$

The definition of accumulated support in equation (2.75) shows that this is the accumulated support, after n pairs have been observed, for the claim that the probability of a match is 0.35. If the true probability of a match is p, the expected value of the expression in (3.27) is

$$n\left(p\log\left(\frac{7}{5}\right) + (1-p)\log\left(\frac{13}{15}\right)\right). \tag{3.28}$$

3.4.5 Hypothesis Testing Using $X_{\max}$ as Test Statistic

Let $X_1, X_2, \ldots, X_n$ be independent normal random variables, each having variance 1. Suppose the null hypothesis states that the mean of each of these random variables is 0, while the alternative hypothesis claims that one of these random variables has mean μ greater than 0, the remaining variables having mean 0. It is not, however, stated by the alternative hypothesis which of the random variables has mean μ. An example where this situation arises in practice is discussed in Example 2 of Section 8.4.2.

This is a different alternative hypothesis from the one considered in Example 1 of Section 3.4.3. The statistical principles given in Chapter 8 that lead to $\bar{X}$ as optimal test statistic in that example lead in this case to $X_{\max}$, the maximum of $X_1, X_2, \ldots, X_n$, as optimal test statistic. If the desired Type I error is α and K is computed so that $\text{Prob}(X_{\max} \geq K \,|\, \text{null hypothesis true}) = \alpha$, the null hypothesis is rejected if $X_{\max} \geq K$. Equivalently, if $x_{\max}$ is the observed value of $X_{\max}$, the null hypothesis is rejected if the P-value $\text{Prob}(X_{\max} \geq x_{\max} \,|\, \text{null hypothesis true})$ is less than α.

The calculation of the value of K is found in principle from the null hypothesis probability distribution of the maximum $X_{\max}$ of n NID$(0,1)$ random variables. No simple exact form is available for this distribution. Various approximations can be found in the literature, usually based on the asymptotic density function (2.122) for the maximum of n iid continuous random variables. However, a direct calculation of P-values is possible using normal distribution tables and equation (2.91). For example, suppose that $n = 100$ and $x_{\max} = 3.6$. The probability that an $N(0,1)$ random variable is less than 3.6 is approximately 0.9998409, and equation (2.91) then shows that the P-value associated with the observed value 3.6 of $X_{\max}$ is approximately $1 - (0.9998409)^{100} \approx 0.0158$, which is significant if the Type I error is 0.05. If $n = 500$ and $x_{\max} = 3.6$, the approximate P-value is $1 - (0.9998409)^{500} \approx 0.0765$, which is not significant if the Type I error is 0.05.

3.5   Nonparametric Alternatives to the Two-Sample t-Test

The hypothesis testing methods described above may be summarized briefly as follows. We are concerned with random variables having some probability distribution that depends on one or more unknown parameters. Apart from the unknown parameters, the form of the distribution is assumed known, for example normal. The aim is to test hypotheses about one or more of the unknown parameters, using the observed value of some test statistic. The test statistic is often determined by theoretical and optimality arguments. The null hypothesis is rejected if this observed value is sufficiently extreme, in some well-defined sense.

The two-sample t-test, introduced in Example 2 of Section 3.4.3, is an example of such a test. In this test we consider a set $X_{11}, X_{12}, \ldots, X_{1m}$ of $N(\mu_1, \sigma^2)$ random variables and a second set $X_{21}, X_{22}, \ldots, X_{2n}$ of $N(\mu_2, \sigma^2)$ random variables. All $m + n$ random variables are independent. The observed values of these random variables are used to calculate the observed value of the two-sample t-test statistic defined in (3.19). The null hypothesis claims that $\mu_1 = \mu_2$, and if, for example, the alternative hypothesis is $\mu_1 > \mu_2$, the null hypothesis is rejected if the observed value of t is greater than or equal to a significance point determined so that the Type I error is at the chosen value.

These significance points are provided by t tables. The values in these tables are calculated under the assumption that the random variables X_{jk} have normal distributions that have equal means and variances. If either the normal distribution assumption or the equal variance assumption is not justified, the use of significance points given by the t tables is not valid. In Example 2 of Section 3.4.3 it is possible in practice that neither assumption is justified.

This difficulty leads to hypothesis testing methods that do not rely on any specific assumption about the form of the probability distribution of the X_{jk}. Such methods are called *nonparametric*, or (perhaps more accurately) *distribution-free*. We now describe two nonparametric alternatives to the two-sample t-test, the *Mann–Whitney* test and the *permutation* test.

In both tests it is assumed, under the null hypothesis, that all $m + n$ random variables involved not only have the same mean, but also have the same (continuous) probability distribution. However no statement is made about the form of this distribution. The alternative hypothesis is that the (common) mean of the X_{1j}'s is different from the (common) mean of the X_{2j}'s, in some specified way, for example that it exceeds this mean.

The Mann–Whitney Test

In the Mann-Whitney test the observed values $X_{11}, X_{12}, \ldots, X_{1m}$ and $X_{21}, X_{22}, \ldots, X_{2n}$, namely $x_{11}, x_{12}, \ldots, x_{1m}$ and $x_{21}, x_{22}, \ldots, x_{2n}$, are listed in increasing order, and each observation is associated with its rank in this list. Thus each observation is associated with one of the numbers $1, 2, \ldots, m+n$. (This statement assumes that no two observations take the same value. A generalization of the procedure is available when ties exist.) The test statistic is the sum of the ranks of the observations in the first group. The sum of the ranks of all $m+n$ observations is $(m+n)(m+n+1)/2$ (see Section B.18). The observations in the first group provide a fraction $m/(m + n)$ of all observations, and proportionality arguments then show that when the null hypothesis is true, the expected value of the sum of the ranks of the observations in the first group is the fraction $m/(m + n)$ of the sum $(m + n)(m + n + 1)/2$ of all ranks. Thus this expected value is $m(m + n + 1)/2$. A more advanced calculation (see Problem 3.6) shows that the null hypothesis variance of the sum of the ranks for the first group in the sample is $mn(m + n + 1)/12$. It is also possible to show that this sum has very close to a normal distribution. The Mann–Whitney test thus becomes a test for the value of the mean of a normal distribution, whose null-hypothesis variance is known. This test will be either one-sided, or two-sided, depending on the context, and is carried out as in Example 1 of Section 3.4.3.

The Permutation Test

The permutation test is carried out as follows. When the null hypothesis is true, all possible $\binom{m+n}{m}$ distinct permutations of the data, for which m randomly chosen data values are taken as the "observations" for the first group and the remaining n taken as the "observations" for the second group, are equally likely. For each such permutation we calculate the value of some test statistic. There is no clear-cut guidance for the choice of test statistic, and often the classical t statistic given in equation (3.19) is used. If this is done, one of the values of t will be that arising for the permutation corresponding to the observed data. If the alternative hypothesis claims that the mean of the first group exceeds that of the second group, then using a Type I error α, the null hypothesis is rejected if the observed value of t is among the largest positive $100\alpha\%$ permutation values.

This procedure has several appealing properties. First, it falls under the general heading of permutation tests, about which there exists substantial theory. Second, it is not necessary to find the probability distribution of the test statistic used, a procedure that can be quite difficult for parametric tests. Third, if t is chosen as the test statistic, it is not necessary to compute the value of t for each of the permutations, and a simpler equivalent procedure requires only the calculation of the difference $\bar{x}_1 - \bar{x}_2$ of the averages of the "observations" in the two groups for each permutation. Indeed, it is necessary only to calculate the average of the "observations" in the first group. This follows because the sum of the observations in the two groups is invariant under permutation, and the average of the "observations" in the second group, and hence $\bar{x}_1 - \bar{x}_2$, is determined by the average of the "observations" in the first group. For comparison with the bootstrap procedure discussed in Section 12.3, it is convenient to use the t statistic as the test statistic.

A final appealing feature is that when the data do have a normal distribution, use of the permutation procedure gives results close to those obtained using the t-test.

Clearly the permutation procedure, when applicable, will involve substantial computation unless both n and m are small, since the number of different permutations is extremely large even for moderately large values of n and m. If this number is too large to allow calculation of all possible values of $\bar{x}_1 - \bar{x}_2$, a random sample of a large number R of permutations might be taken, and the null hypothesis rejected if the observed value is among the largest $100\alpha\%$ positive values so found. Current computing power now make a direct computational procedure such as this increasingly more feasible for hypothesis testing.

The permutation method and the Mann–Whitney procedure suffer one disadvantage. The null hypothesis assumption that the distribution of an observation in the first group is the same as that of an observation in the second group is stronger than the hypothesis of interest, namely that the means of the two groups differ. If for example the means of the two groups are the same, but the variances differ, the symmetry assumptions on which the method is based break down. The bootstrap alternative discussed in Section 12.3 overcomes this problem.

There are several other nonparametric alternatives to the two-sample t-test. The reason for the existence of several tests is that in the nonparametric case the choice of test statistic is made on common-sense grounds rather than by theory or optimality principles, so that several reasonable choices of test statistic are usually possible. This matter is discussed further in the context of BLAST in Section 9.7.

3.6   The Bayesian Approach to Hypothesis Testing

Let H be some hypothesis and let $\text{Prob}(H)$ be the probability that H is true, before any data are seen. This is called the *prior* probability of H, and some of the controversies referred to in Section 3.2 refer to the meaningfulness of such a probability or to its numerical value if meaningful. We proceed here assuming that this probability is meaningful and that its numerical value is specified.

After the data D are obtained, the conditional probability formula (1.95) gives the conditional probability of the hypothesis H, given the data D, in terms of the probability of D given H. Specifically, the *posterior* probability $\text{Prob}(H \mid D)$ of the hypothesis H, given the data D, is

$$\text{Prob}(H \mid D) = \frac{\text{Prob}(H)\,\text{Prob}(D \mid H)}{\text{Prob}(D)}. \tag{3.29}$$

This posterior probability reflects the change in the value of the prior probability $\text{Prob}(H)$ due to the information in the data D.

The way in which equation (3.29) is used when there is a finite number of hypotheses differs from that when there is an infinite number of hypotheses. An example of each case is given below.

The Finite Prior Distribution Case

Suppose that there are $h + 1$ different hypotheses, $H_0, H_1, \ldots, H_h$, with respective prior probabilities $\pi_0, \pi_1, \ldots, \pi_h$, $\sum_j \pi_j = 1$. Equation (3.29) shows that given D, the posterior probability of H_i is

$$\text{Prob}(H_i \mid D) = \frac{\text{Prob}(H_i)\,\text{Prob}(D, \mid H_i)}{\text{Prob}(D)}. \tag{3.30}$$

By (1.96), the denominator can be written as

$$\sum_j \pi_j \,\text{Prob}(D \mid H_j). \tag{3.31}$$

Thus the posterior probability that hypothesis H_i is true becomes

$$\text{Prob}(H_i \mid D) = \frac{\pi_i \,\text{Prob}(D \mid H_i)}{\sum_j \pi_j \,\text{Prob}(D \mid H_j)}. \tag{3.32}$$

The preferred hypothesis is then the one that maximizes $\text{Prob}(H_i \mid D)$. Since the denominator in (3.32) does not depend on i, this is the same hypothesis that maximizes the numerator of (3.32).

Example. A discrete prior distribution for a binomial parameter p. Consider the sequence-matching example described above. Suppose the prior probability of the null hypothesis H_0: $p = 0.25$ is $\frac{1}{3}$ and the prior probability of the alternative hypothesis H_1: $p = 0.35$ is $\frac{2}{3}$. Suppose that an alignment of length $n = 1{,}000$ yields 290 matches; this number is the datum D. The probability of D is calculated, under both H_0 and H_1, from the binomial distribution (1.8).

The posterior probability that H_0 is true is then calculated from (3.32) as

$$\frac{\frac{1}{3} \cdot \binom{1000}{290}(0.25)^{290}(0.75)^{710}}{\frac{1}{3} \cdot \binom{1000}{290}(0.25)^{290}(0.75)^{710} + \frac{2}{3} \cdot \binom{1000}{290}(0.35)^{290}(0.65)^{710}}, \qquad (3.33)$$

which is very close to 1, and the posterior probability that H_1 is true is very close to 0. Thus the null hypothesis would not be rejected if we adopt the rule to accept the more likely posterior hypothesis. This is the opposite conclusion from that reached by classical methods with the same data and any commonly used Type I error.

The Continuous Prior Distribution Case

Suppose now in the example above that the set of possible hypothesized values for p is "all values of p in $(0, 1)$." This implies that p has a continuous prior density function over $(0, 1)$. As an example, we suppose that the prior density function of p is the beta distribution (1.71), rewritten (since the random variable is now the parameter p) as

$$f(p) = \frac{\Gamma(\alpha + \beta)}{\Gamma(\alpha)\Gamma(\beta)} p^{\alpha-1}(1 - p)^{\beta-1}, \quad 0 < p < 1. \qquad (3.34)$$

Suppose that in a sequence comparison of length n there are y matches. The probability of these data under the iid assumption is given by the binomial distribution (1.8). The numerator in the continuous analogue of (3.32) is then

$$\frac{\Gamma(\alpha + \beta)}{\Gamma(\alpha)\Gamma(\beta)} \binom{n}{y} p^{\alpha+y-1}(1 - p)^{\beta+n-y-1}, \qquad (3.35)$$

while the denominator is

$$\int_0^1 \frac{\Gamma(\alpha + \beta)}{\Gamma(\alpha)\Gamma(\beta)} \binom{n}{y} x^{\alpha+y-1}(1 - x)^{\beta+n-y-1} dx \qquad (3.36)$$

$$= \frac{\Gamma(\alpha + \beta)}{\Gamma(\alpha)\Gamma(\beta)} \binom{n}{y} \frac{\Gamma(\alpha + y)\,\Gamma(\beta + n - y)}{\Gamma(\alpha + \beta + n)}. \qquad (3.37)$$

Thus the posterior probability distribution of the binomial parameter p is

$$f_{\text{post}}(p) = \frac{\Gamma(\alpha + \beta + n)}{\Gamma(\alpha + y)\Gamma(\beta + n - y)} p^{\alpha+y-1}(1 - p)^{\beta+n-y-1}. \qquad (3.38)$$

By analogy with the discrete case, the hypothesized value of p could be taken as the value where this posterior distribution is maximized, namely

$$\frac{\alpha + y - 1}{\alpha + \beta + n - 2}. \qquad (3.39)$$

3.7   The Bayesian Approach to Estimation

There are various ways in which a Bayesian might reasonably estimate an unknown parameter. These methods share the common feature that they are all based on the posterior distribution of that parameter. In this section we describe one such estimation method, namely that of estimating the parameter by the mean of the posterior distribution of that parameter.

In the binomial example of the previous section, the mean of the posterior distribution is given by

$$\int_0^1 p f_{\text{post}}(p) dp = \frac{\alpha + y}{\alpha + \beta + n}, \tag{3.40}$$

and this is the Bayesian estimate of p. It is clearly a mixture of the parameters α and β in the prior distribution and the data values y and n, being heavily influenced by the latter when α and β are small and y and n are large.

The estimate (3.40) would be identical to the classical estimate y/n if $\alpha = \beta = 0$. However, the beta distribution (3.35) with these parameters is improper (i.e., the integral of the density function over the domain $(0, 1)$ is infinite), so that in this sense the Bayesian estimator and the classical estimator are never equal. However, a relation between the estimator (3.40) and the classical estimator can be found through the concept of "pseudo-counts." If we imagined that we observed α successes and β failures apart from the y successes and $n - y$ failures actually seen in the data (so we replace y by $y + \alpha$ and $n - y$ by $n - y + \beta$), then the classical estimator of p would be given by (3.40). The numbers α and β are then called "pseudo-counts": We do not actually observe these successes and failures, but act as though we do and then use classical estimation theory.

Choosing α and β large overcomes the difficulty mentioned at the beginning of this chapter, that it is unreasonable to use the classical estimate of 1 for the probability of a head if a fair-looking coin gives three heads from three tosses. The larger the values of α and β, the closer the Bayesian estimate found from (3.40) is to 0.5. On the other hand, the larger that α and β are chosen, the more the information in the data is ignored. This introduces the problem, pointed out under the classical approach to estimation, of having to make a possibly arbitrary choice of α and β.

There is a direct extension of the pseudocounts concept to the multinomial case, and this extension is used in Section 6.6 for the purpose of sequence segment alignment.

3.8   Multiple Testing

A problem that arises often in the statistical analyses of large data sets is that of multiple testing. We illustrate this issue with an example.

As described in Example 2 on page 114, the two-sample t-test might be used to test for significant evidence of differential expression of genes between two cell types. It has recently become possible to measure the gene expression of thousands of genes simultaneously, using microarray technology. Suppose that we want to test simultaneously for differential expression between two cell types for all genes on an array. Generally, the number of differentially expressed genes between any two cell types is not large. Suppose that 1000 genes are represented on the array, and that of these, 40 are differentially expressed. In order to find these 40 genes we might use the two-sample t-test for each gene, using for each gene a Type I error of 5%. This will produce approximately $(.05) \cdot 960 = 48$ false positives. Thus there will probably be more false positives than true positives. Despite having started with a reasonable Type I error of 5% for each gene, the *experiment-wise* Type I error is not 5%.

This comment requires a more formal definition of the experiment-wise Type I error. We formulate a collection g of null hypotheses, null hypothesis j stating that for gene j there is no differential expression between the two cell types. It is assumed initially that the data used for any one test are independent of the data used for any other test. An experiment-wise Type I error is the probability of rejecting at least one null hypothesis, given that all are true. If 1000 tests are simultaneously conducted, an experiment-wise Type I error of 5% could be achieved by using a Type I error of 0.005% for each t-test. This is an example of the *Bonferroni* correction, in which the Type I error for each of g tests is replaced by α/g, where α is the desired experiment-wise Type I error. This is a conservative procedure: a less conservative procedure is to replace the Type I error for each of the g tests by $K(g, \alpha)$, where $K(g, \alpha)$ is defined in equation (2.157) with n replaced by g. If this is done the experiment-wise Type I error is exactly α. This is the *Šidák* procedure. Both of the Bonferroni and the Šidák procedures are symmetric in the sense that the Type I errors for the various individual t-tests are all equal. They are also very small when g is large, so that both the Bonferroni and the Šidák procedures make quite stringent requirements for the rejection of any one of the null hypotheses. They may lead us to miss most of the truly differentiated genes. This is an example of the *multiple testing problem*. In principle, the problem can be overcome by taking many more observations in each t-test, but the cost of each experiment might make this impractical.

In some applications it might not be desirable to focus on the experiment-wise Type I error. For example, if one is looking for differentially expressed genes then one might tolerate a high Type I error, as long as the percentage of false positives among the predictions is small. Controlling the experiment-wise Type I error controls the probability of having any false positives at all, but at the cost of possibly missing many true positives. The less stringent approach of controlling the percentage of false positives among the actual predictions is taken, for example, by Manduchi et al. (2000).

The multiple testing problem arises in several applications of statistics and has been discussed extensively in the statistical literature; for a good survey see Shaffer (1995), in which many further references to this problem may be found. The multiple testing problem also arises in linkage analysis, where a large number of marker loci are tested simultaneously for linkage with a purported disease locus. We discuss the multiple testing problem further in Section 12.4 in the context of computationally intensive methods, since in practice the multiple testing problem and computationally intensive methods often arise together. In Section 12.4 we also consider the case where, as is likely with an expression array analysis, the data used for any one test are not necessarily independent of the data used for any other test.

Problems

3.1 Suppose that each of n iid random variables has a normal distribution with mean μ, variance 1. Under the null hypothesis $\mu = 0$, while under the alternative hypothesis $\mu > 0$. Find the P-value associated with an observed value of 4.20 for the maximum of the observed values of these random variables when $n = $ (i) 1,000, (ii) 10,000, (iii) 100,000, (iv) 1,000,000.

3.2 If $\hat{p}$ is the proportion of successes in n independent Bernoulli trials each having probability p of success, show that the mean value of $\hat{p}(1 - \hat{p})/n$ is $p(1 - p)/n - p(1 - p)/n^2$. Comment on the form of this expression when $n = 1$.

3.3 Suppose that $g(Y)$ is an unbiased estimator of p^{-1} in the binomial distribution (1.8). Then it follows that

$$\sum_{y=0}^{n} g(y)\, p \binom{n}{y} p^y (1 - p)^{n-y} = 1.$$

The left-hand side is a Taylor series in p. Use the result of Section B.13 to show that this equation cannot be solved for $g(y)$ uniformly for a continuum of values of p. This conclusion shows that there can be no unbiased estimator of p^{-1}.

3.4 *Continuation.* Find an unbiased estimator of p^2 and an unbiased estimator of p^3. What functions of p do you think admit unbiased estimation?

3.5 Prove equation (3.15).

3.6 (This example indicates a use of indicator functions in statistics.) The sum of the ranks randomly allocated to group 1 in the Mann–Whitney test of Section 3.5 can be written as $\sum_{j=1}^{m+n} jI_j$, where $I_j = 1$ if the jth ranked observation is allocated to group 1, and $I_j = 0$ otherwise. Prove $E(I_j) = m/(m+n)$ and $E(I_jI_k) = m(m-1)/(m+n)(m+n-1)$, $(j \neq k)$, and use these results to show that the variance of the sum of the ranks randomly allocated to group 1 is $mn(m+n+1)/12$.

3.7 Suppose that $X_1, X_2, \ldots, X_n$ are NID(μ, σ^2). Suppose that the prior distribution of μ is normal with mean μ_0 and variance σ_0^2. Show that the posterior distribution of μ is normal, with mean which is a linear combination of $\bar{X}$ and μ_0. Show that the variance of this distribution is the reciprocal of the sum of the reciprocals of σ_0^2 and the variance of $\bar{X}$.

3.8 Show that when the Šidák value (2.157) is used the experiment-wise Type I error is exactly α.

3.9 Consider the permutation two-sample t-test, based on observations $x_{11}, x_{12}, \ldots, x_{1n}$ and $x_{21}, x_{22}, \ldots, x_{2m}$. By writing the numerator of the variance estimator (used in the calculation of the t statistic) as

$$\sum_{j=1}^{m} x_{1j}^2 + \sum_{j=1}^{n} x_{2j}^2 - (m+n)\left(\frac{\sum_{j=1}^{m} x_{1j} + \sum_{j=1}^{n} x_{2j}}{m+n}\right)^2 - \frac{mn}{n+m}(\bar{x}_1 - \bar{x}_2)^2,$$

show that, under permutation, use of t as the test statistic is equivalent to use of $\bar{x}_1 - \bar{x}_2$.

3.10 *Continuation.* Find the mean and variance under permutation of the average $\bar{x}_1$ of the "observations" in the first group. (The answer is a function of $x_{11}, x_{12}, \ldots, x_{1m}$ and $x_{21}, x_{22}, \ldots, x_{2n}$.)

4
Stochastic Processes (i): Poisson Processes and Markov Chains

4.1 The Homogeneous Poisson Process and the Poisson Distribution

In this section we state the fundamental properties that define a Poisson process, and from these properties we derive the Poisson distribution, introduced in Section 1.3.6.

Suppose that a sequence of random events occur during some time interval. These events form a homogeneous Poisson process if the following two conditions are met:

(1) The occurrence of any event in the time interval (a, b) is independent of the occurrence of any event in the time interval (c, d), where (a, b) and (c, d) do not overlap.

(2) There is a constant $\lambda > 0$ such that for any sufficiently small time interval $(t, t + h)$, $h > 0$, the probability that one event occurs in $(t, t + h)$ is independent of t, and is $\lambda h + o(h)$ (the $o(h)$ notation is discussed in Section B.8), and the probability that more than one event occurs in the interval $(t, t + h)$ is $o(h)$.

Condition 2 has two implications. The first is *time homogeneity*: The probability of an event in the time interval $(t, t+h)$ is independent of t. Second, this condition means that the probability of an event occurring in a small time interval is (up to a small order term) proportional to the length of the interval (with fixed proportionality constant λ). Thus the probability of no events in the interval $(t, t + h)$ is

$$1 - \lambda h + o(h), \tag{4.1}$$

and the probability of one *or more* events in the interval $(t, t + h)$ is

$$\lambda h + o(h).$$

Various naturally occurring phenomena follow, or very nearly follow, these two conditions. Continuing the example in Section 1.13, suppose a cellular protein degrades spontaneously, and the quantity of this protein in the cell is maintained at a constant level by the continual generation of new proteins at approximately the degradation rate. The number of proteins that degrade in any given time interval approximately satisfies conditions 1 and 2. The justification that condition 1 can be assumed in the model is that the number of proteins in the cell is essentially constant and that the spontaneous nature of the degradation process makes the independence assumption reasonable. The discussion leading to (1.106) and that following equation (2.84) makes condition 2 reasonable for spontaneously occurring phenomena. This condition also follows from the same logic discussed in the example on page 9 that when np is small, the probability of at least one success in n Bernoulli trials is approximately np. For a precise treatment of this issue, see Feller (1968).

We now show that under conditions 1 and 2, the number N of events that occur up to any arbitrary time t has a Poisson distribution with parameter λt.

At time 0 the value of N is necessarily 0, and at any later time t, the possible values of N are 0, 1, 2, 3, ... We denote the probability that $N = j$ at any given time t by $P_j(t)$. Note that this is a departure from our standard notational convention, which would be $P_N(j)$ with t an implicit parameter. This notational change is made because the main interest here is in assessing how $P_j(t)$ behaves as a function of j and t.

The event that $N = 0$ at time $t + h$ occurs only if no events occur in $(0, t)$ and also no events occur in $(t, t + h)$. Thus for small h,

$$P_0(t + h) = P_0(t)(1 - \lambda h + o(h)) = P_0(t)(1 - \lambda h) + o(h). \tag{4.2}$$

The first equality follows from conditions 1 and 2.

The event that $N = 1$ at time $t + h$ can occur in two ways. The first is that $N = 1$ at time t and that no event occurs in the time interval $(t, t + h)$, the second is that $N = 0$ at time t and that exactly one event occurs in the time interval $(t, t + h)$. This gives

$$P_1(t + h) = P_0(t)(\lambda h) + P_1(t)(1 - \lambda h) + o(h), \tag{4.3}$$

where the term $o(h)$ is the sum of two terms, both of which are $o(h)$. Finally, for $j = 2, 3, \ldots$, the event that $N = j$ at time $t + h$ can occur in three different ways. The first is that $N = j$ at time t and that no event occurs in the time interval $(t, t+h)$. The second is that $N = j-1$ at time t and that exactly one event occurs in $(t, t+h)$. The final possibility is that $N \leq j-2$ at time t and that two or more events occur in $(t, t+h)$. Thus for $j = 2, 3, \ldots$,

$$P_j(t + h) = P_{j-1}(t)(\lambda h) + P_j(t)(1 - \lambda h) + o(h). \qquad (4.4)$$

Equations (4.3) and (4.4) look identical, and the difference between them relates only to terms of order $o(h)$. Therefore, we can take (4.4) to hold for all $j \geq 1$. Subtracting $P_j(t)$ ($j \geq 1$) from both sides of equation (4.4) and $P_0(t)$ from both sides of equation (4.2), and then dividing through by h, we get

$$\frac{P_0(t + h) - P_0(t)}{h} = -\frac{P_0(t)(\lambda h) + o(h)}{h}$$

$$\frac{P_j(t + h) - P_j(t)}{h} = \frac{P_{j-1}(t)(\lambda h) - P_j(t)(\lambda h) + o(h)}{h},$$

$j = 1, 2, 3, \ldots$. Letting $h \to 0$, we get

$$\frac{d}{dt} P_0(t) = -\lambda P_0(t), \qquad (4.5)$$

and

$$\frac{d}{dt} P_j(t) = \lambda P_{j-1}(t) - \lambda P_j(t), \quad j = 1, 2, 3, \ldots. \qquad (4.6)$$

The $P_j(t)$ are subject to the conditions

$$P_0(0) = 1, \quad P_j(0) = 0, \quad j = 1, 2, 3, \ldots. \qquad (4.7)$$

Equation (4.5) is one of the most fundamental of differential equations, and has the solution

$$P_0(t) = Ce^{-\lambda t}. \qquad (4.8)$$

The condition $P_0(0) = 1$ implies $C = 1$, leading to

$$P_0(t) = e^{-\lambda t}. \qquad (4.9)$$

Given this, we now show that the set of equations (4.6) has the solution

$$P_j(t) = \frac{e^{-\lambda t}(\lambda t)^j}{j!}, \quad j = 1, 2, 3, \ldots. \qquad (4.10)$$

The method for solving these equations follows the induction procedure described in Section B.18. Equation (4.9) shows (4.10) is true for $j = 0$. It must next be shown that the assumption that (4.10) holds for $j-1$ implies that it holds for j. Assuming that (4.10) is true for $j - 1$, equation (4.6) gives

$$\frac{d}{dt} P_j(t) = \frac{\lambda e^{-\lambda t}(\lambda t)^{j-1}}{(j-1)!} - \lambda P_j(t).$$

From this,

$$e^{\lambda t}\left(\frac{d}{dt} P_j(t) + \lambda P_j(t)\right) = \frac{\lambda(\lambda t)^{j-1}}{(j-1)!}.$$

This equation may be rewritten as

$$\frac{d}{dt}\left(P_j(t)e^{\lambda t}\right) = \frac{\lambda(\lambda t)^{j-1}}{(j-1)!},$$

and integration of both sides of this equation gives

$$P_j(t)e^{\lambda t} = \frac{(\lambda t)^j}{j!} + C,$$

for some constant C. From (4.7) it follows that $C = 0$. Thus

$$P_j(t) = \frac{e^{-\lambda t}(\lambda t)^j}{j!}, \quad j = 0, 1, 2, \ldots. \tag{4.11}$$

This completes the induction, showing that at time t the random variable N has a Poisson distribution with parameter λt.

Conditions 1 and 2 are often taken as giving a mathematical definition of the concept of "randomness," and since many calculations in bioinformatics, some of which are described later in this book, make the randomness assumption, the Poisson distribution arises often.

4.2   The Poisson and the Binomial Distributions

An informal statement concerning the way in which the Poisson distribution arises as a limiting case of the binomial was made in Section 1.3.6. A more formally correct version of this statement is as follows. If in the binomial distribution (1.8) we let $n \to +\infty$, $p \to 0$, with the product np held constant at λ, then for any y, the binomial probability in (1.8) approaches the Poisson probability in (1.15). This may be proved by writing the binomial probability (1.8) as

$$\frac{1}{y!}(np)((n-1)p)\cdots((n-y+1)p)\left(1 - \frac{\lambda}{n}\right)^n\left(1 - \frac{\lambda}{n}\right)^{-y}. \tag{4.12}$$

Fix y and λ and write $p = \lambda/n$. Then as $n \to \infty$, each term in the above product has a finite limit as $n \to +\infty$: Terms of the form $(n - i)\lambda/n$

approach λ for any i, and

$$\left(1 - \frac{\lambda}{n}\right)^n \to e^{-\lambda},$$

(see (B.3)), and finally,

$$\left(1 - \frac{\lambda}{n}\right)^{-y} \to 1.$$

Therefore, the expression in (4.12) approaches

$$\lambda^y e^{-\lambda}/y! \tag{4.13}$$

as $n \to +\infty$, and this is the Poisson probability (1.15).

4.3 The Poisson and the Gamma Distributions

There is an intimate connection, implied by equation (4.9), between the Poisson distribution and the exponential distribution. The (random) time until the first event occurs in a Poisson process with parameter λ is given by the exponential distribution with parameter λ. To see this, let $F(t)$ be the probability that the first event occurs before time t. Then the density function for the time until the first occurrence is the derivative $\frac{d}{dt} F(t)$. From (4.9), $F(t) = 1 - P_0(t) = 1 - e^{-\lambda t}$. Therefore, $\frac{d}{dt} F(t) = \lambda e^{-\lambda t}$. This is the exponential distribution (1.59), with notation changed from x to t.

It can also be shown that the distribution of the time between successive events is given by the exponential distribution. Thus the (random) time until the kth event occurs is the sum of k independent exponentially distributed times. The material surrounding (2.25) shows that this sum has the gamma distribution (1.68). Let t_0 be some fixed value of t. Then if the time until the kth event occurs exceeds t_0, the number of events occurring before time t_0 is less than k, and conversely. This means that the probability that $k - 1$ or fewer events occur before time t_0 must be identical to the probability that the time until the kth event occurs exceeds t_0. In other words it must be true that

$$e^{-\lambda t_0} \left(1 + (\lambda t_0) + \frac{(\lambda t_0)^2}{2!} + \cdots + \frac{(\lambda t_0)^{k-1}}{(k-1)!}\right) = \frac{\lambda^k}{\Gamma(k)} \int_{t_0}^{+\infty} x^{k-1} e^{-\lambda x} dx.$$

$$\tag{4.14}$$

This equation can also be established by repeated integration by parts of the right-hand side.

4.4 Introduction to Finite Markov Chains

In this section we give a brief outline of the theory of discrete-time finite Markov chains. The focus is on material needed to discuss the construction of PAM matrices as described in Section 6.5.3. Further developments of Markov chain theory suitable for other applications, in particular for the evolutionary applications discussed in Chapter 13, are given in Chapter 10.

We introduce discrete-time finite Markov chains in abstract terms as follows. Consider some finite discrete set S of possible "states," labeled $\{E_1, E_2, \ldots, E_s\}$. At each of the unit time points $t = 1, 2, 3, \ldots$, a Markov chain process occupies one of these states. In each time step t to $t + 1$, the process either stays in the same state or moves to some other state in S. Further, it does this in a probabilistic, or stochastic, way rather than in a deterministic way. That is, if at time t the process is in state E_j, then at time $t + 1$ it either stays in this state or moves to some other state E_k according to some well-defined probabilistic rule described in more detail below. The process is called Markovian, and follows the requirements of a Markov chain if it has the following distinguishing Markov characteristics.

(i) The *memoryless* property. If at some time t the process is in state E_j, the probability that one time unit later it is in state E_k depends only on E_j, and not on the past history of the states it was in before time t. That is, the current state is all that matters in determining the probabilities for the states that the process will occupy in the future.

(ii) The *time homogeneity* property. Given that at time t the process is in state E_j, the probability that one time unit later it is in state E_k is independent of t.

More general Markov processes relax one or both properties.

The concept of "time" used above is appropriate if, for example, we consider the evolution through time of the nucleotide at a given site in some population. Aspects of this process are discussed later in this book. However, the concept of time is sometimes replaced by that of "space." As an example, we may consider a DNA sequence read from left to right. Here there would be a Markov dependence between nucleotides if the nucleotide type at some site depended in some way on the type at the site immediately to its left. Aspects of the Markov chains describing this process are also discussed later in this book. Because Markov chains are widely applicable to many different situations it is useful to describe the properties of these chains in abstract terms.

In many cases the Markov chain process describes the behavior of a random variable changing through time. For example, in reading a DNA sequence from left to right this random variable might be the excess of purines over pyrimidines so far observed at any point. Because of this it is often convenient to adopt a different terminology and to say that the value of the random variable is j rather than saying that the state occupied by the process is E_j. We use both forms of expression below, and also, when no confusion should arise, we abuse terminology by using expressions like "the random variable is in state E_j."

4.5 Transition Probabilities and the Transition Probability Matrix

Suppose that at time t a Markovian random variable is in state E_j. We denote the probability that at time $t + 1$ it is in state E_k by p_{jk}, called the *transition probability* from E_j to E_k. In writing this probability in this form we are already using the two Markov assumptions described above: First, no mention is made in the notation p_{jk} of the states that the random variable was in before time t (the memoryless property), and second, t does not occur in the notation p_{jk} (the time homogeneity property).

It is convenient to group the transition probabilities p_{jk} into the so-called *transition probability matrix*, or more simply the transition matrix, of the Markov chain. We denote this matrix by P, and write it as

$$
P = \begin{array}{c} \\ (\text{from } E_1) \\ (\text{from } E_2) \\ \vdots \\ (\text{from } E_s) \end{array}
\begin{array}{ccccc} (\text{to } E_1) & (\text{to } E_2) & (\text{to } E_3) & \cdots & (\text{to } E_s) \end{array}
\left[\begin{array}{ccccc}
p_{11} & p_{12} & p_{13} & \cdots & p_{1s} \\
p_{21} & p_{22} & p_{23} & \cdots & p_{2s} \\
\vdots & \vdots & \vdots & \ddots & \vdots \\
p_{s1} & p_{s2} & p_{s3} & \cdots & p_{ss}
\end{array} \right]. \tag{4.15}
$$

The rows and columns of P are in correspondence with the states E_1, $E_2, \ldots, E_s$, so these states being understood, P is usually written in the simpler form

$$
P = \left[\begin{array}{ccccc}
p_{11} & p_{12} & p_{13} & \cdots & p_{1s} \\
p_{21} & p_{22} & p_{23} & \cdots & p_{2s} \\
\vdots & \vdots & \vdots & \ddots & \vdots \\
p_{s1} & p_{s2} & p_{s3} & \cdots & p_{ss}
\end{array} \right]. \tag{4.16}
$$

Any row in the matrix corresponds to the state *from* which the transition is made, and any column in the matrix corresponds to the state *to* which the transition is made. Thus the probabilities in any particular row in the transition matrix must sum to 1. However, the probabilities in any given column do not have to sum to anything in particular.

It is also assumed that there is some *initial* probability distribution for the various states in the Markov chain. That is, it is assumed that there is some probability π_i that at the initial time point the Markovian random variable is in state E_i. A particular case of such an initial distribution arises when it is known that the random variable starts in state E_i: In this case $\pi_i = 1$, while $\pi_j = 0$ for $j \neq i$. In principle the initial probability distribution and the transition matrix P jointly determine the probability for any event of interest in the entire process.

Example. Random walks. Random walk theory underpins BLAST theory. The classic description of a random walk involves a gambler, with initial capital i dollars, $i > 0$, and his adversary, with initial capital $(s-i)$ dollars, $s-i > 0$. At unit time points a coin with probability p for heads is tossed, and if the coin lands heads up, the adversary gives the gambler 1 dollar, while if the coin lands tails up, the gambler gives his adversary 1 dollar. The game continues until either the gambler or his adversary has no money left. The random variable of interest is the current fortune of the gambler. This random variable undergoes what is called a simple random walk, taking at any time one of the values $0, 1, 2, \ldots, s$.

If at any time this random variable takes the value j ($1 \leq j \leq s-1$), it moves one time unit later to the value $j+1$ with probability p or to the value $j-1$ with probability $q = 1-p$. This continues until the random variable reaches either 0 or s. When one of these values is reached no further changes in the gambler's fortune occur: His fortune is stopped at 0 or s.

The above implies that the evolution of the random variable is described by a Markov chain. The transition matrix P of this chain is

$$
P = \begin{bmatrix}
1 & 0 & 0 & 0 & \cdots & 0 & 0 & 0 & 0 \\
q & 0 & p & 0 & \cdots & 0 & 0 & 0 & 0 \\
0 & q & 0 & p & \cdots & 0 & 0 & 0 & 0 \\
\vdots & \vdots & \vdots & \vdots & \ddots & \vdots & \vdots & \vdots & \vdots \\
0 & 0 & 0 & 0 & \cdots & q & 0 & p & 0 \\
0 & 0 & 0 & 0 & \cdots & 0 & q & 0 & p \\
0 & 0 & 0 & 0 & \cdots & 0 & 0 & 0 & 1
\end{bmatrix}. \tag{4.17}
$$

The appearance of the 1's in the top left and bottom right entries reflects the fact that the gambler's fortune remains unchanged when it reaches either 0 or s: The probability of a transition from 0 to 0 or from s to s is 1.

The probability that the Markov chain process moves from state E_i to state E_j after two steps can be found by matrix multiplication. It is this fact that makes much of Markov chain theory an application of linear algebra. The argument is as follows.

Let $p_{ij}^{(2)}$ be the probability that if the Markovian random variable is in state E_i at time t, then it is in state E_j at time $t+2$. We can this a *two-step* transition probability. Since the random variable must be in some state k at the intermediate time $t+1$, equation (1.83) gives

$$
p_{ij}^{(2)} = \sum_k p_{ik} p_{kj}.
$$

The right-hand side in this equation is the (i, j) element in the matrix P^2. Thus if we define the matrix $P^{(2)}$ as the matrix whose (i, j) element is $p_{ij}^{(2)}$, then the (i, j) element in $P^{(2)}$ is equal to the (i, j) element in P^2. From this we get the fundamental identity

$$P^{(2)} = P^2.$$

Extension of this argument to an arbitrary number n of steps gives

$$P^{(n)} = P^n. \tag{4.18}$$

That is, the "n-step" transition probabilities are given by the entries in the nth power of P.

4.6   Markov Chains with Absorbing States

The random walk example described by (4.17) is a case of a Markov chain with absorbing states. These can be recognized by the appearance of one or more 1's on the main diagonal of the transition matrix. If there are no 1's on the main diagonal, then there are no absorbing states. For the Markov chains with absorbing states that we consider, sooner or later some absorbing state will be entered, never thereafter to be left. The two most questions we are interested in for these Markov chains are:

(i) If there are two or more absorbing states, what is the probability that a specified absorbing state is the one eventually entered?

(ii) What is the mean time until one or another absorbing state is eventually entered?

We will address these questions in detail in Chapter 10. In the remainder of this chapter we discuss only certain aspects of the theory of Markov chains with no absorbing states, focusing on the theory needed for the construction of substitution matrices, to be discussed in more detail in Chapter 6.

4.7   Markov Chains with No Absorbing States

The questions of interest about a Markov chain with no absorbing state are quite different from those asked when there are absorbing states.

In order to simplify the discussion, we assume in the remainder of this chapter that all Markov chains discussed are *finite, aperiodic,* and *irreducible.*

Finiteness means that there is a finite number of possible states. The aperiodicity assumption is that there is no state such that a return to that state is possible only at $t_0, 2t_0, 3t_0, \ldots$ transitions later, where t_0 is an integer exceeding 1. If the transition matrix of a Markov chain with states E_1, E_2, E_3, E_4 is, for example,

$$P = \begin{bmatrix} 0 & 0 & 0.6 & 0.4 \\ 0 & 0 & 0.3 & 0.7 \\ 0.5 & 0.5 & 0 & 0 \\ 0.2 & 0.8 & 0 & 0 \end{bmatrix}, \tag{4.19}$$

then the Markov chain is periodic. If the Markovian random variable starts (at time 0) in E_1, then at time 1 it must be either in E_3 or E_4, at time 2 it must be in either E_1 or E_2, and in general it can visit only E_1 at times $2, 4, 6, \ldots$. It is therefore periodic. The aperiodicity assumption holds for essentially all applications of Markov chains in bioinformatics, and we often take aperiodicity for granted without any explicit statement being made.

The irreducibility assumption implies that any state can eventually be reached from any other state, if not in one step then after several steps. Except for the case of Markov chains with absorbing states, the irreducibility assumption also holds for essentially all applications in bioinformatics.

4.7.1 Stationary Distributions

Suppose that a Markov chain has transition matrix P and that at time t the probability that the process is in state E_j is φ_j, $j = 1, 2, \ldots, s$. This implies that the probability that at time $t + 1$ the process is in state j is $\sum_{k=1}^{s} \varphi_k p_{kj}$. Suppose that for every j these two probabilities are equal, so that

$$\varphi_j = \sum_{k=1}^{s} \varphi_k p_{kj}, \quad j = 1, 2, \ldots, s. \tag{4.20}$$

In this case we say that the probability distribution $(\varphi_1, \varphi_2, \ldots, \varphi_s)$ is *stationary*; that is, it has not changed between times t and $t+1$, and therefore will never change. It will be shown in Chapter 10 that for finite aperiodic irreducible Markov chains there is a unique such stationary distribution.

If the row vector φ' is defined by

$$\varphi' = (\varphi_1, \varphi_2, \ldots, \varphi_s), \tag{4.21}$$

then in matrix and vector notation, the set of equations in (4.20) becomes

$$\varphi' = \varphi' P. \tag{4.22}$$

The prime here is used to indicate the transposition of the row vector into a column vector. The vector $(\varphi_1, \varphi_2, \ldots, \varphi_s)$ must also satisfy the equation $\sum_k \varphi_k = 1$. In vector notation, this is the equation

$$\varphi' \mathbf{1} = 1, \tag{4.23}$$

where $\mathbf{1} = (1, 1, \ldots, 1)'$.

Equations (4.22) and (4.23) can be used to find the stationary distribution. An example is given in the next section.

In Chapter 10 we will show that if the Markov chain is finite, aperiodic, and irreducible, then as n increases, $P^{(n)}$ approaches the matrix

$$\begin{bmatrix} \varphi_1 & \varphi_2 & \cdots & \varphi_s \\ \varphi_1 & \varphi_2 & \cdots & \varphi_s \\ \varphi_1 & \varphi_2 & \cdots & \varphi_s \\ \vdots & \vdots & \ddots & \vdots \\ \varphi_1 & \varphi_2 & \cdots & \varphi_s \end{bmatrix}, \tag{4.24}$$

where $(\varphi_1, \varphi_2, \ldots, \varphi_s)$ is the stationary distribution of the Markov chain.

The form of this matrix shows that no matter what the starting state was, or what was the initial probability distribution of the starting state, the probability that n time units later the process is in state j is increasingly closely approximated, as $n \to \infty$, by the value φ_j.

There is another implication, relating to long-term averages, of the calculations above. That is, if a Markov chain is observed for a very long time, then the proportion of times that it is observed to be in state E_j is approximately φ_j, for all j.

4.7.2 Example

Consider the Markov chain with transition probability matrix given by

$$P = \begin{bmatrix} 0.6 & 0.1 & 0.2 & 0.1 \\ 0.1 & 0.7 & 0.1 & 0.1 \\ 0.2 & 0.2 & 0.5 & 0.1 \\ 0.1 & 0.3 & 0.1 & 0.5 \end{bmatrix}. \tag{4.25}$$

For this example the vector equation (4.22) consists of four separate linear equations in four unknowns. However, they form a redundant set of equations and any one of them can be discarded. The remaining three equations, together with (4.23), yield four linear equations in the four unknowns which can be solved for a unique solution. Discarding the last equation in (4.22), we get

$$0.6\varphi_1 + 0.1\varphi_2 + 0.2\varphi_3 + 0.1\varphi_4 = \varphi_1,$$
$$0.1\varphi_1 + 0.7\varphi_2 + 0.2\varphi_3 + 0.3\varphi_4 = \varphi_2,$$
$$0.2\varphi_1 + 0.1\varphi_2 + 0.5\varphi_3 + 0.1\varphi_4 = \varphi_3,$$
$$\varphi_1 + \varphi_2 + \varphi_3 + \varphi_4 = 1.$$

To four decimal place accuracy, these four simultaneous equations have the solution

$$\varphi' = (0.2414, 0.3851, 0.2069, 0.1667). \tag{4.26}$$

This is the stationary distribution corresponding to the matrix P given in (4.25). In informal terms, from the point of view of long-term averages, over a long time period the random variable should spend about 24.14% of the time in state E_1, about 38.51% of the time in state E_2, and so on.

The rate at which the rows in $P^{(n)}$ approach this stationary distribution can be assessed from the following values:

$$P^{(2)} = \begin{bmatrix} 0.42 & 0.20 & 0.24 & 0.14 \\ 0.16 & 0.55 & 0.15 & 0.14 \\ 0.25 & 0.29 & 0.32 & 0.14 \\ 0.16 & 0.39 & 0.15 & 0.30 \end{bmatrix}, \tag{4.27}$$

$$P^{(4)} \approx \begin{bmatrix} 0.2908 & 0.3182 & 0.2286 & 0.1624 \\ 0.2151 & 0.4326 & 0.1899 & 0.1624 \\ 0.2538 & 0.3569 & 0.2269 & 0.1624 \\ 0.2151 & 0.4070 & 0.1899 & 0.1880 \end{bmatrix}, \tag{4.28}$$

$$P^{(8)} \approx \begin{bmatrix} 0.24596 & 0.37787 & 0.20961 & 0.16656 \\ 0.23873 & 0.38946 & 0.20525 & 0.16656 \\ 0.24309 & 0.38223 & 0.20812 & 0.16656 \\ 0.23873 & 0.38880 & 0.20525 & 0.16721 \end{bmatrix}, \tag{4.29}$$

$$P^{(16)} \approx \begin{bmatrix} 0.24142 & 0.38494 & 0.20692 & 0.16667 \\ 0.24135 & 0.38510 & 0.20688 & 0.16667 \\ 0.24140 & 0.38503 & 0.20691 & 0.16667 \\ 0.24135 & 0.38510 & 0.20688 & 0.16667 \end{bmatrix}. \tag{4.30}$$

After 16 time units, the stationary distribution has, for most purposes, been reached. The discussion in Chapter 10 shows how the rate at which this convergence occurs can be calculated in a more informative manner.

4.8   The Graphical Representation of a Markov Chain

It is often convenient to represent a Markov chain by a directed graph. A directed graph is a set of "nodes" and a set of "edges" connecting these nodes. The edges are "directed," that is, they are marked with arrows giving each edge an orientation from one node to another.

We represent a Markov chain by identifying the states with nodes and the transition probabilities with edges. Consider, for example, the Markov chain with states $E_1, E_2,$ and E_3 and with probability transition matrix

$$\begin{bmatrix} .2 & .1 & .7 \\ .5 & .3 & .2 \\ .6 & 0 & .4 \end{bmatrix}.$$

This Markov chain is represented by the following graph:

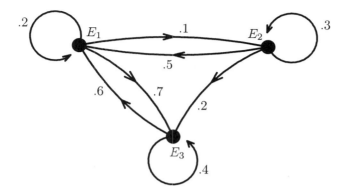

Notice that we do not draw the edge if its corresponding transition probability is known to be zero, as is the case in this example with the transition from E_3 to E_2.

A graph helps us capture information at a glance that might not be so apparent from the transition matrix itself. Sometimes it is also convenient to include a *start state*; this is a dummy state that is visited only once, at the beginning. Therefore, all transition probabilities into the start state are zero. The transition probabilities out of the start state are given by the initial distribution of the Markov chain. If the Markov chain starts at time $t = 0$, we can think of the start state as being visited in time $t = -1$. We can further have an *end state*, which stops the Markov chain when visited.

We refer to the graph structure, without any probabilities, as the *topology* of the graph. Sometimes the topology of a model is known, but the various probabilities are unknown.

We will use these definitions when we discuss hidden Markov models in Chapter 11.

4.9   Modeling

There are many applications of the homogeneous Poisson process in bioinformatics. However, the two key assumptions made in the derivation of the Poisson distribution formula (4.10), namely homogeneity and independence, do not always hold in practice. Similarly, there are many applications of Markov chains in the literature, in particular in the evolutionary processes discussed in Chapter 13. Many of these applications also make assumptions, specifically the two Markov assumptions stated in Section 4.4. The modeling assumptions made in the evolutionary context are discussed further in Section 14.9.

In the context of DNA or protein sequence analysis, where time is replaced by position along the sequence, it is very likely that neither of these two Markov assumptions is correct. The data already available from chromosome 22 in humans (Dunham et al. (1999)) makes it apparent that the probability that the nucleotide a is next followed by g depends to some extent on the current location in the chromosome. Further, it is likely that the Markov chain memoryless assumption does not hold: The probability that the nucleotide a is next followed by g might well depend on the nucleotide (or nucleotides) immediately preceding a. Tests for this possibility are discussed in Section 5.2: Nevertheless these tests are often not applied, and the memoryless Markov chain theory is often assumed when its applicability is uncertain.

The construction of phylogenetic trees, both by algorithmic methods and by methods involving Markov chain evolutionary models, involves many assumptions, both explicit and implicit. An example of quite different phylogenetic trees arising from different models is given in Section 14.8. A discussion of various statistical tests for appropriate evolutionary models is discussed in Section 14.9.

Given that modeling assumptions made for both Poisson processes and Markov chains often do not hold exactly, one might ask why they are made and why there is such an extensive Poisson process and Markov chain literature. Mathematical models generally make simplifying assumptions about properties of the phenomena being modeled. This concern opens up the question of why we model natural phenomena with mathematics if we cannot do so with complete accuracy. In fact, it is not necessarily desirable that we attempt to make an extremely accurate model of reality. The more closely any phenomenon is modeled, the more complicated the model becomes. If a mathematical model becomes too complicated, then solving the equations necessary to find answers to the questions we wish to ask can become intractable. Therefore, we are almost always faced with the task of finding a middle ground between tractable simple models and intractable complex models. The key point is that a model need only capture enough of the true complexity of a situation to serve our purposes, whatever they might be.

Finding this middle ground, however, is not an easy task: Being able to extract the essence of a complex reality in a simplified model that then allows a successful mathematical analysis requires some skill and experience.

In biology it is not always possible to evaluate a model's efficacy directly. Rather, a model is often tested on how well it performs its job. Sometimes benchmarks can be well defined, but often efficacy is not easy to verify empirically, and subjectivity can enter in. This is an unfortunate but usually unavoidable problem.

To illustrate this we consider the early versions of the widely used BLAST procedure discussed in more detail in Chapter 9. One of the simplifying assumptions used is that nucleotides (or amino acids) are identically and independently distributed along a DNA (or protein) sequence. Current data show that this assumption is false. Nevertheless, this simple BLAST procedure does work, in that the model captures enough of biological reality to be effective.

A further aspect of the modeling process is that we do not expect any model to be the final one used. Any given process is often initially modeled using several simplifying assumptions, and then more refined models are introduced as time goes on. Indeed, applications of the simple models often indicate those areas in which more precise modeling is needed. Various updates of the BLAST procedure exemplify this. Recent versions of BLAST remove some of the simplifying assumptions made in earlier versions and provide an example of the joint evolution of models and data analysis. Unfortunately, the mathematical theory involved in these more sophisticated versions is far more complicated than that in the simpler BLAST theory, and we will not consider it in this book.

Not every problem we might wish to solve with a model has a happy middle ground where our assumptions find a workable balance between tractability and reality. Thus while we should be willing to accept simplifying assumptions, we should always be on the lookout for oversimplifications, especially those that are not sufficiently backed up by testing for the efficacy of the model used. Model testing is an active area of statistical research in bioinformatics, and aspects of model testing, especially in the evolutionary and phylogenetic tree contexts, are discussed further in Section 14.9.

Problems

4.1. Prove (4.14) by repeated integration by parts of the right-hand side.

4.2. Events occur in a Poisson process with parameter λ. Given that 10 events occur in the time period $[0,2]$, what is the probability that 6 of these events occur in the time period $[0,1]$? Given that 6 of these events occur in the time period $[0,1]$, what is the probability that 10 of these events occur in the time period $[0,2]$?

4.3. ("Competing Poissons.") Suppose that events occur as described in Section 4.1, but that now each event is of one of k types, labeled types $1, 2, \ldots, k$. The probability that any event is of type i is p_i. Equation (4.10) continues to give the probability that exactly j events occur in the time period $(0,t)$. Assuming this,

(i) Find the (marginal) probability that j_i events of type i occur in the time period $(0,t)$.

(ii) Find the probability that j_1 events of type 1 occur, $\ldots$, j_k events of type k occur in the time period $(0,t)$.

(iii) Find the conditional probability that j_1 events of type 1 occur, ..., j_k events of type k occur in the time period $(0, t)$, given that j events altogether occur in this time period. Relate your answer to expression (2.42).

4.4. The transition matrix of a Markov chain is

$$\begin{bmatrix} .7 & .3 \\ .4 & .6 \end{bmatrix}.$$

Find the stationary distribution of this Markov chain.

4.5. *Continuation.* If the initial probability distribution (at time 0) is $(.8, .2)$, what is the probability that at time 3 the state occupied is E_1?

4.6. The transition matrix of a Markov chain is

$$\begin{bmatrix} 1 - a & a \\ b & 1 - b \end{bmatrix}.$$

Find the stationary distribution of this Markov chain in terms of a and b, and interpret your result.

4.7. Use equations (4.22) and (4.18) to show that the stationary distribution φ' satisfies the equation

$$\varphi' = \varphi' P^{(n)}, \tag{4.31}$$

for any positive integer n. For the numerical example in Section 4.7.2, use the expression for $P^{(2)}$ given in equation (4.27) and the expression for φ' given in (4.26) to check this claim for the case $n = 2$.

Why does equation (4.31) "make sense"?

4.8. Show that if the transition matrix of an irreducible, aperiodic, finite Markov chain is symmetric, then the stationary distribution is a (discrete) uniform distribution.

4.9. Show that if the transition matrix P of a Markov chain is of the circulant form

$$P = \begin{bmatrix} a_1 & a_2 & a_3 & a_4 & \cdots & a_{s-3} & a_{s-2} & a_{s-1} & a_s \\ a_s & a_1 & a_2 & a_3 & \cdots & a_{s-4} & a_{s-3} & a_{s-2} & a_{s-1} \\ a_{s-1} & a_s & a_1 & a_2 & \cdots & a_{s-5} & a_{s-4} & a_{s-3} & a_{s-2} \\ \vdots & \vdots & \vdots & \vdots & \ddots & \vdots & \vdots & \vdots & \vdots \\ a_4 & a_5 & a_6 & a_7 & \cdots & a_s & a_1 & a_2 & a_3 \\ a_3 & a_4 & a_5 & a_6 & \cdots & a_{s-1} & a_s & a_1 & a_2 \\ a_2 & a_3 & a_4 & a_5 & \cdots & a_{s-2} & a_{s-1} & a_s & a_1 \end{bmatrix}, \tag{4.32}$$

where $a_j > 0$ for all j, then the stationary distribution is a (discrete) uniform distribution.

4.10. Suppose that in the gambling game described in Section 4.5, each player gives his opponent 1 dollar if the opponent has no money, so that the transition matrix of the resultant Markov chain is

$$P = \begin{bmatrix} 0 & 1 & 0 & 0 & \cdots & 0 & 0 & 0 & 0 \\ q & 0 & p & 0 & \cdots & 0 & 0 & 0 & 0 \\ 0 & q & 0 & p & \cdots & 0 & 0 & 0 & 0 \\ \vdots & \vdots & \vdots & \vdots & \ddots & \vdots & \vdots & \vdots & \vdots \\ 0 & 0 & 0 & 0 & \cdots & q & 0 & p & 0 \\ 0 & 0 & 0 & 0 & \cdots & 0 & q & 0 & p \\ 0 & 0 & 0 & 0 & \cdots & 0 & 0 & 1 & 0 \end{bmatrix}. \qquad (4.33)$$

Show this Markov chain is periodic. Despite this fact, solve equations (4.22) and (4.23) for the case $p = q$. Is this a stationary distribution?

4.11. Suppose that a transition matrix P is of size $2s \times 2s$ and can be written in the partitioned form

$$P = \begin{bmatrix} 0 & A \\ \hline B & 0 \end{bmatrix},$$

where A and B are both $s \times s$ matrices. Use this expression to find formulae for (i) $P^{(2n)}$, (ii) $P^{(2n+1)}$ in terms of the matrices A and B, and interpret your results.

4.12. Suppose that a finite Markov chain is irreducible and that there exists at least one state E_i such that $p_{ii} > 0$. Show that the Markov chain is aperiodic.

4.13. (More difficult). Any $s \times s$ matrix of nonnegative numbers for which all rows sum to 1 can be regarded as the transition matrix of some Markov chain. Is it true that any such matrix can be the two-step transition matrix of some Markov chain? *Hint:* Consider the case $s = 2$. Write down the general form of a 2×2 Markov chain matrix and find when the sum of the diagonal terms is minimized.

5

The Analysis of One DNA Sequence

5.1 Shotgun Sequencing

Before any analysis of a DNA sequence can take place it is first necessary
to determine the actual sequence itself, at least as accurately as is reasonably possible. Unfortunately, technical considerations make it impossible
to sequence very long pieces of DNA all at once. Instead, many overlapping small pieces are sequenced, each on the order of 500 bases. After this
is done the problem arises of assembling these fragments into one long
"contig." One difficulty is that the locations of the fragments within the
genome and with respect to each other are not generally known. However,
if enough fragments are sequenced so that there will be many overlaps between them, the fragments can be matched up and assembled. This method
is called "shotgun sequencing."

It is customary to say that *n-times coverage* (or nX coverage) is obtained
if, when the length of the original (long) sequence is G, the total length of
the fragments sequenced is nG. It is shown below that to expect to cover
99% of a long sequence it is necessary to have at least 4.6X coverage.

Two strategies have been used to sequence the entire human genome
(International Human Genome Sequencing Consortium (2001), Venter et
al. (2001)). Under one strategy the genome is partitioned into pieces whose
lengths are on the order of 100,000 nucleotides and whose locations in the
genome are known. Then shotgun sequencing is performed on each piece,
with high coverage, in the 8X range. The greater coverage also helps elimi-
nate errors occurring in sequencing the fragments. A competing strategy is

to adopt a "whole genome shotgun" approach, in which the entire genome is broken into small fragments. The assembly problem is much more difficult under this approach, and thus should require high coverage. It is a matter of much debate as to which approach is to be preferred. The human genome has been sequenced using a combination of the two methods.

In this section we address several probabilistic issues arising in the reconstruction of long DNA sequences from comparatively shorter sequences, or fragments. Before proceeding it is important to note the remarks about modeling given in Section 4.9. The probabilistic models described below are simple and do not closely reflect all of the properties of chromosomes as revealed by some of the extraordinary characteristics of (for example) chromosome 22 in Dunham et al. (1999). Less simplified models are referred to at the end of the section. Aspects of the biology are reviewed by Slessinger (1990). Interesting reviews of the assembly of human chromosomes 21 and 22, of *C. elegans* and of *D. melanogaster* are given respectively by Hattori et al. (2000), Dunham et al. (1999), the *C. elegans* Sequencing Consortium (1998), and Adams et al. (2000).

Figure 5.1 shows a collection of $N = 17$ fragments with their locations above the original DNA sequence. It is assumed that overlapping fragments can be recognized and used to determine a collection of "contigs" (thick black lines), of which there are 7 in the example shown. These contigs are then taken as the best possible reconstruction of the original DNA sequence.

FIGURE 5.1.

We assume initially that there are N fragments, each of length L, and that the original full-length DNA sequence, which we call here the genome, is of length G. Therefore, the coverage a is given by

$$a = NL/G.$$

The length G is assumed to be much larger than L, so that end effects are ignored in the calculations below.

The fragments are assumed to be taken at random from the original full-length sequence, so that, ignoring end effects, the position of the left-hand end of any fragment is uniformly distributed in $(0, G)$, and thus falls in an interval $(x, x + h)$ on the original sequence with probability h/G. The number of fragments whose left-hand end falls in this interval has a binomial distribution with mean Nh/G. If N is large and h is small, the discussion of Section 4.2 shows that this distribution is approximately Poisson with mean Nh/G. We use this Poisson approximation throughout.

The Poisson approximation implies that the number of fragments whose left-hand end is located within an interval of length L to the left of a randomly chosen point has a Poisson distribution with mean a, so that the probability that at least one fragment arises in this interval is $1 - e^{-a}$. This calculation, together with other properties of homogeneous Poisson processes, is enough to provide the answer to three basic questions: What is the mean proportion of the genome covered by contigs? What is the mean number of contigs? What is the mean contig size?

The mean proportion of the genome covered by one or more contigs is the probability that a point chosen at random is covered by at least one contig. This is the probability that at least one fragment has its left-hand end in the interval of length L immediately to the left of this point, which is $1 - e^{-a}$ as given above. In order for this probability to be 0.99 it is necessary to have $a = 4.6$, so that the sum of the fragment lengths is then not quite five times the genome length. In order for this probability to be 0.999 it is necessary to have $a = 6.9$. Since the human genome is approximately 3×10^9 nucleotides, 6.9X coverage will still miss approximately 3,000,000 of them.

Each contig has a unique rightmost fragment, so that the formula for the mean of a binomial distribution (see Table 1.1) shows that the mean number of contigs is the number of fragments N multiplied by the probability that a fragment is the rightmost member of a contig. This latter probability is the probability that no other fragment has its left-hand end point on the fragment in question. From the calculation in the previous paragraph, this probability is e^{-a}. Thus

$$\text{mean number of contigs} = Ne^{-a} = Ne^{-NL/G}. \qquad (5.1)$$

Table 5.1 gives some values for this mean for different coverages in the case $G = 100{,}000$, $L = 500$. The mean number of contigs, as a function

a	0.5	0.75	1	1.5	2	3	4	5	6	7
Mean number of contigs	60.7	70.8	73.6	66.9	54.1	29.9	14.7	6.7	3.0	1.3

TABLE 5.1. The mean number of contigs for different levels of coverage, with $G = 100{,}000$ and $L = 500$.

of a, increases and then decreases. The reason for this is that if there is a small number of fragments, there must be a small number of contigs, while a large number of fragments tend to form a small number of large contigs. The mean number of contigs is maximized when $N = G/L$; or equivalently, when $a = 1$; that is, with 1X coverage. In the example in Table 5.1, one contig is expected after 8X coverage. However, the expression (5.1) for the mean number of contigs becomes inaccurate in this range, since with 8X coverage, end effects (which are ignored in deriving (5.1)) become important. While the numbers G and L in this example are realistic, the assumptions of the model are simplified and unrealistic, and in practice the problem of achieving high coverage is more difficult than is implied by the above calculations. For example, the existence of repeated sequences in the junk DNA can cause fragments to appear to overlap when they do not. Furthermore, since some stretches of DNA are technically much more difficult to clone and sequence than others.

We next calculate the mean contig size. The mean contig size is found by considering the left-hand ends of a succession of fragments, starting with the initial left-hand fragment on a given contig. The distance from the left-hand end of the first fragment to the left-hand end of the second fragment has a geometric distribution, which is closely approximated by the exponential distribution (1.59) with parameter $\lambda = N/G$. This second fragment will overlap the first one if this distance is less than the length L of the first fragment. This occurs with probability

$$\int_0^L \lambda e^{-\lambda x} dx = 1 - e^{-a}.$$

A further overlap occurs if the next fragment to the right of the second fragment overlaps that second fragment. The contig is built up in this way until such an overlap fails to occur.

The number of successively overlapping fragments, before such a nonoverlap, has a geometric distribution whose mean (from Table 1.1 with $p = 1 - e^{-a}$) is $e^a - 1$. If n fragments form a contig, the total length of the contig is the length L of the final fragment together with the sum of the $n - 1$ random distances between the left-hand end of any given fragment and the left-hand end of the next fragment to the right. Equation (2.49) shows that the mean of these random distances is

$$\frac{1}{\lambda} - \frac{L}{e^a - 1}.$$

Equation (2.78), referring to the mean of a sum of a random number of random variables, shows that the mean total of these distances is

$$(e^a - 1)\left(\frac{1}{\lambda} - \frac{L}{e^a - 1}\right) = \frac{e^a - 1}{\lambda} - L.$$

Upon adding the length L of the last contig to this, the mean contig size is found to be

$$\frac{e^a - 1}{\lambda} = L\frac{e^a - 1}{a}. \tag{5.2}$$

In the case $L = 500$, $a = 2$, for example, the mean contig size is about 1600.

In the above it was assumed that each fragment had length L. In reality, fragments are obtained in lengths roughly between 400 and 600 bases. Many of the above results can be generalized to the case where the length of any fragment is a random variable L with density function $f_L(\ell)$. We consider only the problem of finding the mean proportion of the genome covered by a contig.

Let P be a given point in the genome. Assuming that end effects and terms of order $o(h)$ can be ignored, the probability that some fragment has its left-hand end in the interval $(P - x, P - x + h)$ and overlaps P is the mean number of fragments having left-hand end point in this interval (λh) multiplied by the probability that the length of a fragment exceeds x (and thus covers the point P). Thus this probability is

$$\lambda h \int_x^{+\infty} f_L(\ell)\, d\ell = \lambda h \left(1 - F_L(x)\right).$$

From this, the mean number of fragments covering the point P is

$$\lambda \int_0^{+\infty} 1 - F_L(x)\, dx,$$

and from equation (1.48) this is

$$\lambda E(L) = \frac{NE(L)}{G}. \tag{5.3}$$

It follows that the probability that a random point is covered by a contig is $1 - e^{-NE(L)/G}$, and this is then also the mean proportion of the genome covered by a contig.

Further aspects of this generalization have been discussed by Lander and Waterman (1998), who, among a variety of other interesting results, show that the mean number of contigs is somewhat larger than the value $Ne^{-NE(L)/G}$, the value that might be obtained intuitively by replacing L by $E(L)$ in equation (5.1).

Short sequences in the DNA that are unique in the genome, and whose locations are known, provide markers that allow us to locate the fragments in the genome that happen to contain these markers. In such a case we say that the fragments are "anchored." Other concepts of an anchor are possible, but this is the simplest. For our present purposes an anchor can be considered simply as a point located on the genome, anchoring and thus locating any fragment lying over it.

An anchored fragment is a fragment with at least one anchor on it, and an anchored contig is a collection of fragments "stapled together" by anchors. Figure 5.2 provides an example where there are 16 fragments, 9 anchors, and 6 anchored contigs, shown as thick black lines. Contigs can overlap, as

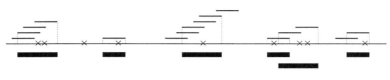

FIGURE 5.2.

shown in this figure, but the overlap is not recognized if there is no anchor on the overlapping section. The genome reconstruction is then carried out using anchored contigs.

We first calculate the expected proportion of the genome covered by anchored contigs in the case where all fragments are of fixed length L. The number of anchors is denoted by M, and it is assumed that anchors are to be placed according to a homogeneous Poisson process. Then the mean number of anchors in any length L of the DNA sequence is $b = ML/G$, and the number of anchors falling in any segment of length L has a Poisson distribution with mean b. The Poisson process determining the location of the anchors is assumed to be independent of the Poisson process locating the fragments.

The mean proportion of the DNA segment covered by anchored contigs is the probability that a point P chosen at random is covered by an anchored contig. It is convenient to consider the complementary probability that P is not covered by an anchored contig. A point can fail to be covered by an anchored contig for one of three mutually exclusive reasons. First, there might be no fragment covering this point. This event has probability e^{-a} (recall that $a = NL/G$). Second, there might be exactly one fragment covering this point (probability ae^{-a}) but no anchor on this fragment (probability e^{-b}), leading to a probability $ae^{-(a+b)}$ for this event. Finally, there might be $k \geq 2$ fragments covering this point, with no anchor on the span of these fragments. The calculations relating to this third possibility are more complicated and are carried out as follows.

Consider the case typified by that shown in Figure 5.3, where P is the point in question. The k ($k \geq 2$) fragments covering P have total span $X + Y$. Each of these k fragments has a "right-projection," namely the length of that part of the fragment to the right of P. Since the fragments are assumed to be placed at random, this length is uniformly distributed in $(0, L)$. The projection X is the maximum of these lengths. Similarly, each of the k fragments covering P has a "left-projection," namely the length of that part of the fragment to the left of P, also uniformly distributed in $(0, L)$. The projection Y is the maximum of these lengths. Further, $Y = L - W$, where W is the length of the smallest right-projection length.

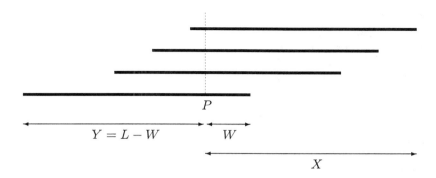

FIGURE 5.3.

For any fixed $k \geq 2$, the joint density function of W and X is given in (2.152) (with $x_{(1)} = W$ and $x_{(n)} = X$). With this joint density function in hand, transformation methods can be used to find the density function of the span $Z = X + Y$, given that the number of fragments crossing P is k. It can be shown (see Problem 5.1) that the density function of Z is

$$f_Z(z; k) = k(k-1)(2L-z)(z-L)^{k-2}/L^k, \quad L < z < 2L, \qquad (5.4)$$

where we use the notation $f_Z(z; k)$ to indicate the dependence of this distribution on k. The (Poisson distribution) probability that exactly k fragments cover P is $e^{-a}a^k/k!$, so that for small h, the probability of the event "$k \geq 2$ and the span is of length between z and $z + h$" is, ignoring small-order terms,

$$h \sum_{k=2}^{\infty} \frac{e^{-a}a^k}{k!} f_Z(z; k) = ha^2 e^{-2a} L^{-2}(2z-a)e^{az/L}.$$

We write the right-hand side in this expression as $h\, j(z)$. The probability that no anchor lies on a span of length z is $e^{-zb/L}$. The probability that $k \geq 2$ and that the point P is not covered by an anchored contig is then

$$\int_{L}^{2L} j(z)e^{-zb/L}dz = \frac{a^2}{(a-b)^2}e^{-2b} + \frac{a^2(b-a-1)}{(a-b)^2}e^{-a-b}.$$

The total probability that the point P is not covered by an anchored contig, that is, the mean proportion of the DNA not covered by an anchored contig, is found by gathering terms such as

$$e^{-a} + \frac{a(b^2 - ab - a)e^{-(a+b)}}{(b-a)^2} + \frac{a^2}{(b-a)^2}e^{-2b}. \qquad (5.5)$$

A further interesting calculation concerns the mean number of anchored contigs. Each anchored contig has a unique rightmost fragment, so this mean is the mean number of rightmost fragments on an anchored contig, and is thus found by first computing the probability that a given fragment C is the rightmost fragment on an anchored contig.

The first way in which fragment C can be such a rightmost fragment is if (i) it is anchored, with probability $(1 - e^{-b})$, and (ii) no other fragment overlaps it on the right, with probability e^{-a}. This event has total probability $(1 - e^{-b})e^{-a}$.

The second way in which fragment C will be the rightmost fragment on an anchored contig is as follows. If fragment C has k $(k \geq 1)$ fragments overlapping it to the right, there will be some leftmost member of these fragments whose left-hand end is a distance W to the right of the left-hand end of fragment C. Fragment C will be the rightmost fragment of an anchored contig if there is at least one anchor on the leftmost length W of fragment C, with probability $1 - e^{-bW/L}$, but no anchor on the remaining rightmost length $L-W$ of fragment C, with probability $e^{-(L-W)b/L}$. Given k, the length W is the minimum of k iid uniform random variables with range $[0, L]$, each having density function $f_X(x) = \frac{1}{L}$, $0 \leq x \leq L$. The probability distribution of this minimum is given in equation (2.85) as $k(L - x)^{k-1}/L^k$.

The probability that fragment C is overlapped on the right by at least one other fragment, but is nevertheless the rightmost fragment on an anchored contig, is then found to be

$$\sum_{k=1}^{+\infty} \left(e^{-a}a^k/k!\right) \int_0^L k(L - w)^{k-1}(1 - e^{-wb/L}))(e^{-(L-w)b/L})/L^k \, dw. \quad (5.6)$$

Evaluating this expression and adding the probability $(1 - e^{-b})e^{-a}$, we arrive at the probability

$$b\frac{(e^{-a} - e^{-b})}{(b - a)}$$

that fragment C is the rightmost fragment of an anchored contig. Since there are N fragments, the mean number of anchored contigs is

$$Nb\frac{(e^{-a} - e^{-b})}{(b - a)}. \quad (5.7)$$

The calculations for the mean contig size are more complicated and are not given here.

The calculations described above rely on several simplifying assumptions, for example that both the fragment and the anchor Poisson processes are homogeneous and that all fragments are of the same length. A wide variety of calculations applying in more realistic cases, removing these and other restrictions, is discussed by Arratia et al. (1991), Schbath (1997), and Schbath et al. (2000).

5.2   Modeling DNA

On its most elementary level, the structure of DNA can be thought of as long sequences of nucleotides. These sequences are organized into coding sequences, or genes, that are separated by long intergenic regions of noncoding sequence. Most eukaryotic genes have one further level of organization: Within each gene, the coding sequences (exons) are usually interrupted by stretches of noncoding sequences (introns). During processing of DNA into messenger RNA (mRNA), these intron sequences are spliced out and thus do not appear in the final mature mRNA which is translated into protein sequence.

Intergenic regions and introns have different statistical properties from those of exons. By capturing these properties in a model we can construct statistical procedures for testing whether or not an uncharacterized piece of DNA is part of the coding region of a gene. The model is based on a set of "training" data taken from already characterized sequences. In the simplest model it is assumed that the nucleotides at the various sites are independent and identically distributed (iid). If this is the case, the difference between coding and noncoding DNA could be captured in the difference between the frequencies of the four nucleotides in the two different cases (intron vs. exon). These distributions can then be estimated from the training data.

The accuracy of this procedure depends on the accuracy of the assumptions made. Thus it is important, for example, to develop tests of the (null) hypothesis of independence. One such test is based on a Markov chain analysis. In the use of Markov chains in sequence analysis, the concept of "time" is naturally replaced with "position along the sequence." Under a Markov chain alternative hypothesis the probability that a given nucleotide is present at any site depends on the nucleotide at the preceding site. The null hypothesis of independence corresponds to a Markov chain in which all rows in the transition probability matrix are identical, so that the probabilities for the various nucleotides at any site are independent of the nucleotide at the preceding site. It is possible that neither hypothesis is correct, but it is of interest to assess whether the Markov-dependence model describes reality significantly better than the independence model and thus might raise the accuracy of our predictive procedures. Furthermore, if it does, then models with even more complex dependence may be considered.

The statistical test for Markov independence is an association test in a 4×4 table such as in Example 4 of Section 3.4.3. In this case $r = c = 4$, and Y_{jk} is the number of times, in the DNA sequence of interest, that a nucleotide of type "j" is followed by a nucleotide of type "k." The data would then appear as shown in Table 5.2. In this case the null hypothesis of independence becomes the null hypothesis of no association in the table.

Tests of this type show that the nucleotides at adjoining DNA sites are often dependent, and a first-order Markov model fits real data significantly

		\multicolumn{5}{c}{nucleotide at site $i+1$}
		a	g	c	t	Total
nucleotide at site i	a	Y_{11}	Y_{12}	Y_{13}	Y_{14}	$y_{1\cdot}$
	g	Y_{21}	Y_{22}	Y_{23}	Y_{24}	$y_{2\cdot}$
	c	Y_{31}	Y_{32}	Y_{33}	Y_{34}	$y_{3\cdot}$
	t	Y_{41}	Y_{42}	Y_{43}	Y_{44}	$y_{4\cdot}$
	Total	$y_{\cdot 1}$	$y_{\cdot 2}$	$y_{\cdot 3}$	$y_{\cdot 4}$	y

TABLE 5.2. Table of numbers of times a nucleotide of one specified type follows one of another specified type.

better than the independence model both in introns and in exons. Higher-order homogeneous Markov models and nonhomogeneous models often provide even better fits. We discuss tests of higher-order Markov dependence in Section 10.3.

5.3 Modeling Signals in DNA

5.3.1 Introduction

Genes contain "signals" in the DNA to indicate the start and end of the transcribed region, the exon/intron boundaries, as well as other features. The cell machinery uses these signals to recognize the gene, to edit it correctly, and to translate it appropriately into protein.

A signal is a short sequence of DNA having a specific purpose, for example indicating an exon/intron boundary. If nature were kind, each such signal would consist of a unique DNA sequence that did not appear anywhere in the DNA except where it serves its specific purpose. In reality, there may be many DNA sequences that perform the same signal function; we call these "members" of that signal. Furthermore, members of signals also appear randomly in the nonfunctional DNA, making it hard to sort out the functional from nonfunctional signals.

In practice, some but not all members of a signal are known. Our aim is to use known members to assess the probability that a new uncharacterized DNA sequence is also a member of the signal. Our assumption is that the different members arose from common ancestors via stochastic processes. Therefore, it is reasonable to construct a statistical model of the data. Some signals require only simple models; others require complex ones. When the model required is complex and the data are limited, we must make carefully chosen simplifying assumptions in the model in order to utilize the data in the most efficient manner. In Chapter 11 we describe a complex model for modeling an entire gene. That application will use all of the different signal models described below.

We assume that all members of the signal of interest have the same length, which we denote by n. This assumption is not too restrictive, since the members of many signals do have the same lengths, and for those that do not we can capture portions of them with this model, which is often sufficient. To model the properties of any signal we must have a set of "training data": These are extensive data in which members of the signal are known and recognized. For example, in investigating signals in human DNA we might use a set of known members of signals taken from known human genes in public databases.

We now consider some of the basic signal models that are used in bioinformatics. For a more exhaustive discussion of this subject, see Burge (1998).

5.3.2   Weight Matrices: Independence

The test of whether a given uncharacterized DNA sequence is a member of a given signal is most easily carried out in the case where the nucleotide in any position in the signal is independent of the nucleotides in any other position in the signal.

It is therefore necessary to form a test of independence, and we consider here a test of whether the nucleotide at position a in a signal is independent of the nucleotide at position b. This can be done in various ways. It is natural to generalize the test described in connection with the data in Table 5.2, which tests for independence at adjacent sites. In this generalization "site i" is replaced by "position a in the signal" and "site $i+1$" is replaced by "position b in the signal," and Y_{jk} is interpreted as the number of times in the training data that nucleotide j occurs at position a in the signal and nucleotide k occurs at position b in the signal. This test is then performed for all pairs a and b. The individual tests are then corrected for multiple testing, either using a Bonferroni or Šidák correction, as discussed in Section 3.8.

Suppose that as a result of this or some other testing procedure independence can be assumed. A $4 \times n$ matrix is then constructed whose (i, j) entry is the proportion of cases in the training data for which the ith nucleotide occurs at the jth position of a signal. This is referred to as a *weight matrix* and is denoted by M. The set of training data is assumed to be sufficiently large that the proportions in this matrix can be taken as probabilities. An example for the case $n = 5$ is given in Table 5.3. The entries in column j in this matrix give (estimated) probabilities for the four nucleotides at position j in the signal.

The matrix M defines a probability $\mathrm{Prob}(s \mid M)$ for any sequence s of length n. Weight matrices are used in many contexts. We will use them as a component of a gene-finding algorithm given in Chapter 11.

		\multicolumn{5}{c}{position}				
		1	2	3	4	5
	a	0.33	0.34	0.19	0.20	0.21
nucleotide	g	0.22	0.27	0.23	0.24	0.21
	c	0.31	0.18	0.34	0.30	0.25
	t	0.14	0.21	0.24	0.26	0.33
Total		1	1	1	1	1

TABLE 5.3. Signal probabilities.

5.3.3 Markov Dependencies

If the nucleotides at the sites in a signal are not independent, a possibility is that there are dependencies of the first order Markov type. Under this hypothesis the first position of the signal has a probability distribution determined by overall nucleotide frequencies. Nucleotides at each subsequent position have a probability distribution depending on the nucleotide at the previous position as determined by a 4×4 Markov chain transition probability matrix whose (i, j) element is the probability of the jth nucleotide in position k, given the ith nucleotide in position $k - 1$. The elements in this matrix are estimated from the training data.

More generally, higher-order Markov dependencies such as those described in Section 10.3 may be modeled. However, as discussed in Section 10.3, these transition matrices become large as the order of the Markov dependency increases, so that for a high-order dependence the amount of data needed for satisfactory estimation might be excessive. When the set of training data is limited we must economize and capture only the most informative dependencies in our model. A method for doing this is discussed in the next section.

5.3.4 Maximal Dependence Decomposition

As discussed in Section 5.3.3, it may be impossible, due to limited data, to obtain satisfactory estimates of all dependencies in the sequences in a given signal. This motivates us to search for a method that captures those dependencies that are most informative. We now describe the approach, known as *maximal dependence decomposition* (MDD) (Burge (1997)), to this problem.

Suppose we wish to model a signal of length n. The first step is to find one position which has the greatest influence on the others. To do this we construct an $n \times n$ table whose (i, j) entry is the observed value of the chi-square statistic obtained from a 4×4 table such as Table 5.2 but that compares the nucleotides at a fixed position i with those at a fixed position j (instead of specifically position $i+1$). If the hypothesis of independence is true, we expect to find about one significant value out of 20 just by chance if the Type I error is 5%. Thus if only a few of the observed chi-square values are significant but not highly significant, we might conclude that the positions in the signal can be taken as independent. If more than a few are significant, or if there are values with high levels of significance, then we would conclude that the nucleotides at the various positions are not independent.

	1	2	3	4	5	total
1		34.2*	7.1	37.2*	2.8	81.3
2	34.2*		.4	72.4*	4.5	111.5
3	7.1	.4		15.3	98.3*	121.1
4	37.2*	72.4*	15.3		14.2	139.1
5	2.8	4.5	98.3*	14.2		119.8

TABLE 5.4.

Table 5.4 provides an example with $n = 5$ in which many of the chi-square values are significant at the 5% level (indicated by asterisks). The row with the greatest sum gives an indication of which position has the greatest influence on the other $n - 1$ positions. In the above example position 4 has the greatest row total, so we take it as the position with greatest influence and try to assess this influence further. In this case we say that we *split* on position 4.

We illustrate first by constructing a model that determines distributions for positions 1, 2, 3, and 5 conditional on position 4. This is done as follows. We divide the set of test sequences into four sets, each set being determined by the nucleotide in position 4, so that for each nucleotide x we have a set T_x consisting of those sequences where there is an x in position 4. We then calculate $p_x = n_x/d$, for each $x = a, g, c, t$, where n_x is the number of members of the training data having nucleotide x in position 4, and $d = \sum n_x$. Next, for each $x = a, g, c, t$, we calculate a 4×4 weight matrix M_x from the sequences in T_x, for positions 1, 2, 3, and 5.

The model M then consists of the distribution $\{p_a, p_g, p_c, p_t\}$ together with the four weight matrices $\{M_a, M_g, M_c, M_t\}$. For any sequence s of length 5 we calculate $\text{Prob}(s \mid M)$ as follows.

If nucleotide x occurs in position 4 of s, the weight matrix M_x is used to assign a probability p_k to positions $k = 1, 2, 3$, and 5 of s. $\text{Prob}(s \mid M)$ then equals $p_x \cdot p_1 \cdot p_2 \cdot p_3 \cdot p_5$.

In general, some or all of the sets T_x might be large enough that we can repeat the entire process recursively, splitting T_x further on one of the positions 1, 2, 3, 5. In this case, for each sufficiently large T_x we make a new table similar to Table 5.4, this time 4×4, to find the position of these four that has the greatest influence on the other three. We then decompose T_x into T_{xy} for $y = a, g, c$, and t, and continue to T_{xyz}, and so on, as long as there are enough data. A rule of thumb is to stop whenever the set $T_{xyz\cdots}$

has fewer than 100 sequences. When such a limit is reached, the remaining positions are modeled with a weight matrix.

5.4   Long Repeats

Suppose that we are interested in one specific nucleotide, say a, and ask whether there is significant evidence of long repeated sequences of this nucleotide. More specifically, we wish to test the null hypothesis that nucleotides occur at random against the alternative that there is a tendency for long repeats of a. We have discussed one test for Markov dependence of nucleotides in Section 5.2 and will discuss more detailed tests in Section 10.3. In this section we discuss a test of randomness tailored to the "repeats" alternative hypothesis.

Consider a long DNA sequence of length N, where N is assumed to be so large that end effects can be ignored in our calculations. Suppose that the null hypothesis probability that the nucleotide a occurs at any site in this sequence is some known value p. We call the occurrence of a at any site a "success" and the occurrence of any other nucleotide at any site a "failure."

Suppose that the DNA sequence is scanned from left to right, and consider the situation immediately following a failure. There will then be a sequence of successes, possibly of length 0 (if the next nucleotide is c, g, or t). Assuming that the null hypothesis is true, equation (1.10) shows that the length Y of this sequence has the geometric distribution, which is reproduced here for convenience:

$$P_Y(y) = \text{Prob}(Y = y) = (1-p)p^y, \quad y = 0, 1, 2, \ldots. \tag{5.8}$$

Equation (2.107) shows that if n independent such sequences are given, and $Y_{\max}$ is the length of the longest of these, then

$$\text{Prob}(Y_{\max} \geq y) = 1 - (1 - p^y)^n. \tag{5.9}$$

If $Y_{\max}$ is used as test statistic, the P-value for any observed value $y_{\max}$ of $Y_{\max}$ can be calculated directly from equation (5.9) once an appropriate value of n is chosen. Now, any sequence of successes (of length 0 or more) must be preceded by a failure. Under the null hypothesis there will be approximately $(1-p)N$ such failures, so that there will also be approximately $(1-p)N$ sequences of successes of length 0 or more. This is the value we use for n. This implies that

$$P\text{-value} \approx 1 - (1 - p^{y_{\max}})^{(1-p)N} \tag{5.10}$$

If $N = 100{,}000$ and $p = \frac{1}{4}$ and the observed length $y_{\max}$ of the longest repeated sequence of a is 10, the approximate P-value calculated from

(5.10) is 0.0690272. Thus given an observed value of 10 for $Y_{\max}$ the null hypothesis would not be rejected with any conventionally used Type I error. On the other hand, if the observed longest length were 12, the P-value would be approximately 0.45%, and this would often be taken as being sufficiently small for rejection of the null hypothesis.

The discussion following the exponential approximation in (B.3) shows that the expression (5.10) is well approximated, when N is large and $(1 - p)Np^{y_{\max}} \leq 1$, by

$$P\text{-value} \approx 1 - e^{-(1-p)Np^{y_{\max}}}. \qquad (5.11)$$

In the case $N = 100,000$, $p = \frac{1}{4}$, the approximation (5.11) gives 0.0690275, essentially identical to the value found from (5.10). Expressions of the form (5.11) arise later in connection with BLAST.

There is one approximation used in these calculations that should be checked. It was assumed in the calculations that the number of failures is $(1-p)N$. More exactly, this number is a random variable having a binomial distribution with parameters N and $1 - p$. Thus when $N = 100,000$ and $1 - p = \frac{3}{4}$, a precise calculation for the null hypothesis probability that $Y_{\max} \geq 10$ is, from equations (1.45) and (1.96),

$$\sum_{j=0}^{100,000} \binom{100,000}{j} \left(\frac{3}{4}\right)^j \left(\frac{1}{4}\right)^{100,000-j} \left(1 - \left(1 - \left(\frac{1}{4}\right)^{10}\right)^j\right). \qquad (5.12)$$

Equation (B.1) shows that this expression reduces to

$$1 - \left(1 - \frac{3}{4}\left(\frac{1}{4}\right)^{10}\right)^{100,000} \approx 0.0690276.$$

This agrees to seven decimal place accuracy with the approximate value found from (5.11). This close agreement is typical of the values for other cases, and we conclude that the approximation of regarding the number of sequences as fixed at its mean value is sufficiently accurate for all practical purposes.

It was shown in Table 2.5 that for the geometric distribution (the $k = 0$ case in Table 2.5), the expression in (2.139) gives a very accurate approximation for P-values, essentially identical to the exact calculation given in (2.107), provided that one uses the expressions $\mu_{\max}$ and $\sigma^2_{\max}$ for the mean and variance approximations for $Y_{\max}$ given in equations (2.119) and (2.121). These expressions are repeated here for convenience, with a slightly altered notation:

$$\mu_{\max} \approx \frac{\gamma + \log n}{-\log p} - \frac{1}{2}, \quad \sigma^2_{\max} \approx \frac{\pi^2}{6(\log p)^2} + \frac{1}{12}. \qquad (5.13)$$

A more difficult question is to ask whether there is significant evidence of long repeated sequences of any nucleotide, not only a. Here we consider a collection of repeated sequences, some of which might be of a, some of g, some of c, and some of t. Karlin et al. (1983) found the mean and variance of the length of the longest such repeated sequence. Clearly, this mean must be larger than the mean given in (5.13), since now repeated sequences of any nucleotide, not only a, are allowed. Perhaps surprisingly, this mean and variance are rather similar to those given in the approximation (5.13), being

$$\mu_{\max} \approx \frac{0.6359 + 2\log n + \log(1 - P)}{-\log P} - 1, \quad \sigma^2_{\max} \approx \frac{\pi^2}{6(\log P)^2}, \quad (5.14)$$

where P is the sum of the squares of the four nucleotide frequencies. These expressions can be used in conjunction with (2.139) to find the approximate P-value corresponding to the length of the longest repeated sequence.

If all nucleotides are equally frequent, this mean is approximately twice that given in (5.13) when n is large. This indicates the size of the increase in the mean when repeats of *all* nucleotides are of interest.

5.5 r-Scans

We define a very short DNA sequence, such as *gaga*, as a *word*. Suppose that we are interested in some specific word and observe in a long segment of DNA of length L that this word occurs n times. The word may be assumed to be so short relative to L that we may regard the positions of the word as points, and it is convenient to normalize lengths so that the long DNA sequence is then taken to be of length 1.

We wish to test whether the locations of the occurrences of this word appear to be random in the long DNA segment. Karlin and Macken (1991a,b) derive a variety of tests to assess heterogeneity of the location of various restriction sites and other markers in genetic sequence data. We give examples of their analysis in this section. Various asymptotic results for r-scans not discussed here are given by Reinert et al. (2000).

The problem to be addressed is that of testing the null hypothesis that the points where the word appears are iid uniformly distributed random variables with range $[0, L]$. The alternative hypothesis of interest is, in practice, either that the points occur in a clumped fashion or that they occur in an overly dispersed fashion.

The alternative hypothesis is not sufficiently precise to determine an unambiguous test statistic through the methods to be described in Chapter 8. Here we examine properties of a collection of test statistics arrived at on reasonable commonsense grounds, culminating in the r-scan statistics of Karlin and Macken (1991a,b). These test statistics are all based on the sizes $U_1, U_2, \ldots, U_{n+1}$ of the subintervals of $(0, 1)$ that are defined by the n points at which the word of interest occurs. Properties of $U_1, U_2, \ldots, U_{n+1}$

were examined extensively in Section 2.15.2 as a preliminary to the r-scan analysis discussed here.

Suppose first that the alternative hypothesis is that the points occur in an overly dispersed fashion. Here we might use as test statistic the maximum $U_{\max}$ of $U_1, U_2, \ldots, U_{n+1}$, and we will accept the alternative hypothesis if this maximum is too large. To assess how large a value is needed it is necessary to find the null hypothesis distribution of this maximum.

For any given number u, $0 < u < 1$, the probability that at least one length U_j exceeds u is given by equation (1.90). This is also the probability that the maximum length exceeds u. This probability is then the P-value associated with any observed value u of $U_{\max}$. Often the first term on the right-hand side of (1.90) supplies a close approximation to this P-value.

An approximate approach to finding this P-value is based on the "conditioned exponentials" calculation of Section 2.15.2. Each U_j has mean $1/(n+1)$ (by symmetry, since $\sum_j U_j = 1$ and the U_j are identically distributed). It may also be shown that if $X_1, X_2, \ldots, X_{n+1}$ are iid random variables, each having the exponential distribution (1.59) with the parameter λ arbitrary, then U_j has the same probability distribution as the random variable

$$\frac{X_j}{\sum_{k=1}^{n+1} X_k}. \tag{5.15}$$

If λ is chosen to be $n+1$, the mean of the denominator in (5.15) is 1. Approximating the denominator by this mean allows the mean and variance of U_j to be approximated by those of X_j, namely $1/(n+1)$ and $1/(n+1)^2$, respectively.

Equations (2.97) and (2.99) then show that the mean of the maximum of $n+1$ such variables is

$$\frac{1}{n+1} \left(\frac{1}{n+1} + \frac{1}{n} + \cdots + \frac{1}{2} + \frac{1}{1} \right), \tag{5.16}$$

and the variance is approximately

$$\frac{\pi^2}{6(n+1)^2}. \tag{5.17}$$

The expression (5.16) is in fact not an approximation but is the exact mean of $U_{\max}$ given in equation (2.168).

Strictly speaking, one should not use the calculations in Section 2.11.3 to approximate the complete distribution of $U_{\max}$, since these calculations require a domain of the form $(A, +\infty)$ for the random variables whose maximum is of interest, and this domain does not apply for $U_{\max}$. However, for large n this requirement becomes increasingly unimportant, and we then use (2.126) to find approximate P-values for $U_{\max}$.

Using the mean given in (2.168) and the variance (5.17) in the approximation (2.126), we get, as $n \to \infty$,

$$\text{Prob}(U_{\max} \geq u) \sim 1 - e^{-(n+1)e^{-(n+1)u}}. \tag{5.18}$$

If $(n+1)e^{-(n+1)u}$ is small, this expression may in turn be approximated by

$$\text{Prob}(U_{\max} \geq u) \sim S(n+1)e^{-(n+1)u}. \tag{5.19}$$

This approximation is essentially the same as that found by using only the first term on the right-hand side of (1.90).

In view of the various approximations involved in reaching (5.18) and (5.19), it is useful to calculate a selection of P-values exactly, using (1.90), and to compare these with the values found from (5.18) and (5.19). If $u = 0.01$ and $n + 1 = 1000$, the exact P-value found from (1.90) is 0.0482. The value found from the approximation (5.18) is 0.0444 and that found from (5.19) is 0.0454. When $u = 0.000017$ and $n + 1 = 1,000,000$, the exact P-value is 0.0406. The value found from the approximation (5.18) is also 0.0406, while that found from (5.19) is 0.0414. This demonstrates the accuracy of (5.18), at least for large n.

The approximation (5.18) may be written in the equivalent form

$$\text{Prob}\left(U_{\max} \geq \frac{\log(n+1) + u}{n+1}\right) \sim 1 - e^{-e^{-u}}. \tag{5.20}$$

The advantage of this formulation is that it can be generalized to apply to a wide variety of "r-scan" test statistics. Instead of using $U_{\max}$ as test statistic, it might be thought to be more reasonable to use the maximum $R_{\max(r)}$ of the "r-fragment lengths" $R_i(r)$ ($i = 1, 2, \ldots, n - r + 1$) where

$$R_i(r) = \sum_{j=i}^{i+r-1} U_j. \tag{5.21}$$

Karlin and Macken (1991a,b) show that

$$\text{Prob}\left(R_{\max(r)} \geq \frac{\log(n+1) + (r-1)\log(\log(n+1)) + u}{n+1}\right)$$
$$\sim 1 - e^{-e^{-u}/(r-1)!}. \tag{5.22}$$

This is a direct generalization of (5.20), to which it reduces when $r = 1$. This probability calculation allows us to calculate approximate P-values for any observed value of $R_{\max(r)}$. A further generalization allows the use of the kth largest of the $R_i(r)$ values as test statistic.

In order to assess clumping of the word of interest we might decide to use the minimum $U_{\min}$ of the U_j values, sufficiently small values of this statistic being significant. An exact expression for the probability that $U_{\min}$ is less than u is given in (2.87). An approximation for this probability analogous to (5.19) is

$$\text{Prob}(U_{\min} \leq u) \sim 1 - e^{-u(n+1)^2}. \tag{5.23}$$

These expressions allow exact and approximate P-value calculations associated with any observed value u of $U_{\min}$.

More generally, we can use the smallest $R_{\min(r)}$ of the r-fragment lengths as test statistic. Here P-values are found from the approximation

$$\text{Prob}\big(R_{\min(r)} \leq u\big) \sim 1 - e^{-u^r(n+1)^{r+1}/r!}, \tag{5.24}$$

which generalizes (5.23). A further generalization allows us to use the kth smallest of the r-fragment lengths as test statistic.

5.6   The Analysis of Patterns

The discussion in this section, and in Sections 5.7, 5.8, and 5.9, is self-contained and is not needed for the material that follows. This material serves to illustrate some unintuitive results about patterns in sequences that arise even in simple cases involving independence.

Suppose we are interested in some word, for example *gaga*. We ask the following two questions of an iid DNA sequence of length N: "What is the mean number of times that this word arises in a segment of length N?" and "What is the mean length between one occurrence of this word and the next?" We will call these two questions, and (later) the parallel questions about variances, "number of occurrences" and "distance between occurrences" questions respectively.

The analysis of nucleotide sequences leads to a focus on an "alphabet" of size 4. We use this example throughout, although generalizations to arbitrary alphabet sizes are straightforward. Words other than *gaga* and words of arbitrary length will also be considered later. The analysis in this chapter assumes that the nucleotide types at different sites are independent. This assumption is made to introduce some of the unexpected features of word pattern properties in a simple setting. In view of the fact that dependencies appear to exist between nucleotides at adjoining sites, the analysis is extended to the Markov-dependent case in Section 10.4.

There are various reasons why these questions might be asked. One is that there might be some a priori reason to suspect that the word *gaga* occurs significantly more often than it should if the nucleotides were generated in an iid fashion. To test for this it is necessary to discuss probabilistic aspects of the frequency of this word under the iid assumption. Perhaps surprisingly, probabilistic aspects of the frequencies of other words such as *gggg*, *gaag*, and *gagc* are often different from those of the word *gaga*, even under the iid assumption.

A second reason has been discussed by Bussemaker et al. (2000). Here the aim is to discover promoter signals by looking for DNA patterns common to upstream regions of genes. This is done by creating a dictionary of words of different lengths, each with an assigned probability. Any word that occurs more frequently than expected in these regions is a candidate for such a sequence. Here the analysis becomes very complicated, since there is no word that is of a priori interest, and indeed no word length that can be defined in advance as being specifically sought. Difficult statistical problems of multiple testing then arise: These are beyond the level of this book.

Examples of words that appear with unusually large frequencies at various types of sites have been investigated by Biaudet et al. (1998), Chedin et al. (1998), and Karlin et al. (1992). Leung et al. (1996) investigate under and overrepresented words in specific genomes. A general review of the theory of word occurrences is given by Reinert et al. (2000).

5.7 Overlaps Counted

5.7.1 General Comments

Consider some word such as *gaga*. It is necessary to address the question of how to count the number of times that this word, or any other word, occurs in a given DNA sequence. If overlapping occurrences of *gaga* are all relevant and counted, then in the sequence

$$t \ a \ t \ g \ a \ g \ a \ g \ a \ t \ c \ c \ g \ a \ g \ a \qquad (5.25)$$

the word *gaga* is counted as occurring in positions 4–7, 6–9, and 13–16, thus a total of three times. Even though two of these words overlap in positions 6 and 7, both are counted. If second and higher overlapping words are not counted, the word *gaga* is counted as occurring only twice, namely in positions 4–7 and 13–16. A more precise definition of the nonoverlap accounting is given later, but it is already clear that the number of times that any given word is counted depends on which of the two possible overlap counting conventions is adopted. The case where overlaps are counted is discussed in this section, and the case where they are not counted is discussed in Section 5.8.

5.7.2 Number of Occurrences

Suppose first that overlaps are counted. We continue to focus on the word *gaga*, and define $Y_1(N)$ as the number of times that the word *gaga* occurs in a nucleotide sequence of length N. We assume throughout this section that nucleotides in the sequence are generated independently, with any given nucleotide arising at any site with probability 0.25. Our first aim is to find the mean and variance of $Y_1(N)$ under these assumptions.

Define the indicator variable I_j by $I_j = 1$ if the word *gaga* finishes in position j, $I_j = 0$ if it does not. Then the total number of times $Y_1(N)$ that the word *gaga* occurs in a sequence of length N can be represented as $I_4 + I_5 + \cdots + I_N$. The mean of $Y_1(N)$ can then be found by applying equation (2.27). The mean of I_j is the probability that *gaga* finishes in position j, namely $1/256$. Equation (2.27) then shows that the mean number of times that the word *gaga* arises in a sequence of length N is 0 if $N < 4$, while for $N \geq 4$,

$$\text{mean of } Y_1(N) = \frac{N-3}{256}. \tag{5.26}$$

The same conclusion holds for any other word of length four.

The variance of $Y_1(N)$ is the variance of $I_4 + I_5 + \cdots + I_N$, namely

$$E(I_4 + I_5 + \cdots + I_N)^2 - \left(\frac{N-3}{256}\right)^2.$$

Now,

$$
\begin{aligned}
E(I_4 + &I_5 + \cdots + I_N)^2 \\
&= E(I_4^2 + I_5^2 + \cdots + I_N^2) \\
&\quad + 2(N-4) \text{ terms of the form } E(I_j I_{j+1}) \\
&\quad + 2(N-5) \text{ terms of the form } E(I_j I_{j+2}) \\
&\quad + 2(N-6) \text{ terms of the form } E(I_j I_{j+3}) \\
&\quad + (N-6)(N-7) \text{ terms of the form } E(I_j I_k),\ |j-k| > 3.
\end{aligned} \tag{5.27}
$$

This calculation assumes that $N \geq 6$, and we make this assumption from now on. The case $N \leq 5$ is easily handled separately.

Since the only possible values of I_j are 0 and 1, it follows that $I_j^2 = I_j$. Thus the first term on the right-hand side of (5.27) is $E(I_4 + I_5 + \cdots + I_N)$. This is the mean number of times that the word *gaga* occurs, namely $(N-3)/256$ (from equation (5.26)).

It is impossible for the word *gaga* to finish at both positions j and $j+1$. Thus one (or both) of I_j and I_{j+1} must be zero, so that $I_j I_{j+1} = 0$. Thus the term $E(I_j I_{j+1})$ in (5.27) must be zero. It is also impossible for *gaga* to finish at both positions j and $j+3$, so that the term $E(I_j I_{j+3})$ in (5.27) must also be zero.

Next, $I_j I_{j+2}$ is zero unless the word *gaga* finishes both in position j and also in position $j+2$, in which case it takes the value 1. This happens if and only if the word *gagaga* occupies positions $j-3, j-2, j-1, j, j+1$, and $j+2$. The probability of this is $1/4096$, so that the third term on the right-hand side of (5.27) contributes an amount $2(N-5)/4096$.

Finally, when $|j - k| > 3$, $I_j I_k$ is zero unless the word *gaga* finishes both in position j and also in position k. These refer to nonoverlapping positions, so for any such (j, k) pair the probability of this event is $(1/256)^2 = 1/65536$. The final term on the right-hand side of (5.27) thus contributes $(N - 6)(N - 7)/65536$.

Addition of these various terms shows that the variance of $Y_1(N)$ is

$$\text{variance of } Y_1(N) = \frac{281N - 895}{65536} \approx 0.004288N - 0.01366. \qquad (5.28)$$

The complete distribution of $Y_1(N)$ is complicated. When N is large, $Y_1(N)$ has an approximate normal distribution. We use here and below various calculations for the case $N = 1{,}000{,}000$, which, although too long for most practical applications, is nevertheless useful for calculation purposes. The mean and variance calculated above can then be used in an approximate test of the assumptions made above, which we take as the null hypothesis. If $N = 1{,}000{,}000$, the mean and variance of $Y_1(N)$ under the null hypothesis are 3906.24 and 4287.71, respectively. If the observed value of $Y_1(N)$ is 4023, the associated z-score (1.55) is $(4022.5 - 3906.24)/\sqrt{4287.71} = 1.775$. This corresponds to a P-value of about 0.038, just small enough to reject the null hypothesis if the Type I error is chosen to be 5%.

Several interesting points can be made about the variance calculation (5.28). First, $Y_1(N)$ does not have a binomial distribution; if *gaga* finishes in position j, for example, it cannot finish in position $j + 1$. Thus the binomial independence requirement does not hold. If $Y_1(N)$ had (incorrectly) been taken to have a binomial distribution with index $N - 3$ and parameter $1/256$, the mean value given in equation (5.26) would continue to hold, but the variance of $Y_1(N)$ would be calculated as $(255N - 765)/65536 \approx 0.0038910N - 0.0117$. This is less than the correct variance given in equation (5.28) whenever $N \geq 6$, the values of N for which equation (5.28) holds. In other words, the number of occurrences of the word *gaga* is more variable than the incorrect binomial formula suggests. This added variability arises because the occurrence of the word *gaga* at position j increases the probability of the occurrence of the word at position $j + 2$, whereas the nonoccurrence of this word at position j decreases the probability of the occurrence of the word at position $j + 2$. This "reinforcement" behavior explains the increased variance.

Second, similar calculations can be made for other words. Under the null hypothesis the mean in equation (5.26) applies for the number of occurrences of any word of four letters. However, the variance of this number depends on the word of interest. This is illustrated by considering the four words *gaga, gggg, gaag,* and *gagc*.

The calculation in equation (5.27) shows that whether or not a term of the form $E(I_j I_{j+k})$ contributes to the variance depends on the nature of the word. We define the constant w_j to be equal to 1 if the first j letters in the word of interest are the same as, and in the same order as, the last j letters, and 0 otherwise. Then the variance of the number of times that this word occurs in a sequence of length N can be written, after some simplification, as

$$\frac{(249 + 128w_3 + 32w_2 + 8w_1)N - (735 + 512w_3 + 160w_2 + 48w_1)}{65536}. \qquad (5.29)$$

For any given word the values of the constants w_j are easily established, and from these this variance follows immediately. The w_j values for the four words $gaga$, $gggg$, $gaag$, and $gagc$ are given in Table 5.5. The value w_4 is always 1 for words of four letters, and is included in the table, since it is used in some future calculations.

word	w_1	w_2	w_3	w_4
$gaga$	0	1	0	1
$gggg$	1	1	1	1
$gaag$	1	0	0	1
$gagc$	0	0	0	1

TABLE 5.5. Values of w_j ($j = 1, 2, 3, 4$) for the words $gaga$, $gggg$, $gaag$, and $gagc$.

The variance expression (5.29) and the values in Table 5.5 jointly show that for the words $gggg$, $gaag$, and $gagc$ the variances of $Y_1(N)$ are, respectively,

$$\frac{417N - 1455}{65536}, \quad \frac{257N - 783}{65536}, \quad \frac{249N - 735}{65536}. \qquad (5.30)$$

As a numerical example, the variances of the numbers of times that the words $gaga$, $gggg$, $gaag$, and $gagc$ occur in a DNA sequence of length $N = 1,000,000$ are, respectively,

$$4284, \quad 6363, \quad 3292, \quad 3799. \qquad (5.31)$$

These all differ from the binomial variance 3890.10. More generally, the expression (5.29) shows that there is no possible choice of the constants w_j leading to the binomial variance $(255N - 765)/65536$.

All the above calculations assume that the probability of the occurrence of any nucleotide at any site is .25, that the word of interest is of length 4 and that the alphabet of interest is of size 4. More general calculations, where these assumptions are relaxed, are given in Section 5.9.

5.7.3   Distance Between Occurrences

Exact Results

Properties of the distance between occurrences of the word of interest have recently been discussed by Robin and Daudin (1999), following the work of Blom (1982), Blom and Thorburn (1982), and Cowan (1991). In the discussion in this section of their results, all the assumptions of the previous section are retained, as are the definitions of the constants w_j.

Any word of length 4, having just occurred at some site i, can next occur at site $i+1$, $i+2$, or $i+3$, that is, at distance y ($y = 1, 2, 3$), only if, for that word, $w_{4-y} = 1$. Define Y_2 to be the distance until the next occurrence of the word after site i, and for notional convenience, let $p(y) = p_{Y_2}(y)$ be the probability that $Y_2 = y$. Then, as explained below,

$$p(y) = w_{4-y}4^{-y} - \sum_{j=1}^{y-1} p(j)w_{4+j-y}4^{j-y} \tag{5.32}$$

for $y = 1, 2, 3$, while if $y \geq 4$,

$$p(y) = 4^{-4} - 4^{-4}\sum_{j=1}^{y-4} p(j) - \sum_{j=y-3}^{y-1} p(j)w_{4+j-y}4^{j-y}. \tag{5.33}$$

The reasoning behind equation (5.33) is as follows. Let E be the event that the word of interest arises at site $i + y$. Let F be the event that the word of interest arises at site $i + y$, and does not arise at any sites between sites i and $i + y$. For each $j = 1, 2, \ldots, y - 1$, let A_j be the event that the word arises at sites $i + j$ and $i + y$, and does not occur anywhere between i and $i + j$. Then

$$E = F \cup A_1 \cup \cdots \cup A_{y-1},$$

and the terms on the right are all disjoint. Therefore,

$$p(y) = \text{Prob}(F) = \text{Prob}(E) - \sum_{j=1}^{y-1} \text{Prob}(A_j).$$

The event (E) has probability 4^{-4}. The probability of (A_j), for $j \leq y - 4$, is $p(j)4^{-4}$, since the events that the word of interest occurs at sites $i + j$ and $i + y$ are independent. Similar arguments show that the final term on the right-hand side of (5.33) gives the probabilities of (A_j) for $j = y - 3$, $y - 2$, and $y - 1$. These calculations lead to (5.33), and equation (5.32) is found by similar arguments.

We now find the pgf $\mathbf{p}(t)$ of the distance between successive occurrences of the word if interest. Multiplying throughout in equations (5.32) and (5.33) by t^y, we get, for example,

$$p(1)t = \tfrac{w_3}{4}t$$

$$p(2)t^2 = \tfrac{w_2}{16}t^2 \qquad\qquad -\tfrac{w_3}{4}p(1)t^2$$

$$p(3)t^3 = \tfrac{w_1}{64}t^3 \qquad\qquad -\tfrac{w_3}{4}p(2)t^3 - \tfrac{w_2}{16}p(1)t^3$$

$$p(4)t^4 = \tfrac{1}{256}t^4 \qquad\qquad -\tfrac{w_3}{4}p(3)t^4 - \tfrac{w_2}{16}p(2)t^4 - \tfrac{w_1}{64}p(1)t^4$$

$$p(5)t^5 = \tfrac{1}{256}t^5 - \tfrac{1}{256}p(1)t^5 \qquad -\tfrac{w_3}{4}p(4)t^5 - \tfrac{w_2}{16}p(3)t^5 - \tfrac{w_1}{64}p(2)t^5$$

$$p(6)t^6 = \tfrac{1}{256}t^6 - \tfrac{1}{256}(p(1) + p(2))\, t^6 - \tfrac{w_3}{4}p(5)t^6 - \tfrac{w_2}{16}p(4)t^6 - \tfrac{w_1}{64}p(3)t^6$$

We now sum these equations over all possible values of y. The sum of the terms on the left-hand sides is, by definition, $\mathbb{p}(t)$. The sum of the terms on the right-hand side is

$$\frac{w_3}{4}t + \frac{w_2}{16}t^2 + \frac{w_1}{64}t^3 + \frac{t^4}{256(1-t)} - \frac{t^4\mathbb{p}(t)}{256(1-t)} - \left(\frac{w_3}{4}t + \frac{w_2}{16}t^2 + \frac{w_1}{64}t^3\right)\mathbb{p}(t).$$

Equating these two sums and solving for $\mathbb{p}(t)$, we find that

$$\mathbb{p}(t) = \frac{t^4 + (1-t)(64w_3t + 16w_2t^2 + 4w_1t^3)}{t^4 + (1-t)(256 + 64w_3t + 16w_2t^2 + 4w_1t^3)}. \tag{5.34}$$

Differentiation of the generating function (5.34) with respect to t, together with use of equation (1.36), shows that the mean of Y_2 is

$$E(Y_2) = 256. \tag{5.35}$$

A second differentiation with respect to t, together with equation (1.37), shows that the variance of Y_2 is

$$\text{Var}(Y_2) = 512\sum_{j=1}^{4} w_j 4^j - 67{,}328. \tag{5.36}$$

While the mean of Y_2 is independent of the word of interest, the variance depends on this word. For the words *gaga*, *gggg*, *gaag*, and *gagc*, the variances given by formula (5.36) are, respectively,

$$71{,}936, \quad 106{,}752, \quad 65{,}792, \quad \text{and} \quad 63{,}744. \tag{5.37}$$

The Geometric Approximation

It is natural to assume that the distance between consecutive occurrences of a given word has the geometric distribution, at least to a close approximation. This assumption has been made on occasion in the biological literature. The assumption of a geometric distribution is, however, incorrect, as a comparison of the pgf of the geometric distribution (given in Problem 1.11) and the pgf (5.34) shows. In some cases the error can be substantial, as we now show.

Whatever the word of interest is, the mean distance between consecutive occurrences is 256. If we choose the parameters of the geometric distribution so that the mean of that distribution takes this value, then from the results of Problems 1.2 and 1.10, the variance of this geometric distribution is 65,792. This differs from three of the variances listed in (5.37), and differs substantially from the variance 106,752 for the case of the word *gggg*.

Beginning at the Origin

In this section we consider the number Y_3 of sites until the first occurrence of a word of interest, starting at the origin. The generating function (5.34) is not appropriate for this case since it was calculated allowing for overlaps of words, whereas overlaps at the origin into "negative values" cannot occur. This observation is confirmed by noting that if $P(y)$ is the probability that the word of interest first occurs at distance y from the origin, then in contrast to the values found from (5.32), $P(1) = P(2) = P(3) = 0$ and $P(4) = 1/256$.

The recurrence relation (5.33), however, continues to hold when $y \geq 5$. From these observations the generating function for the distribution of the distance until the first occurrence of any word can be found. This is

$$\mathbb{p}(t) = \frac{t^4}{t^4 + (1-t)(256 + 64w_3t + 16w_2t^2 + 4w_1t^3)}. \tag{5.38}$$

This differs from the generating function (5.34). The mean and variance of the random variable Y_3, found by differentiating $\mathbb{p}(t)$, are, respectively,

$$E(Y_3) = \mu_3 = \sum_{j=1}^{4} w_j 4^j, \tag{5.39}$$

$$\text{Var}(Y_3) = \mu_3^2 + \mu_3 - 2\sum_{j=1}^{4} jw_j 4^j. \tag{5.40}$$

For the words *gaga, gggg, gaag,* and *gagc,* the respective means of Y_3 are

$$272, \quad 340, \quad 260, \quad 256, \tag{5.41}$$

and the respective variances of Y_3 are

$$72{,}144, \quad 113{,}436, \quad 65{,}804, \quad 63{,}744. \tag{5.42}$$

It might be surprising that both the mean and variance depend on the nature of the word, even though the four words considered are equally likely to occur in any given four consecutive sites. The values for the various means are generally close to those in (5.37), and are identical for the word *gagc*. This identity arises because for this word, no overlaps are possible in successive occurrences of the word. The differences between the two sets of values emphasizes the different properties of "between word" distances and "beginning at the origin" distances. The latter distances will arise again in a different context in Section 5.8.2.

5.8   Overlaps Not Counted

5.8.1   General Comments

Let R_1 be the first occurrence of some word in a sequence, R_2 the next occurrence that does not overlap with the first, R_3 the next that does not overlap with R_2, and so on. This gives a sequence of occurrences R_1, R_2, $R_3, \ldots$, which we refer to as the *recurrences* of the word (with R_1 called the first recurrence). Such recurrences form a special case of *recurrent events*, the theory of which is a major area of probability theory, the definitive account being in Feller (1968). We use some of this theory below.

There are practical reasons why overlapping words should not be counted. For example, restriction endonucleases are enzymes that cut a chromosome whenever a "recognition" sequence (such as *gaga*) specific to that enzyme is encountered in that chromosome. A property of these enzymes is that they do not cut the DNA twice in immediate succession in the sense that if a DNA subsequence is $\ldots attgagagaacc\ldots$, and the recognition sequence of a restriction endonuclease is *gaga*, then the DNA sequence will be cut at position 7 in the subsequence above but not also at position 9. This implies that cuts in the DNA occur as recurrent events.

5.8.2   Distance Between Recurrences

We will say that a recurrence of a word occurs at site n if it finishes in that site.

Let u_n be the probability that a recurrence of a given word occurs at site n and f_n be the probability that the first recurrence arises at site n. The mean number μ of sites between consecutive recurrences is $\mu = \sum_{i=1}^{\infty} i f_i$. It is assumed below that μ is finite. A fundamental result of recurrent event theory concerns the limiting behavior of u_n as $n \to +\infty$. This result is stated as a theorem (without proof).

Theorem 5.1. $u_n \to \mu^{-1}$ as $n \to +\infty$.

This theorem allows direct calculation of μ, and we illustrate this with the word *gaga*. Let E be the event that *gaga* arises at site n, for some $n \geq 6$. Let A be the event that it occurs at site n as one of the recurrences, and let B be the event that it arises at site $n-1$ as one of the recurrences, and that the word *ga* appears at sites $n-1$ and n. Then $E = A \cup B$, and A and B are disjoint. Therefore, $\mathrm{Prob}(E) = \mathrm{Prob}(A) + \mathrm{Prob}(B)$, giving

$$\frac{1}{256} = u_n + \frac{1}{16} u_{n-2}. \tag{5.43}$$

Letting $n \to \infty$ and using Theorem 5.1, we have

$$\frac{1}{256} = \mu^{-1}\left(1 + \frac{1}{16}\right),$$

so that

$$\mu = 272. \tag{5.44}$$

Results such as this will be derived using the theory of Markov chains in Section 10.6.2.

The argument leading to (5.44) shows how the case of an arbitrary four-letter word can be handled, using the notation w_j established in Section 5.7. The arguments leading to equation (5.43) lead to

$$\frac{1}{256} = u_n + \frac{1}{4}w_3 u_{n-1} + \frac{1}{16}w_2 u_{n-2} + \frac{1}{64}w_1 u_{n-3}. \tag{5.45}$$

The limiting process $n \to \infty$ shows that in general,

$$\mu = 4w_1 + 16w_2 + 64w_3 + 256. \tag{5.46}$$

This is identical to formula (5.39) in Section 5.7.3, found in considering distances from the origin to the first occurrence of a specified word. The reason for this identity is that the theory of the distance in counting from the origin to the first occurrence of a word is identical to the theory of the distance between consecutive recurrences of the word, since in the former, overlaps of a word over the origin cannot occur. Thus not only the mean, but the entire distribution of the distance between consecutive counted occurrences of a specified word in the nonoverlapping case follows immediately from the results of Section 5.7.3.

5.8.3 Number of Recurrences

Recurrent event theory shows that in the nonoverlapping case the mean number of recurrences of any word in a long sequence N of trials is asymptotically

$$\text{mean number of recurrences} \sim \frac{N}{\mu}, \tag{5.47}$$

where μ is the mean number of sites between successive recurrences of that word. This equation, together with (5.46), shows that that the respective mean number of recurrences of the words *gaga*, *gggg*, *gaag*, and *gagc* in a DNA sequence of length 1,000,000 are, approximately, 3676.47, 2941.18, 3846.15, and 3906.25. These should be compared to the (unique) value 3906.25 arising when overlapping words are counted, as calculated in Section 5.7. The two means are identical for the word *gagc*, since for this word overlaps of successive occurrences are not possible. For the other three words the mean number of recurrences is less than the mean total number of occurrences, as expected.

The number of recurrences of any word in a sequence of length N has, approximately, a normal distribution with mean (5.47) and asymptotic variance

$$\text{variance} \sim \frac{N\sigma^2}{\mu^3}, \tag{5.48}$$

where μ is the mean, and σ^2 is the variance, of the number of sites between successive recurrences of the word. Exact formulae for μ and σ^2 are given in equations (5.39) and (5.40), respectively.

It is interesting to compare variances calculated from (5.48) with those given in (5.31) for the case where overlapping words are counted. With $N = 1,000,000$, (5.48) gives approximate respective variances

$$3585, \quad 2886, \quad 3744, \quad 3799, \tag{5.49}$$

for the number of recurrences of the words $gaga, gggg, gaag$, and $gagc$. The comparison between the two sets of variances is illuminating. Perhaps the most surprising result is that whereas the total number of occurrences of the word $gggg$ has the largest of the four variances given in (5.31), the number of recurrences of this word has the smallest variance. This fact is largely due to the high mean distance between successive recurrences. The variances for the word $gagc$ are the same in both cases, since overlaps are impossible for this word. The fact that the approximate calculation using (5.48) yields the correct value for this latter variance shows that this formula provides an excellent approximation.

5.9   Generalizations

The formulae derived so far in this chapter all depend on special assumptions, in particular that the probability of the occurrence of any nucleotide at any site is .25 and that the word of interest is of length 4. Much of the analysis of Section 5.7 can be generalized to the case of arbitrary nucleotide probabilities and words of arbitrary length. We consider these cases separately, and do not attempt to generalize them simultaneously because the formulae become quite complicated. However it is clear from the examples given how to generalize to both cases simultaneously.

5.9.1   Arbitrary Nucleotide Probabilities

If the four nucleotides a, g, c, and t occur with respective probabilities p_a, p_g, p_c, and p_t, equation (5.26) when applied to the word $gaga$ generalizes to

$$\text{mean of } Y_1(N) = (N - 3)p_a^2 p_g^2. \tag{5.50}$$

The extension of this formula to other words of length 4, or words of any length, is immediate. The calculation of the variance of $Y_1(N)$ is more complicated. For example, equation (5.27) is relevant only for words of length 4, and a different equation is appropriate for words of other lengths. Even for words of length 4 a general variance formula such as (5.29) no longer exists, and a separate calculation for each word is required.

As an example, if the word of interest is $gaga$, the terms $E(I_j I_{j+1})$ and $E(I_j I_{j+3})$ in (5.27) are zero, while $E(I_j) = p_a^2 p_g^2$, $E(I_j I_{j+2}) = p_a^3 p_g^3$, and $E(I_j I_k) = p_a^4 p_g^4$ when $|j - k| > 3$. This shows that the variance of $Y_1(N)$ is

$$(N - 3)p_a^2 p_g^2 + 2(N - 5)p_a^3 p_g^3 - (7N - 33)p_a^4 p_g^4. \tag{5.51}$$

The variance for the words $gaag$, $gggg$, and $gagc$ differ from this (see Problem 5.11).

5.9.2 Arbitrary Word Length

The analysis of Section 5.7.3 can be generalized to the case where the word size is some arbitrary number k. This generalization is first described under the assumption of equal probabilities for the four nucleotides.

The argument that led to equations (5.32) and (5.33) shows that when k is arbitrary,

$$q(y) = w_{k-y} 4^{-y} - \sum_{j=1}^{y-1} q(j) w_{k+j-y} 4^{j-y} \tag{5.52}$$

for $y = 1, 2, \ldots, k - 1$. If $y \geq k$,

$$q(y) = 4^{-k} - 4^{-k} \sum_{j=1}^{y-k} q(j) - \sum_{j=y-k+1}^{y-1} q(j) w_{k+j-y} 4^{j-y}. \tag{5.53}$$

If the function $w(t, k)$ is defined by

$$w(t, k) = \sum_{j=1}^{k} w_j 4^j t^{k-j}, \tag{5.54}$$

these equations may be used to show that the generating function (5.34) generalizes to

$$\mathbb{P}(t) = \frac{t^k + (1 - t)\left(w(t, k) - 4^k\right)}{t^k + (1 - t)w(t, k)}. \tag{5.55}$$

From this, the mean and variance of Y_2, the number of sites between successive occurrences of the word of interest, are found to be, respectively,

$$E(Y_2) = 4^k, \tag{5.56}$$

$$\mathrm{Var}(Y_2) = 2 \times 4^k w(1, k) - (2k - 1)4^k - 4^{2k}. \tag{5.57}$$

These equations reduce, when $k = 4$, to (5.35) and (5.36).

The corresponding formulae for the first occurrence of the word of interest can be found by similar arguments. In this case the generating function of Y_3 is

$$\mathbb{p}(t) = \frac{t^k}{t^k + (1 - t)w(t, k)}, \tag{5.58}$$

and from this,

$$E(Y_3) = \mu_3 = \sum_{j=1}^{k} w_j 4^j, \tag{5.59}$$

$$\mathrm{Var}(Y_3) = \mu_3^2 + \mu_3 - 2\sum_{j=1}^{k} jw_j 4^j. \tag{5.60}$$

Equations (5.59) and (5.60) reduce respectively to (5.39) and (5.40) when $k = 4$. The discussion below equation (5.46) shows that these formulae apply also to the number of sites between successive recurrences of the word.

Several of the calculations in Section 5.8.2 may be generalized immediately to the case of arbitrary word lengths and arbitrary alphabet sizes, since the pgf (5.58), the mean (5.59), and the variance (5.60) apply to the case where overlaps are not counted.

Equations such as (5.43) and conclusions typified by (5.44) generalize easily to the case where the four nucleotides a, g, c, and t have arbitrary probabilities p_a, p_g, p_c, and p_t, respectively. For example, the argument that led to equation (5.43) shows that the probability u_n that the word atg recurs at site n satisfies the equation

$$p_a p_t p_g = u_n. \tag{5.61}$$

Letting $n \to +\infty$ in this equation and using the fact that $u_n \to \mu^{-1}$, we get

$$\mu = \frac{1}{p_a p_t p_g}. \tag{5.62}$$

If the word of interest is ata, then the equation corresponding to (5.61) is

$$p_a p_t p_a = u_n + p_t p_a u_{n-2}, \tag{5.63}$$

and this leads to

$$\mu = \frac{1 + p_a p_t}{p_a^2 p_t}. \tag{5.64}$$

We will rederive (5.62) and (5.64) in Section 10.6.2, using the theory of Markov chains.

Problems

5.1. Use the joint density function (2.152), together with transformation techniques, to prove equation (5.4).

5.2. Show that if L and G are fixed, the mean number of contigs given in (5.1) is maximized (as a function of N) when $N = G/L$.

5.3. Use the approximation (B.21) to approximate the mean contig size (5.2) when a is small, and interpret your result.

5.4. Write the mean number of anchored contigs (5.7) as
$$\frac{abG}{L} \frac{(e^{-a} - e^{-b})}{(b - a)}.$$

(i) Assume that G, L, and b are fixed, so that this mean number is a function of a only. Suppose that $G/L = 100{,}000$ and $b = 5$. Evaluate this mean for $a = 0.8, 1, 1.2, 1.4, 1.6, 1.8, 2.0$.

(ii) Now do the same calculation when $b = 10$.

(iii) For the cases $b = 5$, $b = 10$, estimate the value of a for which the mean number of contigs is maximized. How do these values compare with the case $b = +\infty$?

5.5. There are four models described below for a signal of length five: iid, weight matrix, first-order Markov, and MDD. For each of the sequences $CCGAT$ and $CATAT$ find the probability of the sequence given the model, for each of the four models (so your answer should consist of eight probabilities).

(i) iid. The probabilities of the four nucleotides are $\{p_a = .2, p_c = .1, p_g = .1, p_t = .6\}$.

(ii) Weight Matrix. The weight matrix (for the nucleotide ordering: a, c, g, t) is
$$\begin{bmatrix} .2 & .3 & .2 & .1 & .1 \\ .1 & .2 & .15 & .6 & .6 \\ .3 & .4 & .6 & .1 & .15 \\ .4 & .1 & .05 & .2 & .15 \end{bmatrix}.$$

(iii) First-Order Markov. The initial distribution is $\{p_a = .2, p_c = .1, p_g = .1, p_t = .6\}$, and the transition matrix (for the nucleotide ordering a, c, g, t) is
$$\begin{bmatrix} .1 & .8 & .05 & .05 \\ .35 & .1 & .1 & .45 \\ .3 & .2 & .2 & .3 \\ .6 & .1 & .25 & .05 \end{bmatrix}.$$

(iv) MDD. The first split is on position 2, with probabilities for this position $\{p_a = .2, p_c = .3, p_g = .1, p_t = .4\}$. For position 2 equal to c, g, or t we model the remaining positions with weight matrices

$$W_c = \begin{bmatrix} .2 & .1 & .2 & .8 \\ .5 & .1 & .2 & .1 \\ .2 & .1 & .3 & .05 \\ .1 & .7 & .3 & .05 \end{bmatrix}, \ W_g = \begin{bmatrix} .4 & .1 & .2 & .2 \\ .3 & .4 & .1 & .3 \\ .2 & .1 & .3 & .2 \\ .1 & .4 & .4 & .3 \end{bmatrix}, \ W_t = \begin{bmatrix} .1 & .1 & .2 & .2 \\ .6 & .6 & .4 & .35 \\ .2 & .1 & .3 & .15 \\ .1 & .2 & .1 & .3 \end{bmatrix}.$$

If position 2 equals a, we split further the other four positions on position 1, with probabilities for this position $\{p_a = .5, p_c = .1, p_g = .1, p_t = .3\}$. The remaining 3 positions are modeled with weight matrices

$$W_a = \begin{bmatrix} .3 & .4 & .2 \\ .4 & .3 & .1 \\ .2 & .2 & .6 \\ .1 & .1 & .1 \end{bmatrix}, \ W_c = \begin{bmatrix} .1 & .5 & .3 \\ .1 & .2 & .1 \\ .05 & .2 & .55 \\ .75 & .1 & .05 \end{bmatrix},$$

$$W_g = \begin{bmatrix} .1 & .2 & .2 \\ .5 & .1 & .1 \\ .1 & .1 & .5 \\ .3 & .6 & .2 \end{bmatrix}, \ W_t = \begin{bmatrix} .1 & .6 & .05 \\ .5 & .2 & .05 \\ .2 & .1 & .4 \\ .2 & .1 & .5 \end{bmatrix}.$$

5.6. For a sequence of DNA of length $N = 6$, find the probability that the word *gaga* occurs 0, 1, and 2 times, and hence verify the variance formula (5.28) for this case.

5.7. What changes are needed in the variance formula (5.28) for the cases $N = 5$, $N = 6$? Find the appropriate variance formulae in these two cases.

5.8. Prove equation (5.29).

5.9. For the case where the probability of each nucleotide at any site is $\frac{1}{4}$, find the mean number of sites between successive "counted" occurrences of the word *gagg*.

5.10. For the case where the probability of each nucleotide at any site is $\frac{1}{4}$, find the variance of the number of sites between successive counted occurrences of the word *gggg*. Use this to find the asymptotic variance of the number of counted occurrences of this word in a DNA sequence of length N. Compare your result with the first result in (5.30).

5.11. Show that the variance formula corresponding to (5.51) for the word *gaag* is

$$(N - 3)p_a^2 p_g^2 + 2(N - 6)p_a^4 p_g^3 - (7N - 33)p_a^4 p_g^4. \tag{5.65}$$

Find the corresponding formulae for the words *gggg* and *gagc*.

6
The Analysis of Multiple DNA or Protein Sequences

6.1 Two Sequences: Frequency Comparisons

In Example 3 of Section 3.4.3 we considered the test of the hypothesis that the probabilities for the four nucleotides in a DNA sequence are equal to a set of prescribed values. In this section we consider the test of the hypothesis that the two sets of probabilities for the four nucleotides in two DNA sequences are equal, no specific claim being made as to what the probabilities are. The data used for this test are as given in Table 6.1.

	nucleotide				
	a	g	c	t	Total
sequence 1	Y_{11}	Y_{12}	Y_{13}	Y_{14}	$Y_{1.}$
sequence 2	Y_{21}	Y_{22}	Y_{23}	Y_{24}	$Y_{2.}$
Total	$Y_{.1}$	$Y_{.2}$	$Y_{.3}$	$Y_{.4}$	Y

TABLE 6.1. Nucleotide counts.

The test is a particular case of the two-way table test discussed in Section 3.4.3, and the test statistic is either (3.23) or (3.24), applied to the data of Table 6.1. When the null hypothesis that the two sequences are drawn from populations with identical nucleotide frequencies is true, both statistics have an approximate chi-square distribution with three degrees of freedom. Further, the numerical values of the two statistics are usually close. For example, with the numerical values given in Table 6.2, the statistic (3.23) and the statistic (3.24) respectively take the values 0.96416 and 0.96426.

The 5% Type I error significance point for chi-square with three degrees of freedom is 7.81, so that the null hypothesis is not rejected using either statistic.

| | nucleotide | | | | |
	a	g	c	t	Total
sequence 1	273	258	233	236	1000
sequence 2	281	244	246	229	1000
Total	554	502	479	465	2000

TABLE 6.2. Nucleotide counts: numerical example.

The statistic (3.24) when applied to the data of Table 6.1 can be written in a different form. If

$$p_{1j} = Y_{1j}/Y_{1\cdot}, \quad p_{2j} = Y_{2j}/Y_{2\cdot}, \quad p_j = Y_{\cdot j}/Y,$$

the statistic (3.24) becomes

$$2Y \left(\frac{Y_{1\cdot}}{Y} \sum p_{1j} \log p_{1j} + \frac{Y_{2\cdot}}{Y} \sum p_{2j} \log p_{2j} - \sum p_j \log p_j \right). \tag{6.1}$$

The entropy-like statistic

$$\frac{Y_{1\cdot}}{Y} \sum p_{1j} \log_2 p_{1j} + \frac{Y_{2\cdot}}{Y} \sum p_{2j} \log_2 p_{2j} - \sum p_j \log_2 p_j \tag{6.2}$$

has been used as a chi-square in the literature. This is not appropriate, since the statistic (6.2) uses logarithms to the base 2 and omits the factor $2Y$ in (6.1).

Bernaola-Galván et al. (2000) take up the much more difficult statistical question of testing whether the nucleotide frequencies change at some undetermined point along a DNA sequence. A DNA sequence of length N may be divided into two subsequences of respective lengths n and $N - n$, and the DNA frequencies in these two subsequences may be compared using the statistic (6.1). For any given value of n, denote this statistic by C_n. If n is varied from a small number a to a number b close to N, a collection of values $C_a, C_{a+1}, \ldots, C_b$ will be obtained. A reasonable test statistic to use for this purpose is $C_{\max}$, the maximum of $C_a, C_{a+1}, \ldots, C_b$. To assess whether the observed value of $C_{\max}$ is significant, it is necessary to find its null hypothesis distribution. This is not easy, since the values of (6.1) are highly correlated and equations such as (2.102) may not be used for this purpose. It appears that the empirical methods discussed in Section 1.16 are necessary to find approximations for this distribution. Bernaola-Galván et al. (2000) conduct simulations to find an approximation to this distribution, and claim from these that a good approximation is found by replacing N by an effective length N_{eff} (approximately $2.45 \log N$), and then using equation (2.102) with N replaced by N_{eff} and y replaced by $.84y$. However, it remains an open matter to find theoretical properties of the probability distribution of $C_{\max}$.

6.2 Alignments

As a segment of genetic material is passed on through the generations in some line of descent in a population, the sequence constituting this material will change through the process of mutation. The simplest mutations are of the form of a switch from one nucleotide to another, or in the form of an insertion or a deletion. Mutations can spread to an entire species, or nearly so, through the process of natural selection or random drift. When a switch in nucleotides spreads throughout most of a species we call it a *substitution*. (When in a population at a given site there does not exist a single nucleotide type, we say that a *polymorphism* exists at that site.) As substitutions, insertions, and deletions get passed along through two independent lines of descent, the two sequences will slowly diverge from each other. For example, the original sequence may have been

cggtatgcca,

whereas the two descendents might be

cgggtatccaa

and

ccctaggtccca.

This divergence will happen at varying rates, depending on the function of the piece of DNA in question and how well that function tolerates substitutions and other changes. For protein coding DNA, the corresponding protein sequences also evolve through time as a result of DNA sequence evolution. Individual genes typically contain stretches that change rapidly and other stretches that remain relatively constant. The latter regions are called "functional domains" since their low tolerance to change suggests that they have a critical functional role in the viability of the organism.

Many problems in bioinformatics relate to the comparison of two (or more) DNA or protein sequences. In order to compare sequences of nucleotides or amino acids, we use *alignments*. The following is an example of an alignment of the above two descendent sequences:

c g g g t a − − t c c a a
c c c − t a g g t c c c a

The symbol "−" is called an *indel*: it represents an assumed insertion or deletion at some point in the evolutionary history leading to the two sequences. A sequence of ℓ consecutive indels is called a *gap* of length ℓ. In the above alignment there are two gaps, one of length 1 and one of length 2.

There are many types of alignments. There are *global* alignments, in which the entire lengths of the sequences are aligned, and there are *local* alignments, which align only subsequences of each sequence. There are *gapped* alignments, in which indels are allowed, and there are *ungapped* alignments, in which indels are not allowed. There are *pairwise* alignments, which are alignments of two sequences, and there are *multiple* alignments, which align more than two sequences. Of course, these can be combined into, for example, "gapped global pairwise" alignments. There is a large literature on alignments, of which we will examine just a few of the basics.

For a particular type of pairwise alignment (for instance ungapped global), there are many possible such alignments between any two sequences. Good alignments of related sequences are ones that better reflect the evolutionary relationship between them. There are several ways to discriminate between good and bad alignments. In the sections that follow we consider this problem in some detail.

6.3   Simple Tests for Significant Similarity in an Alignment

In this section we consider tests of the hypothesis that the two sequences in an ungapped alignment have been generated at random with respect to each other, that is, that there is no significant similarity between them reflected in the alignment.

In later sections we consider other more sophisticated solutions to this problem. We will also consider the more difficult problem of determining whether there is significant similarity between unaligned sequences, and the problem of finding the "best" alignments when there is significant similarity.

Consider two aligned DNA sequences such as those given in (1.1). In this section we wish to assess the significance of the similarity by using one or another "local" criterion. This can be done in various ways of which we here consider two, one requiring a stringent similarity criterion, the other a less stringent requirement.

Exactly-Matching Subsequences

As the most stringent similarity criterion, we consider as similar only those subsequences where the elements match exactly. In the array (1.1) there are five of these: the subsequences 'a' in position 3, 'c' in position 6, '$t\,a\,g$' in positions 9–11, '$c\,a$' in positions 13–14, and '$t\,a\,t$' in positions 23–25. We will generically denote the length of any such subsequence by Y. We denote the length of the longest such subsequence by $Y_{\max}$, and this is the statistic we will use to test for significant similarity between the two sequences. In the array (1.1) the observed value $y_{\max}$ of $Y_{\max}$ is 3. To assess in general whether the observed value of $Y_{\max}$ is significant, we have to find out the probability distribution of $Y_{\max}$ when the null hypothesis that there is no significant similarity between the two sequences, that is, that one is generated independently of the other, is true.

If a match is thought of as a success, the theory of Section 5.4 can be used for this problem if p is defined as the match probability $p_a^2 + p_g^2 + p_c^2 + p_t^2$, where p_a, p_g, p_c, and p_t are the respective frequencies of a, g, c, and t. With this interpretation of p, any subsequence of successes (of length 0 or more) must be preceded by a pair of nonmatching nucleotides, i.e., a failure. Under the null hypothesis there will be approximately $(1-p)N$ such failures, so that there will also be approximately $(1-p)N$ subsequences of successes in two long matched sequences each of length N. As in Section 5.4, the approximation that takes the number of subsequences as fixed at this value is sufficiently accurate for all practical purposes. This claim continues to be true for the "well-matching" subsequences considered in the following section, and we thus make this approximation there also.

Well-Matching Subsequences

The use of the length of the longest exactly matching subsequence as a test statistic is probably unwise. Even if two DNA sequences have a reasonably recent common ancestor, evolutionary changes will usually cause at least a small number of differences between them. Thus it is more appropriate to use a test procedure that focuses on well-matching, rather than exactly matching, subsequences.

One such approach is to consider subsequences where k mismatches (more strictly, because of end-effects, up to k mismatches) are allowed. The length Y of any such subsequence is the number of trials up to but not including the $(k+1)$th failure, so that the probability distribution of Y is the generalized geometric distribution given in equation (1.14). Because this distribution enters naturally into calculations involving well-matching subsequences, we now consider its distribution function at greater length than was done in Section 1.3.5.

The probability that y trials or fewer are required before failure $k+1$ is

$$F_Y(y) = \sum_{j=k}^{y} \binom{j}{k} p^{j-k}(1-p)^{k+1}, \quad y = k, k+1, k+2, \dots . \tag{6.3}$$

We will later wish to calculate $\text{Prob}(Y \le y-1)$. Replacing y by $y-1$ in equation (6.3), we immediately obtain

$$\text{Prob}(Y \le y-1)$$
$$= \sum_{j=k}^{y-1} \binom{j}{k} p^{j-k}(1-p)^{k+1}, \quad y = k+1, k+2, k+3, \dots . \tag{6.4}$$

The expression on the right-hand side of equation (6.4) can be calculated in various ways. One of these is straightforward numerical computation, which is now possible and available in most mathematical computer packages. Two perhaps unexpected ways of calculating this sum numerically are given in Section C. These calculations also show how we can use equation (6.4) to calculate the distribution function for the binomial distribution.

With these calculations in hand we turn to the statistical hypothesis-testing process using $Y_{\max}$ as test statistic. To use $Y_{\max}$ as test statistic we must be able to compute the P-value associated with any observed value y of $Y_{\max}$. We now address the question of this calculation, using as an example the case $N = 100,000$, $p = \frac{1}{4}$. In view of a calculation in Section 5.4, we assume that the number of subsequences is fixed at the value $n = N(1 - p) = 75,000$.

The main difficulty in calculating P-values exactly is that the lengths of the subsequences are not independent when $k > 1$. For example, a single long run of exact matches will lead to long lengths of two or more well-matching subsequences overlapping this run. Unfortunately, the theory for the probability distribution of the maximum of dependent subsequence lengths is difficult, and we are forced to consider alternative approaches.

One possible approach is to ignore this dependence and to use values such as those given in Table 2.5, calculated assuming independence of subsequence lengths. This is potentially dangerous, since as we have seen, P-values change very rapidly as a function of y, and thus the approximation implied by assuming independence is not guaranteed to give accurate P-values.

Fortunately, a second approach is available. Given current computing power, the dependent case is one where a computational approach using simulations is now possible. This approach in based on the discussion in Section 1.16, and allows us to assess the magnitude of any error incurred by making the independence assumption. We now discuss this approach.

A long random DNA sequence can be generated repeatedly by a simple program, and from this an empirical probability distribution for $Y_{\max}$ can be obtained. From this, quite precise estimates of P-values for $Y_{\max}$ can be found. Some representative estimates found in this way are given in Table 6.3. A comparison of the "independence" P-values in Table 2.5 and the simulation values in Table 6.3 shows that the independence assumption does cause somewhat inaccurate P-value approximations, the inaccuracy increasing, as might be expected, as k increases.

A second approach to the approximation of P-values is based on (2.139), which requires knowledge of, or approximations for, the mean $\mu_{\max}$ and the variance $\sigma^2_{\max}$ of $Y_{\max}$. To assess the accuracy of this approach it is first necessary to compare the simulation values for the mean and variance of $Y_{\max}$ with the "independence" values given in Table 2.5. The simulation values are given in Table 6.4, together with their standard deviations. A comparison of the values in the two tables shows that as k increases, the

y	$k=0$	$k=1$	$k=2$	$k=3$	$k=4$	$k=5$
7	0.990					
8	0.683					
9	0.247	0.998				
10	0.068	0.845				
11	0.018	0.405	0.999			
12	0.005	0.133	0.875			
13	0.001	0.038	0.464			
14		0.010	0.169	0.860		
15		0.003	0.052	0.469		
16			0.005	0.179	0.823	
17			0.001	0.059	0.441	
18				0.018	0.172	0.763
19				0.006	0.058	0.395
20				0.002	0.019	0.158
21						0.056
22						0.019

TABLE 6.3. Approximate P-values for the length of the largest subsequence allowing k mismatches, in a DNA segment of length 100,000, found by simulation. Estimates have approximate standard deviation 0.001. $p = \frac{1}{4}$.

k	0	1	2	3	4	5
Mean	8.013	10.426	12.582	14.592	16.518	18.383
Variance	0.937	1.081	1.201	1.313	1.417	1.520

TABLE 6.4. Simulation estimates of the mean and variance of the length of the largest subsequence allowing k mismatches, in a DNA segment of length 100,000. Estimates of the mean have approximate standard error 0.002. Estimates of the variance have approximate standard deviation 0.01. $p = \frac{1}{4}$.

"independence" mean and variance in Table 2.5 become somewhat inaccurate. Even though this inaccuracy is small, the sharp changes in P-values near the mean imply that the "independence" values for the mean and variance are possibly not sufficiently accurate for P-value approximations using (2.139) when k is large. This view is supported by the somewhat inaccurate P-values, as noted above, found by assuming independence of subsequence lengths.

Are there approximations for the mean and variance of $Y_{\max}$ more accurate than those given in Table 2.4? Waterman (1995, page 277) claims that for large n, good approximations for the mean and variance are

$$\mu_{max} = \frac{\log n + \gamma + k \log\left(\frac{\log n}{\lambda}\right) + k \log\left(\frac{1-p}{p}\right) - \log(k!)}{\lambda} - \frac{1}{2} + r_1, \quad (6.5)$$

$$\sigma^2_{max} = \frac{\pi^2}{6\lambda^2} + \frac{1}{12} + r_2, \quad (6.6)$$

where $\lambda = (-\log p)$, $n = N(1-p)$, and, for $p = \frac{1}{4}$, $|r_1| \leq 3.45 \times 10^{-4}$ and $|r_2| \leq 2.64 \times 10^{-2}$. These approximations were first calculated assuming independence of subsequence lengths, and later shown not to change significantly in the dependent case. The approximations (6.5) and (6.6) are direct generalizations of the approximations (2.119) and (2.121), respectively, to the case where k is a positive integer.

If these values are to be used in conjunction with the approximation (2.139), it is important to assess their accuracy, especially if we recall the great precision necessary for calculations involving the maximum of generalized geometric random variables, as discussed in Section 2.12.3. To make this assessment, we give in Table 6.5 the approximate values of the mean μ_{max} as calculated by equation (6.5). The calculations in this table are for the values $n = 75,000$, $p = \frac{1}{4}$, $k = 0, 1, 2, 3, 4, 5$. The values in this table differ from those in Table 6.4, when $k > 0$, by more than the maximal error 3.45×10^{-4} claimed for the approximation (6.5). Presumably this arises because the asymptotic theory does not yet apply when $n = 75,000$. Thus when n is not extremely large the use of the means in Table 6.5 in conjunction with the approximation (2.139) could lead to some inaccuracies when it is recalled how quickly the P-values change in the neighborhood of the mean.

k	0	1	2	3	4	5
Mean	8.013	10.315	12.116	13.625	14.926	16.066

TABLE 6.5. Approximation for the mean of the maximum of 75,000 iid generalized geometric random variables, for various values of k, calculated from equation (6.5), $p = \frac{1}{4}$.

The corresponding approximate value for the variance σ^2_{max}, given by equation (6.6), is 0.939. When $k > 0$, this also differs from the values in Table 6.4 by more than the maximal error 2.64×10^{-2} claimed for the approximation (6.6). This is also of concern with regard to P-value approximations using (2.139). It is thus necessary to calculate the P-value approximation given by (2.138) and (2.139), using the mean (6.5) and the variance (6.6) in the approximation, to assess the accuracy both of the approximation and of (6.5) and (6.6). Calculations corresponding to those in Table 2.5 are given in Table 6.6. It is clear that the P-value approximations

y	$k=0$	$k=1$	$k=2$	$k=3$	$k=4$	$k=5$
7	0.990					
8	0.682					
9	0.249	0.999				
10	0.069	0.824				
11	0.018	0.353	0.995			
12	0.004	0.102	0.733			
13	0.001	0.028	0.281			
14		0.007	0.079	0.487		
15		0.002	0.020	0.154		
16			0.005	0.041	0.224	
17			0.002	0.010	0.061	
18				0.003	0.016	0.074
19				0.001	0.004	0.019
20				0.000	0.001	0.005
21						0.001
22						0.000

TABLE 6.6. Approximate P-values for the maximum of 75,000 iid generalized geometric random variables, for various values of k and y, calculated using equations (2.139), (6.5) and (6.6). $p = \frac{1}{4}$.

based on equations (6.5) and (6.6) are somewhat in error, at least when k is 2 or more.

BLAST provides a second (and more frequently used) approach to significance testing based on "well-matching" rather than exactly matching subsequences. Since BLAST calculations depend on an analysis of random walks, we defer its consideration until the relevant random walk theory has been described.

The analysis of patterns in DNA and protein sequences is a large and growing area of bioinformatics. The theory, however, rapidly becomes difficult, and we do not pursue further topics here. Recent research results may be found, for example, in Bailey and Gribskov (1998), Jonassen, Collins, and Higgins (1995), Karlin and Brendel (1992), Neuwald and Green (1994), and Rigoutsos and Floratos (1998).

6.4   Alignment Algorithms for Two Sequences

6.4.1   Introduction

One way to discriminate between good and bad alignments is to use a scoring scheme. A simple example of a scoring scheme is

(the number of matches) − (the number of mismatches and indels).   (6.7)

Scoring schemes used for aligning DNA are often not much different from this simple scheme. For protein sequences, however, a more complex scoring scheme is appropriate. Commonly used scoring schemes are developed using statistical analysis of existing data, and we will discuss the statistical theory behind these scoring schemes in Section 6.5. For now, we assume that we have assigned a score to each alignment in a meaningful way that reflects the likelihood that this alignment was produced as a consequence of divergence from a common ancestor. Then we can consider the alignments with the "best" score, and we can define the score of the sequence pair to be this best score. What "best" means here depends on whether high scores in the scoring scheme are more indicative of relatedness (so the "best" score is the maximum over all alignments), or whether low scores are more indicative (so "best" is the minimum).

This mathematical framework allows a statistical analysis where we make inferences about the relatedness of the sequences. We can investigate the hypothesis that they did indeed diverge from a common ancestor by considering the probability of the observed score (or one more extreme) arising by chance, under some appropriate model of evolution. If the two sequences are judged to be related, we can use their alignment to discover common patterns in the sequences. This is useful in particular for finding functional domains. Finally, by comparing scores among several different species we can get information to help reconstruct the phylogenetic tree that relates them all.

Scores of alignments consist of two main types: *similarity scores* and *distance scores* (also commonly called *distance measures*). In similarity scores the higher the score, the more closely related are the two aligned sequences; in the distance measures the opposite is the case. In the remainder of this section we use similarity scores. These are usually computed as the sum of individual scores, one for each aligned pair of residues, together with a score for each gap. We will denote by $s(X, Y)$ the score assigned to the aligned pair consisting of the residues X and Y. This score reflects how conservative the substitution represented by the alignment of X with Y is. For example, it is much less likely that the amino acid W (tryptophan) will be substituted for V (valine) in a functional domain than it is that W will be substituted for R (arginine). (This is not only an empirically observed fact, but also makes sense in terms of the chemical properties involved.) Thus the score $s(W, V)$ assigned to an alignment of the two symbols W and V is lower than $s(W, R)$, the score assigned to an alignment of the two symbols W and R. The score assigned to a gap of length ℓ is usually a function of ℓ, which we denote by $\delta(\ell)$. It represents the cost of having a gap of length ℓ and is therefore zero or negative. The simplest gap penalty model is a *linear gap model*, where $\delta(\ell) = -\ell d$ for some nonnegative constant d, called the *linear gap penalty*. Therefore, in the linear gap model, each indel in a gap is weighted in the same way, namely by a penalty of d.

Thus if the alphabet has size N ($N = 4$ for nucleotides and $N = 20$ for amino acids), a scoring scheme consists of an $N \times N$ matrix S and a gap cost function δ. The matrix S is called a *substitution matrix* and the entry in its ith row and jth column is the score of the alignment of the ith and jth symbols in the alphabet.

Example. Consider the comparison of two nucleotide sequences with a simple scoring scheme that assigns $+1$ to each match, -1 to each mismatch, and a linear gap score with $d = 2$. Then the score for the following alignment of the two sequences *cttagg* and *catgagaa* is $1 - 1 + 1 - 2 + 1 - 2 + 1 - 4 = -5$:

$$
\begin{array}{ccccccccc}
c & t & t & a & g & - & g & - & - \\
c & a & t & - & g & a & g & a & a
\end{array}
$$

One of the main aims of the statistical theory is to find for nucleotides, and more importantly for amino acids, what an optimal scoring scheme should be. This matter is taken up in detail in Section 6.5 and Chapter 9.

6.4.2 Gapped Global Comparisons and Dynamic Programming Algorithms

Suppose that we are given a scoring scheme made up of a substitution matrix and a linear gap penalty. Our aim is to find, of the possible global alignments of two sequences (with gaps allowed), that one (or those ones) with the highest score. One method in principle for doing this is to list exhaustively all possible alignments and their scores, and then note the highest-scoring alignment(s). However, when the sequences are long, this is not computationally feasible, and more efficient algorithms are needed. We describe one such algorithm below, but first we will justify our assertion that the exhaustive search illustrated above is indeed not efficient, by getting a sense of how large the number of global alignments between a sequence $x = X_1 X_2 \ldots X_m$ of length m and a sequence $y = Y_1 Y_2 \ldots Y_n$ of length n is. We will denote this number by $c(m, n)$. Since there is no point in matching two deletions, no alignments of one indel with another are allowed.

Let $g(m, n)$ be the number of groups obtained by grouping together those alignments that have the same combination of aligned residue pairs ignoring the indels. Then $g(m, n) < c(m, n)$, and this provides a lower bound for $c(m, n)$. We can compute $g(m, n)$ as follows.

The number k of aligned residues for two sequences of lengths m and n is between 0 and $\min\{m, n\}$. Moreover, for each such k there are $\binom{m}{k}$ ways of choosing the residues of x that align with residues of y, and $\binom{n}{k}$ ways of choosing the residues of y that align with residues of x. So there are

$\binom{m}{k}\binom{n}{k}$ alignments with k aligned residues. Therefore,

$$g(m,n) = \sum_{k=0}^{\min\{m,n\}} \binom{m}{k}\binom{n}{k}. \tag{6.8}$$

From the result of Problem 6.1 below, it follows that

$$g(m,n) = \binom{m+n}{n}. \tag{6.9}$$

In particular, when $m = n$,

$$g(n,n) = \binom{2n}{n}.$$

This number grows quite fast with n. Stirling's approximation (B.4), and even more directly (B.5), shows that

$$\binom{2n}{n} \sim \frac{2^{2n}}{\sqrt{\pi n}}. \tag{6.10}$$

Thus the number $c(1000, 1000)$ of global alignments between two sequences each of length 1000 satisfies

$$c(1000, 1000) \geq g(1000, 1000) \approx \frac{2^{2000}}{\sqrt{1000\pi}} \approx 10^{600}.$$

This shows why it is not feasible to examine all possible alignments. This motivates the search for algorithms that can compute the best score efficiently and an alignment with this score, without having to examine all possibilities. One such algorithm is the Needleman–Wunsch algorithm (1970), and we discuss a version of this procedure introduced by Gotoh (1982). These are examples of dynamic programming algorithms, and we will use them to illustrate the general concept of dynamic programming.

The input consists of two sequences,

$$\boldsymbol{x} = X_1 X_2 \ldots X_m \text{ and } \boldsymbol{y} = Y_1 Y_2 \ldots Y_n,$$

of lengths m and n, respectively, whose elements belong to some alphabet of N symbols (for DNA or RNA sequences $N = 4$, for proteins $N = 20$). We assume that we are given a substitution matrix S and a linear gap penalty d. The output consists of the highest score over all alignments between $\boldsymbol{x}$ and $\boldsymbol{y}$ and a highest-scoring global alignment between $\boldsymbol{x}$ and $\boldsymbol{y}$.

The broad approach is to break the problem into subproblems of the same kind and build the final solution using the solutions for the subproblems: This is the basic idea behind any dynamic programming algorithm. In this problem we find a highest-scoring alignment using previous solutions for highest-scoring alignments of smaller subsequences of x and y. We denote by $x_{1,i}$ the initial segment of x given by $X_1 X_2 \cdots X_i$ and similarly we denote by $y_{1,j}$ the initial segment of y given by $Y_1 Y_2 \cdots Y_j$. For $i = 1, 2, \ldots, m$ and $j = 1, 2, \ldots, n$, we denote by $B(i,j)$ the score of a highest-scoring alignment between $x_{1,i}$ and $y_{1,j}$. For $i = 1, 2, \ldots, m$, we denote by $B(i,0)$ the score of an alignment where $x_{1,i}$ is aligned to a gap of length i, so $B(i,0) = -id$. Similarly, for $j = 1, 2, \ldots, n$, we denote by $B(0,j)$ the score of an alignment where $y_{1,j}$ is aligned to a gap of length j, so $B(0,j) = -jd$. Finally, we initialize $B(0,0) = 0$. These calculations lead to an $(m+1) \times (n+1)$ matrix B. The entry in the last row and in the last column of B, namely $B(m,n)$, is the score of a highest-scoring alignment between our two sequences x and y, and it is one of the things we want our algorithm to output.

The essence of the procedure is to fill in the elements of the matrix B recursively. We already have the values of B at $(0,0)$, $(i,0)$, and $(0,j)$, for $i = 1, 2, \ldots, m$ and $j = 1, 2, \ldots, n$. Now we proceed from top left to bottom right by noting that a highest-scoring alignment between $x_{1,i}$ and $y_{1,j}$ could terminate in one of three possible ways, namely, with

$$\begin{matrix} X_i \\ Y_j \end{matrix}, \quad \begin{matrix} X_i \\ - \end{matrix}, \quad \text{or} \quad \begin{matrix} - \\ Y_j \end{matrix}.$$

In the first case, $B(i,j)$ is equal to the sum of the score for a highest-scoring alignment between $x_{1,i-1}$ and $y_{1,j-1}$ together with the extra term $s(i,j)$ to account for the match between X_i and Y_j; that is, $B(i,j) = B(i-1,j-1) + s(i,j)$. In the second case, $B(i,j)$ is equal to the sum of the score for a highest-scoring alignment between $x_{1,i-1}$ and $y_{1,j}$ together with an extra term $-d$ to account for the indel to which X_i is aligned, i.e., $B(i,j) = B(i-1,j) - d$. Similarly, in the third case, $B(i,j) = B(i,j-1) - d$. These are all the possible options, and hence $B(i,j)$ is the highest of the three. In other words,

$$B(i,j) = \max\{B(i-1,j-1) + s(i,j), B(i-1,j) - d, B(i,j-1) - d\}. \quad (6.11)$$

In this way we recursively fill in every cell in the matrix B and determine the value of $B(m,n)$, which is the desired maximum score. The running time of this algorithm is clearly $O(mn)$. To find an alignment that has this score we must keep track, at each step of the recursion, of one of the three choices giving the value of the maximum. Although there could be more than one choice giving the maximum, if we are interested in finding only one alignment, we choose one and keep a pointer to it. Once $B(m,n)$ is obtained, by tracing back through the pointers, we can reconstruct an alignment with the highest score. We now illustrate this procedure with an example.

Example. Let $x = gaatct$ and $y = catt$, so that $m = 6$ and $n = 4$. Using the same scoring scheme as in the example in Section 6.4.1, B is given in

Figure 6.1, where we have used arrows to denote, for each cell, where it came from. The best score for an alignment is given by the element in the

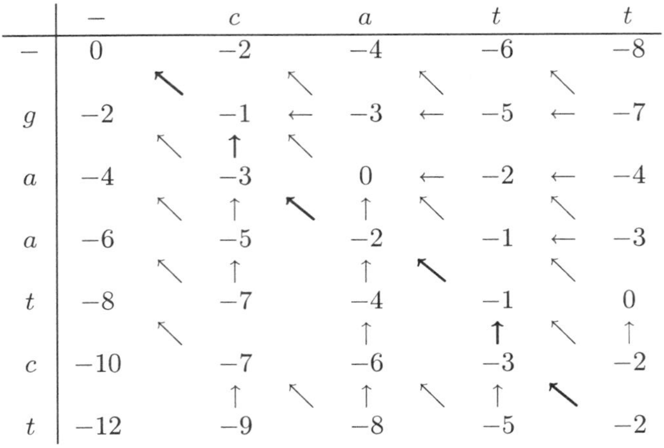

	$-$	c	a	t	t
$-$	0	-2	-4	-6	-8
g	-2	-1	-3	-5	-7
a	-4	-3	0	-2	-4
a	-6	-5	-2	-1	-3
t	-8	-7	-4	-1	0
c	-10	-7	-6	-3	-2
t	-12	-9	-8	-5	-2

FIGURE 6.1.

bottom rightmost cell, which is -2. Tracing back along the bold arrows, we get the highest-scoring alignment

$$\begin{array}{cccccc} g & a & a & t & c & t \\ c & - & a & t & - & t \end{array}.$$

By making different choices of arrows in the traceback procedure we can get the following other alignments, which are also highest-scoring, i.e., which also have a score of -2:

$$\begin{array}{cccccc} g & a & a & t & c & t \\ c & a & - & t & - & t \end{array} \quad \text{and} \quad \begin{array}{cccccc} g & a & a & t & c & t \\ - & c & a & t & - & t \end{array}.$$

We next consider modifications of the Needleman–Wunsch algorithm, which can be used to address other kinds of pairwise alignment problems.

6.4.3 Fitting One Sequence into Another Using a Linear Gap Model

In this section we address the following problem: Given two sequences, a longer and a shorter one, find the subsequence(s) of the longer one that can be best aligned with the shorter sequence, where gaps are allowed. This procedure is relevant when one is interested in locating a specified pattern within a sequence.

Let $x = X_1 X_2 \ldots X_m$ and $y = Y_1 Y_2 \ldots Y_n$ be two sequences with $n \geq m$. For $1 \leq k \leq j \leq n$, denote by $y_{k,j}$ the subsequence of y given by

$Y_k Y_{k+1} \ldots Y_j$. For two sequences $\boldsymbol{u}$ and $\boldsymbol{v}$, denote by $B(\boldsymbol{u}, \boldsymbol{v})$ the score of a highest-scoring (global) alignment between $\boldsymbol{u}$ and $\boldsymbol{v}$. Our aim is to find

$$\max\{B(\boldsymbol{x}, \boldsymbol{y}_{k,j}) : 1 \leq k \leq j \leq n\}. \tag{6.12}$$

For each choice of k and j the running time of the Needleman–Wunsch algorithm, giving the value of $B(\boldsymbol{x}, \boldsymbol{y}_{k,j})$, is $O(m(j-k))$. Thus if we used this algorithm for all possible choices of k and j, and then took the maximum over all such choices, the total running time would be $O(mn^3)$, since there are $\binom{n}{2}$ possible choices for j and k. We now illustrate another approach with a better running time, namely an $O(mn)$ running time.

For $1 \leq i \leq m$ and $1 \leq j \leq n$, let $F(i, j)$ be the maximum of the scores $B(\boldsymbol{x}_{1,i}, \boldsymbol{y}_{k,j})$ over the values of k between 1 and j. That is, of all the possible scores for highest-scoring alignments between the initial segment of $\boldsymbol{x}$ up to x_i and the segments of $\boldsymbol{y}$ ending at y_j and beginning at some k we take $F(i, j)$ to be the greatest of such scores. The value of (6.12) is the maximum of $F(m, j)$ over all values of j between 1 and n. To find this, we initialize $F(i, 0) = -id$ for $1 \leq i \leq m$ and initialize $F(0, j) = 0$ for $0 \leq j \leq n$, since deletions of the beginning of $\boldsymbol{y}$ should clearly be without penalty. Then we fill in the matrix F recursively by

$$F(i, j) = \max\{F(i-1, j-1) + s(i, j), F(i, j-1) - d, F(i-1, j) - d\},$$

where the reasoning behind this formula is analogous to that behind (6.11). Note that there might be more than one value of j giving the maximum score. In order to recover the highest-scoring alignments of $\boldsymbol{x}$ to subsequences of $\boldsymbol{y}$ we can keep pointers, as in the Needleman–Wunsch algorithm.

6.4.4 Local Alignments with a Linear Gap Model

Another interesting alignment problem is to find, given two sequences, which respective subsequences have the highest-scoring alignment(s) (with gaps allowed). This is called a local alignment problem, and it is appropriate when one is seeking common patterns/domains in two sequences.

In what follows we make the assumption that the scoring scheme we use is such that the expected (or mean) score for a random alignment is negative. If this assumption did not hold, then long matches between subsequences could score highly just because of their lengths, so that two long unrelated subsequences could give a highest-scoring alignment. Clearly, we do not want this to occur.

For $1 \leq h \leq i \leq m$ we denote by $\boldsymbol{x}_{h,i}$ the subsequence of $\boldsymbol{x}$ given by $X_h X_{h+1} \ldots X_i$. With the notation as in the previous section, we want to find

$$\max\{B(\boldsymbol{x}_{h,i}, \boldsymbol{y}_{k,j}) : 1 \leq h \leq i \leq m, 1 \leq k \leq j \leq n\}, \tag{6.13}$$

when this is nonnegative. There are $\binom{m}{2}\binom{n}{2}$ pairs of subsequences of $\boldsymbol{x}$ and $\boldsymbol{y}$, one for each choice of h and i among m possible values and of k and j

among n possible values. Thus computing a highest-scoring alignment for each pair, using the Needleman–Wunsch algorithm, requires a total running time of $O(m^3 n^3)$. Clearly, we want to give a more efficient approach to this problem. Such an approach is provided by the Smith–Waterman algorithm (Smith and Waterman (1981)) which computes (6.13) in $O(mn)$ time. The procedure is as follows.

For each $1 \leq i \leq m$ and $1 \leq j \leq n$, we define $L(i,j)$ to be the maximum of 0 and the maximum of all possible scores for alignments between a subsequence of $\boldsymbol{x}$ ending at X_i and one of $\boldsymbol{y}$ ending at Y_j. That is,

$$L(i,j) = \max\{0, B(\boldsymbol{x}_{h,i}, \boldsymbol{y}_{k,j}) : 1 \leq h \leq i, 1 \leq k \leq j\}.$$

The reason we want $L(i,j) = 0$ when the max of the $B(\boldsymbol{x}_{h,i}, \boldsymbol{y}_{k,j})$'s is negative is because it is sensible to always remove the first part of an alignment if this part has a negative score, as it will just decrease the overall score of the alignment. Then the maximum of 0 and (6.13) is the maximum of $L(i,j)$ over all values of i between 1 and m and of j between 1 and n. To determine this maximum we use again dynamic programming, by initializing $L(i,0) = 0 = L(0,j)$ for $0 \leq i \leq m$ and $0 \leq j \leq n$ (since deletions at the beginning or end of our two sequences should not be penalized), and by computing

$$L(i,j) = \max\{0, L(i-1,j-1) + s(i,j), L(i-1,j) - d, L(i,j-1) - d\}.$$

We then calculate the maximum of $L(i,j)$ over all values of i and j. As in the previous maximizing procedures there might be more than one highest-scoring local alignment. To find a highest-scoring alignment, we follow the traceback procedure previously described. However, for this algorithm, we stop this process when we encounter a 0.

Figure 6.2 shows an example of an $L(i,j)$ matrix arising in locally aligning two sequences of lengths 7 and 9. In this example, the score of an optimal local alignment of the two sequences is 28, and there is only one alignment of subsequences giving this score, the one indicated by the bold arrows, which is

$$\begin{array}{ccccc} X_2 & X_3 & - & X_4 & X_5 \\ Y_5 & Y_6 & Y_7 & Y_8 & Y_9 \end{array}.$$

6.4.5 Other Gap Models

There are many variants and extensions of the algorithms discussed above. For example, while the linear gap model used above is appealing for its simplicity, it is often not appropriate for biological sequences, since often it is harder for a gap to open (i.e., start) than it is for it to extend. Thus it is often more appropriate not to penalize additional gap steps as much as the first one, and to use a more complicated gap cost $\delta(\ell)$. This implies that the recurrence relations need to be adjusted. For instance, in (6.11),

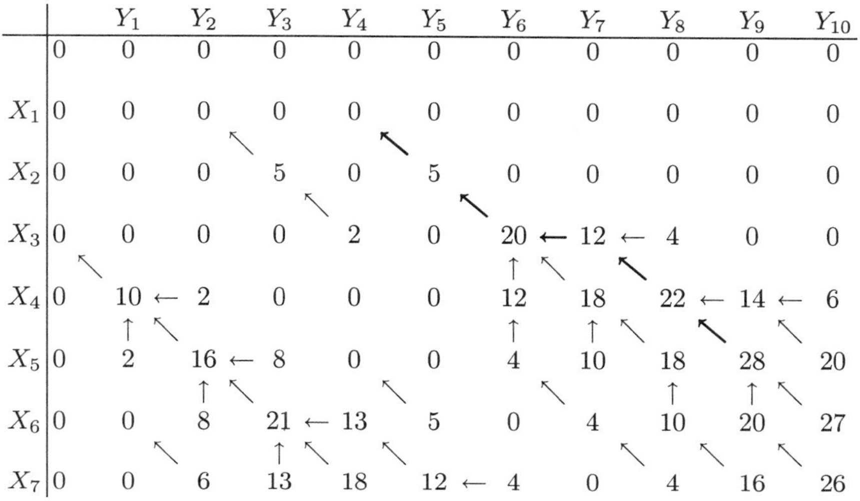

		Y_1	Y_2	Y_3	Y_4	Y_5	Y_6	Y_7	Y_8	Y_9	Y_{10}
	0	0	0	0	0	0	0	0	0	0	0
X_1	0	0	0	0	0	0	0	0	0	0	0
X_2	0	0	0	5	0	5	0	0	0	0	0
X_3	0	0	0	0	2	0	20 ← 12 ← 4			0	0
X_4	0	10 ← 2		0	0	0	12	18	22 ← 14 ← 6		
X_5	0	2	16 ← 8		0	0	4	10	18	28	20
X_6	0	0	8	21 ← 13		5	0	4	10	20	27
X_7	0	0	6	13	18	12 ← 4		0	4	16	26

FIGURE 6.2.

we now have to distinguish among the alignments of $\boldsymbol{x}_{1,i}$ with $\boldsymbol{y}_{1,j}$ that end with X_i aligned to an indel. The score of such an alignment will also depend on how many symbols in $\boldsymbol{x}$ immediately preceding X_i are aligned to indels. Suppose that in such an alignment the last symbol preceding X_i that is not aligned to an indel (hence is aligned to Y_j) is X_k. Then the $i-k$ symbols from X_{k+1} to X_i are aligned to indels, and the score of a highest-scoring alignment is $B(k,j) + \delta(i-k)$. Similar reasoning must be applied to those alignments between $\boldsymbol{x}_{1,i}$ and $\boldsymbol{y}_{1,j}$ that end with Y_j aligned to an indel. Hence (6.11) must be replaced with

$$
\begin{aligned}
B(i,j) = \max\{ & B(i-1,j-1) + s(i,j), \\
& B(k,j) + \delta(i-k) : k = 0,1,\ldots,i-1, \\
& B(i,k) + \delta(j-k) : k = 0,1,\ldots,j-1\},
\end{aligned}
$$

and the initialization is given by $B(0,0) = 0$, $B(i,0) = \delta(i)$ for $i = 1,2,\ldots,m$, and $B(0,j) = \delta(j)$ for $j = 1,2,\ldots,n$. So, in general, finding a highest-scoring alignment between two sequences of lengths m and n takes $O(m^2 n + mn^2)$ operations, as opposed to $O(mn)$ operations needed if the gap penalty model is linear. This is because for each cell of B we now need to consider $i + j + 1$ previous cells, instead of just three.

If, however, $\delta(\ell)$ satisfies certain conditions, there are algorithms that take $O(mn)$ operations. One simple example of this is that of an *affine gap model*, in which $\delta(\ell) = -d - (\ell - 1)e$, for some (nonnegative) d and e. d is called the *gap-open* penalty, and e is called the *gap-extension* penalty. Usually, e is set smaller than d. Thus all gap steps other than the first have the same cost, but each of them is penalized less than the first. We now describe a dynamic programming implementation of a global alignment algorithm for this case whose running time is also $O(mn)$.

Instead of using just one matrix B, the algorithm uses three matrices. Let x and y be the sequences we want to align. We will use the same notation as above for $x_{1,i}$ and $y_{1,j}$. For $i = 1, 2, \ldots, m$ and $j = 1, 2, \ldots, n$, we denote by $S(i, j)$ the score of a highest-scoring alignment between $x_{1,i}$ and $y_{1,j}$, given that the alignment ends with X_i aligned to Y_j. We denote by $I_x(i, j)$ the score of a highest-scoring alignment between $x_{1,i}$ and $y_{1,j}$, given that the alignment ends with X_i aligned to an indel. Finally, we denote by $I_y(i, j)$ the score of a highest-scoring alignment between $x_{1,i}$ and $y_{1,j}$, given that the alignment ends with an indel aligned to Y_j. Then if we assume that a deletion will not be followed directly by an insertion, we have, for $i = 1, 2, \ldots, m$ and $j = 1, 2, \ldots, n$,

$$
\begin{aligned}
&S(i,j) \\
&= \max\{S(i{-}1,j{-}1){+}s(i,j), I_x(i{-}1,j{-}1){+}s(i,j), I_y(i{-}1,j{-}1){+}s(i,j)\},
\end{aligned}
$$

where

$$I_x(i, j) = \max\{S(i - 1, j) - d, I_x(i - 1, j) - e\},$$

and

$$I_y(i, j) = \max\{S(i, j - 1) - d, I_y(i, j - 1) - e\}.$$

These recurrence relations allow us to fill in the matrices S, I_x, and I_y, once we initialize $S(0,0) = I_x(0,0) = I_y(0,0) = 0$, $S(0,j) = I_x(0,j) = -d - (j-1)e$, and $S(i,0) = I_y(i,0) = -d - (i-1)e$, for $i = 1, 2, \ldots, m$ and $j = 1, 2, \ldots, n$. The score of a highest-scoring alignment is then given by $\max\{S(m,n), I_x(m,n), I_y(m,n)\}$.

6.4.6   Limitations of the Dynamic Programming Alignment Algorithms

All the algorithms discussed above yield the exact highest score according to the given scoring scheme. However, when one has to deal with very long sequences, as would occur if one wished to align a given sequence with each of several sequences in a big database, a time complexity of $O(mn)$ might not be good enough for performing the required search in an acceptable amount of time. So various other algorithms have been developed to overcome this difficulty, BLAST being one of them. These algorithms use heuristic techniques to limit the search to a fraction of the possible alignments between two sequences in a way that attempts not to miss the high-scoring alignments. The trade-off is that one might not necessarily find the best possible score. Various aspects of BLAST will be discussed extensively in Chapters 7 and 9.

Another consideration is that of space, since memory usage can also be a limiting factor in dynamic programming. In the Needleman–Wunsch algorithm, for instance, one needs to store the $(m+1) \times (n+1)$ matrix B. If one cares only about the value of the best score, without wanting to find a highest-scoring alignment, then there is no need to store all cells of B, since the value of $B(i,j)$ depends only on entries up to one row back. Thus one can throw away rows of the matrix that are further back. However, usually one wants to find also a highest-scoring alignment, not only its score. There are methods that allow one to do this in $O(m+n)$ space, rather then $O(mn)$, and this with no more than doubling in time, so the time complexity remains at $O(mn)$. For more details see Section 2.6 in Durbin et al. (1998) or Section 9.7 in Waterman (1995).

6.5   Protein Sequences and Substitution Matrices

6.5.1   Introduction

In the study of DNA sequences, simple scoring schemes are usually effective. For protein sequences, however, some substitutions are much more likely than others. The performance of any alignment algorithm is improved when it accounts for this difference. In all cases we will consider, higher scores will represent more likely substitutions.

There are two frequently used approaches to finding substitution matrices. One leads to the PAM (Accepted Point Mutation) family of matrices, and the other to the BLOSUM (BLOcks SUbstitution Matrices) family. Table 6.7 gives an example of a typical BLOSUM substitution matrix (called the BLOSUM50 matrix). In this section we discuss how these substitution matrices are derived.

Any attempt to create a scoring matrix for amino acid substitutions must start from a set of data that can be trusted. The "trusted" data are then used to determine which substitutions are more or less likely. The matrix is then derived from these data, using (as we will see) aspects of statistical hypothesis-testing theory.

Historically, the PAM matrices were developed first (in 1978), but since the derivation of BLOSUM matrices is somewhat simpler than that for PAM matrices, we will start by considering BLOSUM matrices.

	A	R	N	D	C	Q	E	G	H	I	L	K	M	F	P	S	T	W	Y	V
A	5	-2	-1	-2	-1	-1	-1	0	-2	-1	-2	-1	-1	-3	-1	1	0	-3	-2	0
R	-2	7	-1	-2	-4	1	0	-3	0	-4	-3	3	-2	-3	-3	-1	-1	-3	-1	-3
N	-1	-1	7	2	-2	0	0	0	1	-3	-4	0	-2	-4	-2	1	0	-4	-2	-3
D	-2	-2	2	8	-4	0	2	-1	-1	-4	-4	-1	-4	-5	-1	0	-1	-5	-3	-4
C	-1	-4	-2	-4	13	-3	-3	-3	-3	-2	-2	-3	-2	-2	-4	-1	-1	-5	-3	-1
Q	-1	1	0	0	-3	7	2	-2	1	-3	-2	2	0	-4	-1	0	-1	-1	-1	-3
E	-1	0	0	2	-3	2	6	-3	0	-4	-3	1	-2	-3	-1	-1	-1	-3	-2	-3
G	0	-3	0	-1	-3	-2	-3	8	-2	-4	-4	-2	-3	-4	-2	0	-2	-3	-3	-4
H	-2	0	1	-1	-3	1	0	-2	10	-4	-3	0	-1	-1	-2	-1	-2	-3	2	-4
I	-1	-4	-3	-4	-2	-3	-4	-4	-4	5	2	-3	2	0	-3	-3	-1	-3	-1	4
L	-2	-3	-4	-4	-2	-2	-3	-4	-3	2	5	-3	3	1	-4	-3	-1	-2	-1	1
K	-1	3	0	-1	-3	2	1	-2	0	-3	-3	6	-2	-4	-1	0	-1	-3	-2	-3
M	-1	-2	-2	-4	-2	0	-2	-3	-1	2	3	-2	7	0	-3	-2	-1	-1	0	1
F	-3	-3	-4	-5	-2	-4	-3	-4	-1	0	1	-4	0	8	-4	-3	-2	1	4	-1
P	-1	-3	-2	-1	-4	-1	-1	-2	-2	-3	-4	-1	-3	-4	10	-1	-1	-4	-3	-3
S	1	-1	1	0	-1	0	-1	0	-1	-3	-3	0	-2	-3	-1	5	2	-4	-2	-2
T	0	-1	0	-1	-1	-1	-1	-2	-2	-1	-1	-1	-1	-2	-1	2	5	-3	-2	0
W	-3	-3	-4	-5	-5	-1	-3	-3	-3	-3	-2	-3	-1	1	-4	-4	-3	15	2	-3
Y	-2	-1	-2	-3	-3	-1	-2	-3	2	-1	-1	-2	0	4	-3	-2	-2	2	8	-1
V	0	-3	-3	-4	-1	-3	-3	-4	-4	4	1	-3	1	-1	-3	-2	0	-3	-1	5

TABLE 6.7. The BLOSUM50 substitution matrix

6.5.2   BLOSUM Substitution Matrices

The BLOSUM approach was introduced by Henikoff and Henikoff (1992). Henikoff and Henikoff started with a set of protein sequences from public databases that had been grouped into related families. From these sequences they obtained "blocks" of aligned sequences. A block is the *ungapped* alignment of a relatively highly conserved region of a family of proteins. Methods for producing such alignments are given in Section 6.6. These alignments provide the basic data for the BLOSUM approach to constructing substitution matrices. An example of such an alignment leading to four blocks is given in Table 6.8.

Since the algorithms used to construct the aligned blocks employ substitution matrices, there is a circularity involved in the procedure if the aligned blocks are subsequently used to find substitution matrices. Henikoff and Henikoff broke this circularity as follows. They started by using a simple "unitary" substitution matrix where the score is 1 for a match, 0 for a mismatch. Then, using data from suitable groups of proteins, they constructed only those blocks that they could obtain with this simple matrix. This procedure has the effect of generating a conservative set of blocks; that is, it tends to omit blocks with low sequence identity. While this restricted

```
WWYIR   CASILRKIYIYGPV   GVSRLRTAYGGRK   NRG
WFYVR   CASILRHLYHRSPA   GVGSITKIYGGRK   RNG
WYYVR   AAAVARHIYLRKTV   GVGRLRKVHGSTK   NRG
WYFIR   AASICRHLYIRSPA   GIGSFEKIYGGRR   RRG
WYYTR   AASIARKIYLRQGI   GVGGFQKIYGGRQ   RNG
WFYKR   AASVARHIYMRKQV   GVGKLNKLYGGAK   SRG
WFYKR   AASVARHIYMRKQV   GVGKLNKLYGGAK   SRG
WYYVR   TASIARRLYVRSPT   GVDALRLVYGGSK   RRG
WYYVR   TASVARRLYIRSPT   GVGALRRVYGGNK   RRG
WFYTR   AASTARHLYLRGGA   GVGSMTKIYGGRQ   RNG
WFYTR   AASTARHLYLRGGA   GVGSMTKIYGGRQ   RNG
WWYVR   AAALLRRVYIDGPV   GVNSLRTHYGGKK   DRG
```

TABLE 6.8. A set of four blocks from the Blocks database

the number of blocks derived, the blocks obtained were trustworthy and were not biased toward any specific scoring scheme.

Using the blocks so constructed, Henikoff and Henikoff then counted the number of occurrences of each amino acid and the number of occurrences of each pair of amino acids aligned in the same column. Consider a very simplified example, with only three amino acids, A, B, and C, and only one block:

$$
\begin{array}{cccc}
B & A & B & A \\
A & A & A & C \\
A & A & C & C \\
A & A & B & A \\
A & A & C & C \\
A & A & B & C \\
\end{array}
$$

In this block there are 24 amino acids observed, of which 14 are A, 4 are B, and 6 are C. Thus the observed proportions are

amino acid	proportion of times observed	
A	14/24	
B	4/24	(6.14)
C	6/24	

There are $4 \cdot \binom{6}{2} = 60$ *aligned pairs* of amino acids in the block. These 60 pairs occur with proportions as given in the following table:

aligned pair	proportion of times observed	
A to A	26/60	
A to B	8/60	
A to C	10/60	(6.15)
B to B	3/60	
B to C	6/60	
C to C	7/60	

We now compare these observed proportions to the *expected* proportion of times that each amino acid pair is aligned under a random assortment of the amino acids observed, given the observed amino acid frequencies (6.14). In other words, if we choose two sequences of the same length at random with these frequencies (6.14), and put them into alignment, then the expected proportion of pairs in which A is aligned with A is $\frac{14}{24} \cdot \frac{14}{24}$, the mean proportion of pairs in which A is aligned with B is $2 \cdot \frac{14}{24} \cdot \frac{4}{24}$, and so on. (The factor of 2 in the second calculation allows for the two cases where A is in the first sequence and B in the second, and that where B is in the first sequence and A in the second.)

These fractions are now used to calculate "estimated likelihood ratios" (see Section 3.4.4) as shown in the following table:

aligned pair	proportion observed	proportion expected	$2\log_2\left(\frac{\text{proportion observed}}{\text{proportion expected}}\right)$
A to A	26/60	196/576	0.70
A to B	8/60	112/576	-1.09
A to C	10/60	168/576	-1.61
B to B	3/60	16/576	1.70
B to C	6/60	48/576	0.53
C to C	7/60	36/576	1.80

$$(6.16)$$

For each row in this table the ratio of the entries in the second and third columns is an estimate, from the data, of the ratio of the proportion of times that each amino acid combination occurs in any column to the proportion expected under random allocation of amino acids into columns. With one important qualification, described later, the respective elements in the BLOSUM substitution matrix are now found by calculating twice the logarithm (to the base 2) of this ratio (as shown in the final column of the above table), and then rounding the result to the nearest integer. In this simplified example, the substitution matrix would thus be

$$
\begin{array}{cccc}
 & A & B & C \\
A & 1 & -1 & -2 \\
B & -1 & 2 & 1 \\
C & -2 & 1 & 2
\end{array}
$$

In general, the procedure is as follows. For each pair of amino acids x and y, first count the number of times we see x and y in the same column of an aligned block. We denote this number by n_{xy}. We then put

$$p_{xy} = \frac{n_{xy}}{\sum_{u \leq v} n_{uv}},$$

where we take $u \leq v$ to mean that the letter denoting u precedes the letter denoting v in the alphabet. This number p_{xy} is the estimate of the probability of a randomly chosen pair of amino acids chosen from one column of a block to be the pair x and y. Now, for each amino acid x, let p_x be the proportion of times x occurs somewhere in any block. Consider the quantity

$$
e_{xy} = \begin{cases} \frac{2p_x p_y}{p_{xy}} & \text{if } x \neq y, \\ \frac{p_x p_y}{p_{xy}} & \text{if } x = y. \end{cases}
$$

This quantity is the ratio of the likelihood that x and y are aligned by chance, given their frequencies of occurrence in the blocks, to the proportion of times we actually observe x and y aligned in the same column in the blocks. We convert this into a score by taking -2 times its logarithm to the base 2, and rounding to the nearest integer. In this way pairs that are more likely than chance will have positive scores, and those less likely will have negative scores.

While this approach is still rudimentary, it does yield a more useful scoring scheme than the original one that merely scores 1 for a match and 0 for a mismatch. Its main shortcoming is that it overlooks an important factor that can bias the results. The substitution matrix derived will depend significantly on which sequences of each family happen to be in the database used to create the blocks. In particular, if there are many very closely related proteins in one block, and only a few others that are less closely related, then the contribution of that block will be biased toward closely related proteins. For example, suppose the data in one block are as follows:

$$
\begin{array}{cccc}
A & B & A & A \\
A & B & A & A \\
A & B & A & A \\
A & B & A & A \\
A & A & B & D \\
A & C & B & A \\
D & A & B & A
\end{array}
$$

The first four sequences possibly derive from closely related species and the last three from three more distant species. Since A occurs with high frequency in the first four sequences, the observed number of pairings of A with A will be higher than is appropriate if we are comparing more distantly related sequences. Ultimately, we would prefer in each block to have sequences such that any pair have roughly the same amount of "evolutionary distance" between them. The solution to this problem used by Henikoff and Henikoff is to group, or cluster, those sequences in each block that are "sufficiently close" to each other and, in effect, use the resulting cluster as a single sequence. This step requires a definition of "sufficiently close," and this is done by specifying a cut-off proportion, say 85%, and then grouping the sequences in each block into clusters in such a way that each sequence in any cluster has 85% or higher sequence identity to at least one other sequence in the cluster in that block.

We now describe how the counting is done in this case, and after the general method is described, we illustrate it with an example. The count of each amino acid is found by dividing each occurrence by the number of sequences in the cluster containing that occurrence, and summing over all occurrences. After this is done, we count aligned amino acid pairs. Here the rule followed is that if in any block two sequences are in the same cluster, then in that block no counts are taken between amino acids in those two sequences. For any aligned amino acids in sequences in two different clusters in the same block, the count for any amino acid pair is divided by nm, where n and m are the sizes of the two clusters from which the amino acids are taken.

These weighted counts are then used in the same way as before. Consider a simple example with two blocks

$$
\begin{array}{cccc}
B & A & B & A \\
B & A & B & C \\
A & A & C & C
\end{array}
$$

and

$$
\begin{array}{ccc}
C & B & B \\
C & B & B \\
A & B & C \\
A & A & C
\end{array}
$$

Suppose the identity for clustering is taken to be .75. Thus we cluster the first two sequences in each block together. The A's are counted as follows. The first column of the first block has one A, the second column contributes two A's, since the first two sequences are clustered it has $1 + \frac{1}{2} + \frac{1}{2} = 2$ A's. The fourth column contributes $\frac{1}{2}$ A. In the second block there are three A's, since each occurrence occurs in a cluster of size one. So in total there are $13/2$ A's. Now to get the proportion of A's we must divide by 17, since each column of the first block contributes 2 to the counts of the symbols, and each column of the second block contributes 3 to the counts. So the proportion of A's is $(13/2)/17 = 13/34$. We record the proportions for all symbols in the following table:

amino acid	proportion of times observed
A	$13/34$
B	$5/17$
C	$11/34$

(6.17)

To count the A–B pairs, each occurrence in the first column of the first block contributes $\frac{1}{2}$, and in the second column of the second block the contribution is $\frac{1}{2} + \frac{1}{2} + 1$. So the total A–B count is 3. There are a total of 13 pairs in the blocks, four in the first block (each column contributes one pair, or more precisely, two half pairs) and three in the second block.

Thus the proportion of A–B pairs is 3/13. We record the proportions for all pairs of symbols in the following table:

aligned pair	proportion of times observed	
A to A	2/13	
A to B	3/13	
A to C	5/26	(6.18)
B to B	1/13	
B to C	3/13	
C to C	3/26	

The procedure is then carried out as before.

A further refinement was taken by Henikoff and Henikoff (1992). After obtaining a BLOSUM substitution matrix as just described, the matrix obtained is then used instead of the conservative "unitary" matrix to construct a second, less conservative, set of blocks. A new substitution matrix is then obtained from these blocks. Then the process is repeated a third time. From this third set of blocks Henikoff and Henikoff derive the final family of BLOSUM matrices, and it is these whose use is suggested.

If the .85 similarity score criterion is adopted, the final matrix is called a BLOSUM85 matrix. In general if clusters with $X\%$ identity are used, then the resulting matrix is called BLOSUMX. The BLOSUM matrices currently available on the BLAST web page at NCBI are BLOSUM45, BLOSUM62, and BLOSUM80. Note that the larger-numbered matrices correspond to more recent divergence, and the smaller-numbered matrices correspond to more distantly related sequences.

One often has prior knowledge about the evolutionary distance between the sequences of interest that helps one choose which BLOSUM matrix to use. With no information, BLOSUM62 is often used. We explore the implications of the choice of various matrices in Section 9.2.4.

A central feature of the BLOSUM substitution matrix calculation is the use of (estimated) likelihood ratios. We see in the next section that the same is true of PAM matrices. In Section 8.3.1 it is shown that use of likelihood ratios has a statistical optimality property, and this optimality property explains in part their use in the construction of both BLOSUM and PAM matrices.

6.5.3   PAM Substitution Matrices

In this section we outline the Dayhoff et al. (1978) approach to deriving the so-called PAM substitution matrices. Two essential ingredients in the construction of these matrices, as with construction of BLOSUM matrices, are the calculation of an (estimated) likelihood ratio and the use of Markov chain theory as introduced in Section 4.7. We now describe this construction in more detail.

An "accepted point mutation" is a substitution of one amino acid of a protein by another that is "accepted" by evolution, in the sense that within some given species, the mutation has not only arisen but has, over time, spread to essentially the entire species. A PAM1 transition matrix is the Markov chain matrix applying for a time period over which we expect 1% of the amino acids to undergo accepted point mutations within the species of interest.

The construction of PAM matrices starts with ungapped multiple alignments of proteins into blocks for which all pairs of sequences in any block are, as in the BLOSUM procedure, "sufficiently close" to each other. In the original construction of Dayhoff et al. (1978), the requirement was that each sequence in any block be no more than 15% different from any other sequence. This requirement resulted, for their data, in 71 blocks of aligned proteins. Imposing the requirement of close within-block similarity is aimed at minimizing the number of substitutions in the data that may have resulted from two or more consecutive substitutions at the same site. This is important because the initial goal is to create a Markov transition matrix for a short enough time period so that multiple substitutions are very unlikely to happen during this time period. We discuss later how to handle the multiple substitutions that are expected to arise over longer time periods.

The BLOSUM approach uses clustering to achieve two aims. One is to minimize biases in the databases that the sequences were taken from, since without clustering some closely related sequences may be overrepresented. The second is to account for evolutionary deviation of varying time periods. In the PAM approach, the first aim is approached by inferring a separate phylogenetic tree for the data in each aligned block of sequences, eventually using all the inferred trees in an aggregated manner to estimate a Markov chain transition matrix. The second aim is achieved by using Markov chain theory applied to this matrix.

The phylogenetic reconstruction method adopted for the data within any block in the database is the method of maximum parsimony, described in Chapter 13. This algorithm constructs trees with our sequences at the leaves, and with inferred sequences at the internal nodes, such that the total number of substitutions across the tree is minimal. Such a tree is called a *most parsimonious* tree. There are often several most parsimonious trees for any block, in which case all such trees are used and an averaging procedure is employed from the data in each tree, as described below. From now on, "tree" means one or other of the set of most parsimonious trees for one of the blocks.

For any column in any block, the data in each tree are used to obtain counts in the following manner. Suppose that two different but aligned amino acids A and B occur (in any order) in two nodes of a tree joined by a single edge. Then this edge contributes 1 to the "A–B" count. If the same amino acid A occurs aligned in two nodes of the tree joined by a single edge, then this edge contributes 2 to the "A–A" count. The counts for all

"A–A" and all "A–B" amino acid pairs are then totaled over all edges of all trees in each block. If the block has n most parsimonious trees, these total counts are then divided by n. The sum of the resultant counts over all blocks is then calculated.

The following simple example demonstrates the calculations within any one block. Suppose that a given block of three sequences is

$$AA$$
$$AB \; .$$
$$BB$$

There are $n = 5$ most parsimonious trees leading to these three sequences at the leaves of the trees, as shown in Figure 6.3. Among these trees A

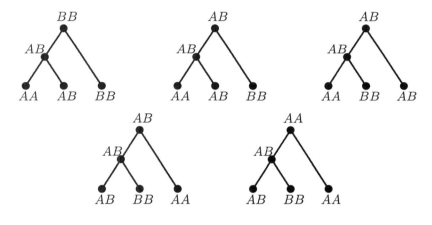

FIGURE 6.3.

is aligned with, and substituted for, B (or conversely) twice in each tree, leading to a total "A–B" count of 10. Division by the number of trees (5) in the block leads to a final contribution from this block of 2 to the A–B count.

Next, A is aligned with A two times in tree 1, three times in tree 2, three times in tree 3, four times in tree 4, and three times in tree 5, leading to a total of 15 A–A alignments. Each A–A alignment leads to a count of 2, so that the total A–A count is 30. Division by the number of trees for this block leads to a final block contribution to the overall A–A count of 6. Similar calculations show that that the contribution to the B–B count from this block is also 6.

If this were the only block in the data set, the final matrix of counts would then be

$$
\begin{array}{c|cc}
 & A & B \\
\hline
A & 6 & \\
B & 2 & 6 \\
\end{array}
$$

In general, there will be more than one block in the data set, and if so, as indicated above, we simply add the counts from the different blocks to obtain an overall count matrix.

Suppose that the amino acids are numbered from 1 to 20 and denote the (j, k)th entry in the overall count matrix by A_{jk}. The next task is to use this count matrix to construct an estimated Markov chain transition matrix. For any j and k (not necessarily distinct), define a_{jk} by

$$
a_{jk} = \frac{A_{jk}}{\sum_m A_{jm}}. \tag{6.19}
$$

For $j \neq k$, let

$$
p_{jk} = c a_{jk}, \tag{6.20}
$$

where c is a positive scaling constant (to be determined later), and let

$$
p_{jj} = 1 - \sum_{k \neq j} c a_{jk}. \tag{6.21}
$$

It follows from these definitions that $\sum_k p_{jk} = 1$. If c is chosen to be sufficiently small so that each p_{jj} is nonnegative, the matrix $P = \{p_{jk}\}$ then has the properties of a Markov chain transition matrix. In this matrix smaller values of c imply larger diagonal entries relative to the nondiagonal entries; however, the relative sizes of the nondiagonal entries are independent of the choice of c. In practice it will always be the case that this matrix is irreducible and aperiodic, so that it has a well-defined stationary distribution.

Although the matrix P is derived from data, and thus any probability derived from it is an estimate, we assume that the data leading to P are sufficiently extensive so that no serious error is incurred by thinking of P not as an estimated transition matrix but as an actual transition matrix. We thus drop the word "estimated" below in discussing the probabilities deriving from this matrix.

The value of c is now chosen so that the weighted expected proportion of amino acid changes, after one step of the Markov chain defined by the transition matrix P, is 0.01, the weights being the various amino acid frequencies. A reasonable estimate for this set of frequencies is the observed distribution found from the data in the original blocks of aligned proteins. Let p_j be the observed such frequency for the jth amino acid. Then the expected proportion of amino acids that change after one step of the Markov chain defined by the transition matrix P is

$$\sum_j p_j \sum_{k \neq j} p_{jk} = c \sum_j \sum_{k \neq j} p_j a_{jk}. \tag{6.22}$$

This implies that if c is defined by the equation

$$c = \frac{.01}{\sum_j \sum_{k \neq j} p_j a_{jk}}, \tag{6.23}$$

the "expected proportion" requirement is satisfied, and the resulting transition matrix then corresponds to an evolutionary distance of 1 PAM. This matrix is often denoted by M_1, with typical element m_{jk}, and we follow this notation here. The matrix corresponding to an evolutionary distance of n PAMs is obtained by raising M_1 to the nth power, in line with the n-step transition probability formula in equation (4.18). This matrix is denoted here by M_n and is called the PAMn matrix. As n gets larger, the matrix M_n gets closer and closer to a matrix all of whose rows are identical to the stationary distribution corresponding to the matrix M_1 (see equation (4.24)). In practice, the element common to all positions in the jth column of this matrix is often close to the background frequency p_j.

The stationary distribution of the matrix M is independent of the choice of the scaling constant c (see Problem 6.6).

The next task is to construct a substitution matrix derived from the probability transition matrix M_n. Let $m_{jk}^{(n)}$ be the (j, k) entry in the matrix M_n, for some extrinsically chosen value of n. Then $m_{jk}^{(n)}$ is the probability, after n steps of the chain defined by the matrix M_1, that the kth amino acid occurs in some specified position, given that initially the jth amino acid occurred in that position. For reasons that will be developed in Section 9.2.4, the typical entry in a PAM substitution matrix is of the form

$$C \cdot \log \left(\frac{m_{jk}^{(n)}}{p_k} \right), \tag{6.24}$$

where C is a positive constant. The choice of C is not crucial: Nevertheless, this also is discussed in Section 9.2.4.

A variant of the expression (6.24) is the following. Denote the joint probability that amino acid j occurs at some nominated position at time 0 and that amino acid k occurs at this position after n steps of the Markov chain whose one-step transition matrix is M_1 by $q(j, k)$. (Note that $q(j, k)$ is a function of n, but in accordance with common practice we suppress this dependence in the notation.) Then

$$q(j, k) = p_j m_{jk}^{(n)}, \tag{6.25}$$

and (6.24) can be written as

$$C \cdot \log \left(\frac{q(j, k)}{p_j p_k} \right). \tag{6.26}$$

The choice of the value n has so far not been discussed. This matter will be taken up in Section 9.6, where the effects of an incorrect choice will be evaluated.

The BLOSUM and PAM procedures differ in one interesting respect: The larger n is for a PAM matrix, the longer is the evolutionary distance, whereas for BLOSUM matrices *smaller* values of n correspond to longer evolutionary distance.

6.5.4   A Simple Symmetric PAM Matrix

In order to elucidate some properties of PAM matrices and to assess the implications of the choice of n in these matrices, it is useful to discuss a simple symmetric example, which, while it does not correspond to any PAM matrix used in practice, has properties that are found easily and that apply broadly to PAM matrices in general.

Suppose that all amino acids are equally frequent (so that $p_j = 0.05$), that all are equally likely to be substituted by some other amino acid in any given time, and that all substitutions are equally likely. Then the matrix M_1 is such that its elements $\{m_{jk}\}$ are given by

$$m_{jj} = 0.99, \quad m_{jk} = 0.01/19, \quad j \neq k. \tag{6.27}$$

The value 0.99 for m_{jj} derives from the fact that we wish to mimic a PAM1 matrix, that is, a matrix for which the probability of an amino acid change in unit time is 0.01. For the simple PAM matrix considered here it can be shown from the spectral theory of Section B.19 that

$$m_{jj}^{(n)} = 0.05 + 0.95(94/95)^n, \tag{6.28}$$

$$m_{jk}^{(n)} = 0.05 - 0.05(94/95)^n, \quad j \neq k. \tag{6.29}$$

These calculations, together with equation (6.24), imply that the typical diagonal entry and the typical off-diagonal entry in the substitution matrix are, respectively,

$$C \cdot \log(1 + 19(94/95)^n), \quad C \cdot \log(1 - (94/95)^n), \tag{6.30}$$

for some positive value of C. The ratio of these is independent of C, being

$$\frac{\log(1 + 19(94/95)^n)}{\log(1 - (94/95)^n)}. \tag{6.31}$$

When $n = 259.0675$, this ratio is very close to -12. A ratio of -12 corresponds to a substitution matrix whose entries are of the form

$$S(j,j) = 12C, \quad S(j,k) = -C. \tag{6.32}$$

For the case $n = 259.0675$,

$$m_{jj}^{(259.0675)} = 0.111251, \quad m_{jk}^{(259.0675)} = 0.046776 \quad j \neq k. \tag{6.33}$$

The definition (6.25) of $q(j, k)$ implies that for this case

$$q(j, j) = 0.0055625, \quad q(j, k) = 0.0023388, \quad j \neq k. \tag{6.34}$$

With these values the probability $20q(j, j)$ of a match at any position is 0.111251, and the probability of a mismatch is 0.888749. Taking the value $C = 1$ in (6.32), the mean score in the substitution matrix is then

$$12(0.111251) - 0.888749 = 0.446. \tag{6.35}$$

We discuss this example further in Section 9.3.3.

6.6 Multiple Sequences

We may have a set of more than two related sequences, all descended from a common ancestor, and in such a case it is often desirable to put them into a multiple alignment. The definition of a multiple alignment is a straightforward generalization of a pairwise alignment. Similarly, the dynamic programming algorithms for constructing the alignments generalize in a straightforward manner. For a small number of sequences, usually fewer than 20, this works well. Problems arise, however, in applying these algorithms to many sequences. The running time for using dynamic programming to do a global multiple alignment of n sequences each of length approximately L is $O((2L)^n)$. For local alignments the situation becomes even worse.

Some algorithms have been developed to find high-scoring alignments quickly, without, however, guaranteeing to find one with the highest score. Perhaps one of the algorithms most commonly used for multiple global alignments is called CLUSTAL W (Thompson et al. (1994)).

In this section we will describe a statistical method for finding ungapped local alignments, introduced by Lawrence et al. (1993). A more general algorithm allowing gapped alignments is given by Zhu et al. (1998). In Section 11.3.2 we give a different method, which constructs gapped multiple alignments.

We describe the process in terms of protein sequences. Label the amino acids in some agreed order as amino acids 1, 2, ..., 20, and suppose that these have respective "background" frequencies $p_1, p_2, \ldots, p_{20}$. Given N protein sequences, of respective lengths $L_1, L_2, \ldots, L_N$, the aim is to find N segments of length W, one in each sequence, that in some sense are most similar to each other. Here the value of W is some chosen fixed number; the choice of W is discussed below. There are $S = \prod_{j=1}^{N}(L_j - W + 1)$ possible choices for the respective locations of these N segments in the N respective sequences, and it is assumed that N and the L_j are so large that a purely algorithmic approach to finding the most similar segments by an extension of the methods discussed in Section 6.4.2 is not computationally feasible.

We describe the Lawrence et al. approach in Markov chain terms. Consider a Markov chain with S "states," each state corresponding to a choice of the locations of the N segments in the N sequences. Each state of the Markov chain is an aligned array of amino acids. This array has N rows and W columns. The aim is to find that array in which the various rows in some sense best align with each other.

The procedure consists of repeated iteration of a basic step, each step consisting of a move from one state of the Markov chain to another, that is, from one array to another. The initial array can be chosen arbitrarily or on the basis of some biological knowledge as an initial guess of the best alignment. In each step one of the various protein sequences is chosen, and the row in the array corresponding to that sequence is allowed to change in a way described below. It is convenient here to assume that the choice of this sequence is made randomly.

Before describing the way in which changes in the array are made, we illustrate in Figure 6.4 the result of the changes made in two consecutive steps of the procedure. In step 1 the third sequence happened to be chosen, so that the third row in the array was allowed to change, and in step 2 the first sequence happened to be chosen, so that the first row was allowed to change.

position	position	position
1 2 3 4 $\cdots$ W	1 2 3 4 $\cdots$ W	1 2 3 4 $\cdots$ W
V Q A L $\cdots$ N	V Q A L $\cdots$ N	C A A N $\cdots$ R
A Q B N $\cdots$ R	A Q B N $\cdots$ R	A Q B N $\cdots$ R
L L C R $\cdots$ N	C Q T N $\cdots$ N	C Q T N $\cdots$ N
W R A A $\cdots$ C	W R A A $\cdots$ C	W R A A $\cdots$ C
S Q C C $\cdots$ T	S Q C C $\cdots$ T	A Q C C $\cdots$ T
S Q T R $\cdots$ C	S Q T R $\cdots$ C	S Q T R $\cdots$ C
$\vdots$	$\vdots$	$\vdots$
G M C R $\cdots$ T	G M C R $\cdots$ T	G M C R $\cdots$ T

step 1 $\longrightarrow$ step 2 $\longrightarrow$

FIGURE 6.4. Full arrays of aligned segments before steps 1, 2, and 3.

We call the array just before any step is taken the "original" array and the array following this step the "new" array. This new array is also the original array for the next step. We now consider the first step in detail.

The segments in all sequences other than randomly chosen sequence 3 define a reduced array of $N - 1$ segments consisting of the original array without row 3. This reduced array is the leftmost array in Figure 6.5. The

$$
\begin{array}{cccccc}
 & \text{position} & & & & \\
1 & 2 & 3 & 4 & \cdots & W \\
\hline
V & Q & A & L & \cdots & N \\
A & Q & B & N & \cdots & R \\
W & R & A & A & \cdots & C \\
S & Q & C & C & \cdots & T \\
S & Q & T & R & \cdots & C \\
 & & \vdots & & & \\
G & M & C & R & \cdots & T \\
\end{array}
\qquad \xrightarrow{\text{step 1}} \qquad
\begin{array}{cccccc}
 & \text{position} & & & & \\
1 & 2 & 3 & 4 & \cdots & W \\
\hline
A & Q & B & N & \cdots & R \\
C & Q & T & N & \cdots & N \\
W & R & A & A & \cdots & C \\
S & Q & C & C & \cdots & T \\
S & Q & T & R & \cdots & C \\
 & & \vdots & & & \\
G & M & C & R & \cdots & T \\
\end{array}
\qquad \xrightarrow{\text{step 2}} \qquad
\begin{array}{cccccc}
 & \text{position} & & & & \\
1 & 2 & 3 & 4 & \cdots & W \\
\hline
C & A & A & N & \cdots & R \\
A & Q & B & N & \cdots & R \\
C & Q & T & N & \cdots & N \\
C & A & A & N & \cdots & R \\
S & Q & T & R & \cdots & C \\
 & & \vdots & & & \\
G & M & C & R & \cdots & T \\
\end{array}
$$

FIGURE 6.5. Partial arrays of aligned segments before steps 1, 2, and 3.

reduced arrays in this figure show that sequence 3 was chosen to change in step 1 and that sequence 1 was chosen to change in step 2, and also shows that sequence 5 was chosen to change in step 3.

Suppose that in the first reduced array, amino acid j $(j = 1, 2, \ldots, 20)$ occurs $c_{i,j}$ times in the ith column. From the $c_{i,j}$ values a probability estimate q_{ij} is calculated, defined by

$$
q_{ij} = \frac{c_{ij} + b_j}{N - 1 + B}. \tag{6.36}
$$

Here the b_j are pseudocounts, as defined in Section 3.7, and $B = \sum_j b_j$. The reason for the introduction of pseudocounts, and the actual choices of the b_j, will be discussed below. For the moment we take the b_j as given.

The aim of the first step is to replace the original segment in the third row by a new segment in a way that tends to increase the overall alignment of the N resulting segments in the new array. There are $L_3 - W$ segments of length W in sequence 3. We call that segment starting in position x in sequence 3 "segment x." Suppose that the amino acids in this segment are $x_1, x_2, \ldots, x_W$. The probability P_x of this ordered set of amino acids under the population amino acid frequencies is $P_x = p_{x_1} p_{x_2} \cdots p_{x_W}$. The estimated probability Q_x of this ordered set of amino acids using the $N-1$ segments in the first reduced array in Figure 6.5 is taken as

$$
Q_x = q_{1,x_1} q_{2,x_2} \cdots q_{W,x_W}.
$$

The likelihood ratio $\mathrm{LR}(x)$ is defined as Q_x/P_x. The numerator may be thought of as the probability of the sequence under the model reflecting only the sequences in the current alignment, while the denominator is the probability of the sequence under background frequencies. The final operation in step 1 is to replace the segment in row 3 of the original array by segment x with probability

$$\frac{\text{LR}(x)}{\sum_{m=1}^{L_3-W} \text{LR}(m)}. \tag{6.37}$$

The reason for this choice will be discussed in Section 10.5.2. The result of this step is to produce a new array (as illustrated in the second array in Figure 6.4). The replacement procedure tends to replace the original segment in row 3 by a new segment more closely aligned with the remaining segments.

In the following step this procedure is repeated, with (in Figure 6.4) the segment from sequence 1 being randomly chosen to change. In the next step the segment from sequence 5 was randomly chosen to change, and so on. As one step follows another the N segments in the array tend to become more similar to each other, or in other words, to align better. This iterative procedure visits various possible alignments according to a random process, and thus does not systematically approach the "best" alignment. However, after many steps the best alignment should tend to arise more and more frequently and hence be recognized.

This procedure is essentially one of Gibbs sampling, and we further consider its properties in the discussion of the Gibbs sampling procedure in Section 10.5.2. In order to introduce the analysis of Section 10.5.2, we adopt here a general notation suitable for the discussion of that section. Suppose that before some specific step is taken, the current array is array s in the collection of S possible arrays, and that after this step is taken the new array is array u. Arrays s and u are identical in all rows other than the one that is changed during this step. Thus the reduced arrays obtained by eliminating the row in which arrays s and u differ lead to identical values of the q_{ij}.

Now consider the amino acids in arrays s and u in the row in which they differ. Suppose that in array s these are denoted by $s(1), s(2), \ldots, s(W)$ and in array u are denoted by $u(1), u(2), \ldots, u(W)$. If the transition probability from array s to array u is denoted by p_{su}, then expression (6.37) shows that

$$\frac{p_{su}}{p_{us}} = \frac{q_{1,u(1)}q_{2,u(2)} \cdots q_{W,u(W)}}{q_{1,s(1)}q_{2,s(2)} \cdots q_{W,s(W)}} \cdot \frac{p_{s(1)}p_{s(2)} \cdots p_{s(W)}}{p_{u(1)}p_{u(2)} \cdots p_{u(W)}}. \tag{6.38}$$

We return to this ratio in Section 10.5.2.

The reason for using pseudocounts in the probability estimate (6.36) is the following. In step 1 above, there might be an excellent alignment of segment x of protein sequence 3 with the $N-1$ segments in the reduced array, except that the amino acid at position i in sequence x is not represented at position i in the reduced array. If pseudocounts were not used and the probability estimate q_{ij} replaced by $c_{ij}/(N-1)$, segment x could not be chosen at this step, since in this case the probability $c_{ij}/(N-1)$

would be 0 for this segment. We therefore assume that each b_j is greater than 0, and then introduction of pseudocounts allows the choice of segment x in the above case.

The choice of the pseudocounts b_j must be made subjectively. Lawrence et al. (1993) make the reasonable suggestion of choosing b_j proportional to the background frequency p_j of amino acid j. The choice of the proportionality constant is less obvious, and Lawrence et al. claim that in practice the constant $\sqrt{N}$ works well.

So far the length W of the segments considered has been assumed to be given. The choice of this length is discussed in detail by Lawrence et al. together with further tactical questions.

It might be asked why the choice of each new segment is random. An alternative procedure in the step described above would be to replace the current segment in sequence one by that segment maximizing the ratio in (6.37). The entire operation is then fully deterministic, and runs the risk of settling on a locally rather than a globally best alignment. The stochastic procedure allows movement from a locally best alignment to a globally best alignment. However, it is not guaranteed that the procedure will escape from the neighborhood of a locally best alignment in reasonable time, and thus may be sensitive to the choice of the initial alignment.

Problems

6.1. Prove that

$$\sum_{k=0}^{\min\{m,n\}} \binom{m}{k}\binom{n}{k} = \binom{m+n}{n} = \binom{m+n}{m}.$$

Hint: Think of having to select n objects from a set with $s = m+n$ objects, of which m are of one kind and n are of another kind.

6.2. The BLOSUM50 substitution matrix given in Table 6.7 is often used to assign a score to each pair of aligned amino acids. Use this matrix and a linear gap penalty of $d = 5$ to find all highest-scoring alignments between $x = EATGHAG$ and $y = EEAWHEAE$.

6.3. Fit $x = cttgac$ into $y = cagtatcgtac$ with the scoring scheme of the example in Section 6.4.1.

6.4. Find the best local alignment(s) of $x = aagtatcgca$ and $y = aagttagttgg$ with the same scoring scheme as in the example in Section 6.4.1.

6.5. The matrix below has been obtained using a dynamic programming algorithm (with a linear gap model) to solve one of the following three problems relative to the two sequences

$$x = X_1 X_2 X_3 X_4$$

and

$$y = Y_1 Y_2 Y_3 Y_4 Y_5 Y_6 Y_7 Y_8 Y_9 Y_{10} Y_{11},$$

(i) global alignment, (ii) fitting x into y, (iii) local alignment.

(1) By considering the matrix it is clear which of the three problems above was the one being tackled. Which was it?

(2) What is the optimal score for the alignment sought?

(3) List all the optimal alignments (in terms of the X_i's and Y_j's) relative to the problem being tackled.

	−	Y_1	Y_2	Y_3	Y_4	Y_5	Y_6	Y_7	Y_8	Y_9	Y_{10}	Y_{11}
−	0	0	0	0	0	0	0	0	0	0	0	0
X_1	−5	−5	−5	5←0		5	5	5←0		5	5←0	
X_2	−10	0←−5		0	10←5←0			0	10←5←0			0
X_3	−15	−5	−5	−5	5	5←0←−5			5	5←0←−5		
X_4	−20	−10	−10	0	0	10	10←5←0			10	10←5	

6.6. Prove that the stationary distribution of the matrix M, defined by equations (6.20) and (6.21), is independent of the choice of the scaling constant c.

6.7. Equations (6.28) and (6.29) can be derived in various ways. One of these is by mathematical induction (see Section B.18). It is easy to see that the equations are true for $n = 1$. Suppose that they are true for some given value n^* of n. Show that they are then true for $n = n^* + 1$, and thus complete a proof by induction that they are true for all positive integers n. *Hint:* Use the facts that

$$m_{jj}^{(n+1)} = 0.99\, m_{jj}^{(n)} + 0.01\, m_{jk}^{(n)}, \quad j \neq k,$$

$$m_{jk}^{(n+1)} = (0.01/19)\, m_{jj}^{(n)} + 0.99\, m_{jk}^{(n)} + (.18/19) m_{jr}^{(n)}, \quad j \neq k \neq r,$$

and that the latter equation can be rewritten as

$$m_{jk}^{(n+1)} = \frac{1}{19}\left(0.01\, m_{jj}^{(n)} + 18.99\, m_{jk}^{(n)}\right).$$

6.8. Use the relation (B.24) to find the limit of the ratio (6.31) as $n \to \infty$.

6.9. Show that the n-step transition probabilities $m_{jk}^{(n)}$ given in equations (6.28) and (6.29) can be found by calculating the spectral expansion of the associated matrix M.

7

Stochastic Processes (ii): Random Walks

7.1 Introduction

Random walks are special cases of Markov chains, and thus can be analyzed by Markov chain methods. However, the special features of random walks allow specific methods of analysis. Some of these methods are introduced in this chapter.

Our main interest in random walk theory is that it supplies the basic probability theory behind BLAST. BLAST is a procedure that searches for high-scoring local alignments between two sequences and then tests for significance of the scores found via P-values. The P-value calculation takes into consideration, as it must, the lengths of the two sequences, since the longer the sequences, the more likely there is to be local homology simply by chance. In practice, BLAST is used to search a database consisting of a number of sequences for similarity to a single "query" sequence. P-values are calculated allowing for the size of the entire database. However, we defer consideration of this matter until Section 9.5, and consider here only the random walk theory underlying the eventual P-value calculations.

The utility of BLAST is that it is able to complete its task very quickly, even when there are many sequences in the database. The efficiency arises for two reasons. First, on the algorithmic side it uses heuristics to avoid searching through all possible ungapped local alignments, of which there is an astronomically large number. Therefore, it can fail to produce exactly the highest-scoring alignments. Its performance, however, is extremely good. Second, the calculation of the P-value uses sophisticated approxi-

mations to achieve extremely fast calculations. The calculation of these
P-values is discussed in Chapter 9.

Consider the simple case of the two aligned DNA sequences given in (1.1)
and repeated here for convenience:

$$\begin{array}{cccccccccccccccccccc}
\downarrow & \downarrow & & \downarrow & \downarrow\downarrow & \downarrow & \downarrow\downarrow & & & & & & & & \downarrow\downarrow\downarrow & \\
g & g & a & g & a & c & t & g & t & a & g & a & c & a & g & c & t & a & a & t & g & c & t & a & t & a & \\
g & a & a & c & g & c & c & c & t & a & g & c & c & a & c & g & a & g & c & c & c & t & t & a & t & c &
\end{array} \qquad (7.1)$$

Suppose we give a score $+1$ if the two nucleotides in corresponding positions
are the same and a score of -1 if they are different. As we compare the
two sequences, starting from the left, the accumulated score performs a
random walk. In the above example, the walk can be depicted graphically
as in Figure 7.1. The filled circles in this figure relate to *ladder points*, that
is to points in the walk lower than any previously reached point.

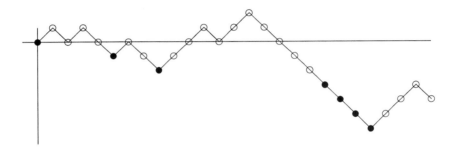

FIGURE 7.1.

The part of the walk from a ladder point until the highest point attained
before the next ladder point is called an excursion. BLAST theory focuses
on the maximum heights achieved by these excursions. In Figure 7.1 these
maximum heights are, respectively, 1, 1, 4, 0, 0, 0, 3. (If the walk moves
from one ladder point immediately to the next, the corresponding height
is taken as 0.)

In practice, BLAST theory relates to cases that are much more complicated than this simple example. It is often applied to the comparison of
two protein sequences and uses scores other than the simple scores $+1$ and
-1 for matches and mismatches. These scores are described by the entries
in a substitution matrix such as those given in the BLOSUM50 substitution matrix shown in Table 6.7. These scores determine the upward or
downward movement of the random walk describing the score for that protein comparison. For example, if the score for any amino acid comparison
is given by the appropriate entry in the BLOSUM50 substitution matrix,

then for the alignment

$$
\begin{array}{cccccccccccccccc}
T & Q & L & A & A & W & C & R & M & T & C & F & E & I & E & C & K & V \\
R & H & L & D & S & W & R & R & A & S & D & D & A & R & I & E & E & G
\end{array}
\qquad (7.2)
$$

the scores are -1, 1, 5, -2, 1, 15, -4, 7, -1, 2, -4, etc..., and therefore the graph of the accumulated score goes through the points

$$
(1, -1), (2, 0), (3, 5), (4, 3), (5, 4), (6, 19), (7, 15), (8, 22), \text{etc.} \qquad (7.3)
$$

The graph of the accumulated score for this walk is depicted in Figure 7.2.

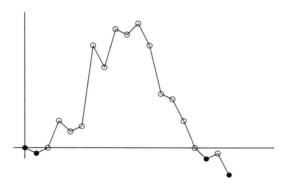

FIGURE 7.2.

To discuss BLAST it is necessary to consider arbitrary scoring schemes and thus aspects of the general theory of random walks. We do this in this chapter. However, we start by analyzing the simple random walk where the only possible step sizes are $+1$ and -1.

For the remainder of this chapter we consider random walks in the abstract, without reference to sequence comparisons, which we will return to in Chapter 9.

7.2   The Simple Random Walk

7.2.1   Introduction

We first consider a process that starts at some arbitrary point h and moves, independently of the previous history of the process, a step down every unit time with probability q or a step up with probability p. (We use the notation "h" throughout to denote the initial position of the walk.) This process will be called a simple random walk. The walk is assumed to be restricted to the interval $[a, b]$, where a and b are integers with $a < h < b$, and stops when it reaches either a or b. We ask, "What is the probability that eventually the walk finishes at b rather than a?" and "What is the mean number of steps taken until the walk stops?"

7.3   The Difference Equation Approach

We begin by discussing the derivation of properties of the simple random walk by using the classical difference equation method, described in detail in Feller (1968). We will find that the difference equation method does not allow ready generalizations to more complicated walks, and in the next section will develop the moment-generating function approach, which does allow these generalizations.

7.3.1   Absorption Probabilities

Let w_h be the probability that the simple random walk eventually finishes at, or is "absorbed" at, the point b rather than at the point a, given that the initial point is h. Then by comparing the situation just before and just after the first step of the walk, we get

$$w_h = pw_{h+1} + qw_{h-1}. \tag{7.4}$$

Furthermore,

$$w_a = 0, \quad w_b = 1. \tag{7.5}$$

Equation (7.4) is a homogeneous difference equation, with boundary conditions (7.5). A solution of (7.4) is a set of values w_h that satisfy it for all integers h. Difference equations are discussed in detail in Feller (1968), and we draw on his results here without extensive discussion because these results will be rederived in Section 7.4. The difference equation methods are given here largely for comparison with the more powerful mgf methods.

One solution to equation (7.4) is of the form

$$w_h = e^{\theta h}$$

for some fixed constant θ. To solve for θ we substitute into (7.4) to get

$$e^{\theta h} = pe^{\theta(h+1)} + qe^{\theta(h-1)}.$$

Multiplying throughout by $e^{\theta - \theta h}$, this becomes

$$pe^{2\theta} - e^{\theta} + q = 0,$$

a quadratic equation in e^{θ}. For the case $p \neq q$, there are two distinct solutions of this equation, namely

$$e^{\theta} = 1, \quad e^{\theta} = \frac{q}{p}, \tag{7.6}$$

or equivalently

$$\theta = 0, \quad \theta = \log\left(\frac{q}{p}\right).$$

Thus when $p \neq q$, we have found two specific solutions of (7.4), namely

$$w_h = 1 \quad \text{and } w_h = e^{\theta^* h},$$

where

$$\theta^* = \log\left(\frac{q}{p}\right). \tag{7.7}$$

The theory of homogeneous difference equations states that the general solution of (7.4) is any linear combination of these two solutions. That is, the general solution of (7.4) is

$$w_h = C_1 + C_2 e^{\theta^* h}, \tag{7.8}$$

for any arbitrary constants C_1 and C_2. We determine C_1 and C_2 by the boundary conditions (7.5), and this leads to

$$w_h = \frac{e^{\theta^* h} - e^{\theta^* a}}{e^{\theta^* b} - e^{\theta^* a}}. \tag{7.9}$$

This is the solution for the case $p \neq q$ only. When $p = q$ $(= \frac{1}{2})$, the two solutions in (7.6) are the same, and the above argument does not hold. In the applications of random walk theory to BLAST we never encounter the symmetric case $p = q$ or its various generalizations (which we call "zero mean" cases). Since consideration of this case usually involves extra work, we will not consider it further.

In some applications interest focuses on the probability u_h that the walk finishes at a rather than at b. The probabilities u_h satisfy the same difference equation (7.4) as w_h, but with the boundary conditions $u_a = 1$, $u_b = 0$. This leads to

$$u_h = \frac{e^{\theta^* b} - e^{\theta^* h}}{e^{\theta^* b} - e^{\theta^* a}}. \tag{7.10}$$

As we expect, $w_h + u_h = 1$.

7.3.2   Mean Number of Steps Taken Until the Walk Stops

We continue to assume that the walk starts at h and ends when one of a or b is reached, $a < h < b$. We define N as the (random) number of steps taken until the walk stops. The probability distribution of N depends on h, a, and b, and in this section we find the mean value m_h of this probability distribution, that is, the mean of N. We do not use the formula (1.16) to find this mean value; in fact, we never compute the complete probability distribution of N. Instead, we use difference equation methods similar to those in the previous section, which lead directly to the value of the mean. These calculations, too, will be repeated in the next section using moment-generating functions.

Considering the situation after one step of the walk in two different ways, we get

$$m_h - 1 = pm_{h+1} + qm_{h-1}. \tag{7.11}$$

This is an inhomogeneous difference equation. Difference equation theory shows that the general solution of (7.11) consists of any particular solution added to arbitrary multiples of solutions of the homogeneous equation

$$m_h = pm_{h+1} + qm_{h-1}. \tag{7.12}$$

A particular solution of (7.11) is $m_h = h/(q-p)$. The homogeneous equation (7.12) is the same as (7.4), and hence has the same general solution (7.8). Therefore, the general solution of (7.11) is

$$m_h = \frac{h}{q-p} + C_1 + C_2 e^{\theta^* h}. \tag{7.13}$$

The constants C_1 and C_2 are determined by boundary conditions. The particular boundary conditions $m_a = m_b = 0$ lead to

$$m_h = \frac{h-a}{q-p} - \frac{b-a}{q-p} \cdot \frac{e^{\theta^* h} - e^{\theta^* a}}{e^{\theta^* b} - e^{\theta^* a}}. \tag{7.14}$$

Using equations (7.9) and (7.10) we can rewrite this as

$$m_h = \frac{w_h(b-h) + u_h(a-h)}{p-q}. \tag{7.15}$$

We will see later that equation (7.15) has a simple intuitive interpretation.

7.4   The Moment-Generating Function Approach

In this section the absorption probabilities w_h and u_h and the value of m_h are rederived using mgf techniques. Various asymptotic results are then derived. Finally, the generalization to an arbitrary random walk will be discussed.

7.4.1   Absorption Probabilities

The mgf of the (random) value, either -1 or $+1$, of any single step taken in the random walk is

$$\mathrm{m}(\theta) = qe^{-\theta} + pe^{\theta}. \tag{7.16}$$

Theorem 1.1 (page 35) shows that for any nonzero-mean random walk for which both positive and negative steps are possible, there exists a unique nonzero value θ^* of θ for which

$$m(\theta^*) = 1. \tag{7.17}$$

The value θ^* is the same as given in equation (7.7). This observation is now used to provide an alternative way of finding the probability that the simple random walk finishes at b rather than at a and also to provide an alternative way of finding the mean number m_h of steps until the walk terminates.

Recall that N is the random number of steps until the walk finishes. The mgf of the total displacement after N steps is, from (2.17), $(qe^{-\theta} + pe^{\theta})^N$. If we choose the value θ^* given in (7.7) for θ, this is

$$(qe^{-\theta^*} + pe^{\theta^*})^N, \tag{7.18}$$

which is equal to 1 for all N. We now find another expression for this mgf.

Consider the situation when the random walk has just finished. At this time the net displacement from the original starting point h is either $b-h$ (if the walk finished at b) or $a-h$, (if the walk finished at a). The probabilities of these two outcomes are w_h and u_h, given respectively by (7.9) and (7.10). The definition (1.73) of the mgf of a discrete random variable shows that the mgf of the net displacement when the walk has just finished is

$$w_h e^{(b-h)\theta} + u_h e^{(a-h)\theta} = w_h e^{(b-h)\theta} + (1 - w_h)e^{(a-h)\theta}. \tag{7.19}$$

With the particular choice $\theta = \theta^*$ this is

$$w_h e^{(b-h)\theta^*} + (1 - w_h)e^{(a-h)\theta^*}. \tag{7.20}$$

But this is another expression for the moment-generation function (7.18), whose value is 1. Thus equating this expression also to 1, we get

$$w_h = \frac{e^{h\theta^*} - e^{a\theta^*}}{e^{b\theta^*} - e^{a\theta^*}}. \tag{7.21}$$

This is identical to the solution (7.9), found by using difference equations. The value of u_h given in (7.10) can be found similarly.

Thus use of the mgf gives us a second approach to finding absorption probabilities of the random walk. For the simple random walk considered in this section the mgf approach does not differ markedly from the approach using difference equations. However, as we see later, the mgf is more readily adapted than the difference equation method for the calculations needed in the general theory used in BLAST.

7.4.2 Mean Number of Steps Until the Walk Stops

We now find m_h using mgf methods. The analysis is not straightforward and relies on Wald's identity (Karlin and Taylor (1975), page 264). We do not prove this identity here, and limit the discussion to conclusions that follow from it.

Consider the random walk discussed above, starting at h and continuing until reaching a or b. Suppose the ith step is S_i ($S_i = \pm 1$). The S_i are iid random variables; let S be a random variable with their common distribution. The mgf $m(\theta)$ of S is given by (7.16).

As above, we denote the (random) number of steps taken until the walk finishes by N, and let $T_N = \sum_{j=1}^{N} S_j$. T_N is either $a - h$, with probability $1 - w_h$, or $b - h$, with probability w_h. With these definitions we have the following:

Theorem 7.1 (Wald's identity).

$$E\left(\text{m}(\theta)^{-N} e^{\theta T_N}\right) = 1 \qquad (7.22)$$

for all θ for which the mgf is defined. (The expected value is taken with respect to the joint distribution of N and T_N.)

The derivative (with respect to θ) of the right-hand side of (7.22) is 0. Since there are infinitely many possible values of N, the left-hand side is an infinite sum of functions of θ. A basic theorem of calculus gives a criterion indicating when the derivative of a sum of functions can be calculated as the sum of the derivatives of the individual functions. This criterion holds in this case, so that

$$\frac{d}{d\theta} E\left(\text{m}(\theta)^{-N} e^{\theta T_N}\right) = E\left(\frac{d}{d\theta}\left(\text{m}(\theta)^{-N} e^{\theta T_N}\right)\right)$$

$$= -N\text{m}(\theta)^{-N-1}\frac{d}{d\theta}\text{m}(\theta)e^{\theta T_N} + \text{m}(\theta)^{-N} T_N e^{\theta T_N}.$$

It follows that

$$E\left(-N\text{m}(\theta)^{-N-1}\frac{d}{d\theta}\text{m}(\theta)e^{\theta T_N} + \text{m}(\theta)^{-N} T_N e^{\theta T_N}\right) = 0.$$

Inserting the specific value $\theta = 0$ into this equation, we get

$$E(-NE(S) + T_N) = -E(N)E(S) + E(T_N) = 0. \qquad (7.23)$$

Thus

$$E(T_N) = E(S)m_h, \qquad (7.24)$$

where $E(S)$ is the mean of the step size. The content of equation (7.24) is intuitively acceptable, namely that the mean value of the final total displacement of the walk, when it stops, is the mean size of each step multiplied by the mean number of steps m_h taken until the walk finishes.

Since the eventual total displacement T_N in the walk is either $b - h$ (with probability w_h given in (7.9)) or $a - h$ (with probability u_h given in (7.10)), the mean value of T_N is

$$E(T_N) = u_h(a - h) + w_h(b - h). \tag{7.25}$$

Since $E(S) = p - q$, equation (7.24) shows that

$$m_h = \frac{u_h(a - h) + w_h(b - h)}{p - q}. \tag{7.26}$$

This is the same as the expression (7.15), found by using difference equations. The present derivation is more flexible than the difference equation approach, and further, it demonstrates where the various terms in equation (7.26) come from.

7.4.3 An Asymptotic Case

The theory of BLAST concerns those walks that start at $h = 0$, where there is a lower boundary at $a = -1$ but no upper boundary, and where $E(S)$, the mean step size, is negative. Such a walk is destined eventually to reach -1. BLAST theory requires the calculation of two quantities:

(i) the probability distribution of the maximum value that the walk ever achieves before reaching -1, and

(ii) the mean number of steps until the walk eventually reaches -1.

BLAST theory generally uses the notation A to denote (ii), so for this particular walk we use this notation rather than m_0.

To discuss the maximum value that the walk ever achieves before eventually reaching -1 we install an artificial stopping boundary at the value y, where $y \geq 1$. Then if $h = 0$ and $a = -1$, equation (7.9) shows that the probability that the unrestricted walk finishes at the artificial boundary y rather than at the value a is

$$\frac{1 - e^{-\theta^*}}{e^{\theta^* y} - e^{-\theta^*}}, \tag{7.27}$$

where θ^* is defined in equation (7.7). Since θ^* is positive, the term $e^{\theta^* y}$ dominates the denominator in (7.27) when y is large, so that the probability (7.27) is asymptotic to

$$(1 - e^{-\theta^*})e^{-\theta^* y}. \tag{7.28}$$

Thus if Y is the maximum height achieved by the walk,

$$\text{Prob}(Y \geq y) \sim Ce^{-\theta^* y} \tag{7.29}$$

as $y \to \infty$, where

$$C = 1 - e^{-\theta^*}. \tag{7.30}$$

This asymptotic relation is in the form of the geometric-like probability displayed in equation (1.63), with the identification $\lambda = \theta^*$.

Turning to (ii), the expression for the mean number A of steps before that walk finishes at -1 or b is found by first making the substitutions $h = 0$, $a = -1$ in equation (7.26), to obtain

$$A = \frac{u_0 - bw_0}{q - p}. \tag{7.31}$$

Equation (7.21) shows that $bw_0 \to 0$ as $b \to \infty$. Since $w_0 \to 0$, $u_0 \to 1$. Taking the limit $b \to \infty$ in equation (7.31) shows that

$$A = \frac{1}{q - p}. \tag{7.32}$$

7.5 General Walks

Suppose generally that the possible step sizes in a random walk are

$$-c, -c + 1, \ldots, 0, \ldots, d - 1, d, \tag{7.33}$$

and that these steps have respective probabilities

$$p_{-c}, p_{-c+1}, \ldots, p_d, \tag{7.34}$$

some of which might be zero. We assume three conditions throughout:

(i) Both $p_{-c} > 0$ and $p_d > 0$.

(ii) This step size is of the "negative mean" type, so that

$$E(S) = \sum_{j=-c}^{d} j p_j < 0.$$

(iii) The greatest common divisor of the step sizes that have nonzero probability is 1.

The mgf of S is

$$\mathrm{m}(\theta) = \sum_{j=-c}^{d} p_j e^{j\theta},$$

and Theorem 1.1 (page 35) states that there exists a unique positive value θ^* of θ for which

$$\sum_{j=-c}^{d} p_j e^{j\theta^*} = 1. \tag{7.35}$$

Our aim is to obtain asymptotic results generalizing those in Section 7.4.3 applying for walks that start at 0 and have a stopping boundary at -1 and no upper boundary. To do this we impose an artificial barrier at y, where $y > 0$. The walk finishes either when it reaches the point -1 or a point less than -1, or when it reaches the point y or a point exceeding y. The possible points where the walk can finish are now

$$-c, -c+1, \ldots, -1, y, \ldots, y+d-1.$$

Let P_k be the probability that the walk finishes at the point k. From equation (7.22) (which holds also for the case of general step sizes),

$$E\left(e^{\theta^* T_N}\right) = 1,$$

where T_N is the total displacement from 0 when the walk stops. Thus

$$\sum_{k=-c}^{-1} P_k e^{k\theta^*} + \sum_{k=y}^{y+d-1} P_k e^{k\theta^*} = 1. \tag{7.36}$$

We consider some implications of this equation in the following two examples.

Example 1. In the simple random walk, equation (7.36) becomes

$$P_{-1}e^{-\theta^*} + P_y e^{y\theta^*} = 1.$$

For this walk $P_{-1} = 1 - P_y$, and by inserting this into the above equation the value of P_y in (7.27) is obtained. Note that in this case, $P_{-1} = u_0$ and $P_y = w_0$.

Example 2. In this example we consider a walk that at each move takes *two* steps down (with probability q) or one step up (with probability p). We assume that the mean step size $p - 2q$ is negative. The mgf of the step size is $qe^{-2\theta} + pe^{\theta}$, and the unique positive value θ^* of θ for which this takes the value 1 is

$$\theta^* = \log\left(\frac{q + \sqrt{4pq + q^2}}{2p}\right). \tag{7.37}$$

With θ^* defined in equation (7.37), equation (7.36) yields

$$P_{-2}e^{-2\theta^*} + P_{-1}e^{-\theta^*} + P_y e^{y\theta^*} = 1. \tag{7.38}$$

Let $R_{-1} = \lim_{y \to +\infty} P_{-1}$ and $R_{-2} = \lim_{y \to +\infty} P_{-2}$. Then

$$P_y \sim (1 - R_{-1}e^{-\theta^*} + R_{-2}e^{-2\theta^*})e^{-y\theta^*}. \tag{7.39}$$

This implies that if there is no upper barrier to the walk and Y is the maximum height achieved by the walk,

$$P_y = \mathrm{Prob}(Y \geq y) \sim Ce^{-y\theta^*}, \tag{7.40}$$

where

$$C = 1 - R_{-1}e^{-\theta^*} + R_{-2}e^{-2\theta^*}. \tag{7.41}$$

Thus Y has the geometric-like distribution. We show in the following section that this conclusion is true for the general walk described above, and we will also find expressions for the values of C and θ^* applying for any walk.

Returning to the general case, we next find a general expression for A, the mean number of steps that the walk takes before stopping at one of the points $-c, -c+1, \ldots, -1$ when no stopping boundary at y is imposed. Wald's identity (7.22) and the calculation (7.24) that follows from it make no assumption about the possible step sizes, so that (7.24) may be used in the case of a general random walk. The mean net displacement when the walk stops at one of these points is $\sum_{j=1}^{c} -jR_{-j}$, where R_{-j} is the probability that the walk finishes at $-j$. The mean step size $E(S)$ is $\sum_{j=-c}^{d} jp_j$, assumed to be negative. Thus from equation (7.24),

$$A = \frac{\sum_{j=1}^{c} jR_{-j}}{-\sum_{j=-c}^{d} jp_j}. \tag{7.42}$$

For BLAST calculations it is necessary to calculate the values of θ^*, C, and A for any random walk. The calculation of θ^* from equation (7.35) for any random walk is straightforward, using numerical methods if necessary. The calculation of C is far less straightforward and is considered in the following section. The calculation of A from (7.42) depends on the calculation of the R_{-j} values, a matter discussed in Section 9.2.3.

7.6   General Walks: Asymptotic Theory

7.6.1   Introduction

In this section we further develop the theory for the general random walk, focusing on the asymptotic $y \to +\infty$ case discussed in the previous section. Our analysis is based on that of Karlin and Dembo (1992). We will show for any walk starting at $h = 0$ and with lower boundary $a = -1$ that

$$\mathrm{Prob}(Y \geq y) \sim Ce^{-\theta^* y}, \tag{7.43}$$

where θ^* is found from equation (7.35) and C is a constant specific to the walk in question. This implies that the maximum height achieved by the walk has a geometric-like distribution whose parameter λ is found from the mgf equation (7.35), replacing θ^* in this equation by λ. We will also find an explicit formula for the constant C in this distribution.

7.6.2 The Renewal Theorem

We start with a technical result.

Theorem 7.2 (The Renewal Theorem).

Suppose three sequences $(b_0, b_1, \ldots)$, $(f_0, f_1, \ldots)$, and $(u_0, u_1, \ldots)$ of non-negative constants satisfy the equation

$$u_y = b_y + (u_y f_0 + u_{y-1} f_1 + u_{y-2} f_2 + \cdots + u_1 f_{y-1} + u_0 f_y), \qquad (7.44)$$

for all y. Suppose further that $B = \sum_i b_i < +\infty$, $\sum_i f_i = 1$, $\mu = \sum_i i f_i < +\infty$, and that the greatest common divisor of $f_0, f_1, f_3, \ldots$ is 1. Then

$$u_y \to B\mu^{-1} \text{ as } y \to +\infty. \qquad (7.45)$$

The proof of this theorem is beyond the scope of this book. The interested reader is referred to Karlin and Taylor (1981).

7.6.3 Unrestricted Walks

We assume that the walk starts at 0. Suppose as in Section 7.5 that the possible step sizes in a random walk are

$$-c, -c+1, \ldots, 0, \ldots, d-1, d,$$

that these steps have respective probabilities $p_{-c}, p_{-c+1}, \ldots, p_d$, and that the three assumptions given on page 228 all hold.

In this section we consider the case where no boundaries are placed on the walk, so that the concept of stopping at a boundary point no longer applies. Because the mean step size is negative, the walk eventually drifts down to $-\infty$. Before doing so, however, it might visit various positive values. The first aim is to find an equation satisfied by the probabilities Q_k, where Q_k is the probability that the walk visits the positive value k before reaching any other positive value. The largest positive step size is d, so that $Q_k = 0$ for all $k > d$. It will be convenient to also define $Q_0 = 0$. Since it is possible that the walk never visits any positive value, we have $\sum_{k=1}^{d} Q_k < 1$.

Although the focus in this section is on unrestricted walks, for the moment we impose an artificial boundary at $+1$ and another at $-L$, where L

is large and positive. Then equation (7.36) becomes

$$\sum_{k=-L-c+1}^{-L} Q_k(L)e^{k\theta^*} + \sum_{k=1}^{d} Q_k(L)e^{k\theta^*} = 1, \tag{7.46}$$

where $Q_k(L)$ is the probability that the walk stops at the value k and the notation recognizes the dependence of the probabilities $Q_k(L)$ on L. Since $+1$ is a boundary, if the walk stops at $k > 0$, then this k is the first positive value reached. Further,

$$\lim_{L\to\infty} Q_k(L) = Q_k.$$

Since θ^* is positive and $Q_k(L) \le 1$ for all k and L, the terms in the first sum in (7.46) all approach zero as $L \to \infty$. Thus letting $L \to \infty$ in (7.46) we have

$$\sum_{k=1}^{d} Q_k e^{k\theta^*} = 1. \tag{7.47}$$

The next aim is to find an expression for $F_{Y_{\mathrm{unr}}}(y)$, the probability that in the unrestricted walk the maximum upward excursion is y or less, for any positive value of y. This is done as follows.

The event that in the unrestricted walk the maximum upward excursion is y or less is the union of several nonoverlapping events. The first of these is the event that the maximum excursion never reaches positive values (probability $\bar{Q} = 1 - Q_1 - Q_2 - Q_3 - \cdots - Q_d$). The remaining events are that the first positive value achieved by the excursion is k, $k = 1, 2, \ldots, y$ (with probability Q_k), and then, starting from this first positive value, that the walk never achieves a further height exceeding $y - k$ (probability $F_{Y_{\mathrm{unr}}}(y - k)$). These various possible events imply that

$$F_{Y_{\mathrm{unr}}}(y) = \bar{Q} + \sum_{k=0}^{y} Q_k F_{Y_{\mathrm{unr}}}(y - k) \tag{7.48}$$

(recall that $Q_0 = 0$).

This is in the form of the renewal equation (7.44), with u_k replaced by $F_{Y_{\mathrm{unr}}}(y - k)$, b_y replaced by $\bar{Q}$, and f_k replaced by Q_k. However, we cannot immediately use the result of Theorem 7.2 to develop properties of $F_{Y_{\mathrm{unr}}}(y)$. This is for two reasons. First, Theorem 7.2 requires that $\sum_k f_k = 1$, whereas here $\sum Q_k < 1$. Second, Theorem 7.2 requires $\sum b_y < \infty$, and this requirement does not hold when $b_y = \bar{Q}$. However, we can use Theorem 7.2 by introducing the quantity $V(y)$, defined by

$$V(y) = \left(1 - F_{Y_{\mathrm{unr}}}(y)\right)e^{y\theta^*}, \tag{7.49}$$

where θ^* is defined by equation (7.35). From (7.49) it follows that implies that

$$F_{Y_{\mathrm{unr}}}(y) = 1 - V(y)e^{-y\theta^*},$$

and equation (7.48) can then be written as

$$1 - V(y)e^{-y\theta^*} = \bar{Q} + \sum_{k=0}^{y} Q_k \left(1 - V(y-k)e^{-(y-k)\theta^*}\right).$$

Elementary reorganization of this equation leads to

$$V(y) = e^{y\theta^*}\left(Q_{y+1} + Q_{y+2} + \cdots + Q_d\right) + \sum_{k=0}^{y}(Q_k e^{k\theta^*})V(y-k) \qquad (7.50)$$

when $y < d$, and

$$V(y) = \sum_{k=0}^{d}(Q_k e^{k\theta^*})V(y-k) \qquad (7.51)$$

when $y \geq d$. This is again in the form of the renewal equation, this time with $f_k = Q_k e^{k\theta^*}$, and with b_y replaced by $e^{y\theta^*}\left(Q_{y+1} + Q_{y+2} + \cdots + Q_d\right)$ if $y < d$ and $b_y = 0$ if $y \geq d$. To use Theorem 7.2 we must show that $\sum b_y < \infty$ and that

$$\sum_{k=0}^{\infty} Q_k e^{k\theta^*} = 1.$$

The first requirement follows since $b_y = 0$ for $y \geq d$, and the second follows from equation (7.47), recalling that Q_k is zero when $k = 0$ or when $k > d$. We must find

$$B = \sum b_k = \sum_{k=0}^{d} e^{k\theta^*}\left(Q_{k+1} + Q_{k+2} + \cdots + Q_d\right).$$

If the right-hand side of this equation is multiplied by $e^{\theta^*} - 1$, the resulting expression is $\sum_{k=1}^{d} Q_k e^{k\theta^*} - (Q_1 + Q_2 + \cdots)$, which from equation (7.35) is $1 - (Q_1 + Q_2 + \cdots) = \bar{Q}$. Thus

$$B = \frac{\bar{Q}}{e^{\theta^*} - 1}.$$

The statement of Theorem 7.2 then implies that if

$$V = \lim_{y \to +\infty} V(y), \qquad (7.52)$$

then

$$V = \frac{\bar{Q}}{(e^{\theta^*} - 1)(\sum_{k=1}^{d} k Q_k e^{k\theta^*})}. \qquad (7.53)$$

7.6.4 Restricted Walks

We now consider what the unrestricted walk calculations of the previous section imply for a restricted random walk having a stopping boundary at the value -1. We again assume that the walk starts at 0. The calculations are best carried out using the complementary probabilities $F^*_{Y_{\mathrm{unr}}}(y) = 1 - F_{Y_{\mathrm{unr}}}(y)$ and $F^*_Y(y) = 1 - F_Y(y)$, the respective probabilities that the size of an excursion in the unrestricted and the restricted walks exceeds the value y. With these definitions, equation (7.49) implies that

$$F^*_{Y_{\mathrm{unr}}}(y) = V(y)e^{-y\theta^*}. \tag{7.54}$$

The definition of V in (7.52) then implies that

$$F_{Y_{\mathrm{unr}}} \sim Ve^{-y\theta^*}. \tag{7.55}$$

It follows that

$$\lim_{y \to +\infty} \left(F^*_{Y_{\mathrm{unr}}}(y) \right) e^{y\theta^*} = V. \tag{7.56}$$

The size of an excursion of the unrestricted walk can exceed the value y either before or after reaching negative values. In the latter case, the first negative value reached by the walk is one of $-1, -2, \ldots, -c$. If the probability that it is $-j$ is R_{-j}, then

$$F^*_{Y_{\mathrm{unr}}}(y) = F^*_Y(y) + \sum_{j=1}^{c} R_{-j} F^*_{Y_{\mathrm{unr}}}(y + j). \tag{7.57}$$

Multiplying by $e^{y\theta^*}$ throughout in this equation and then letting $y \to +\infty$, we get from (7.56)

$$V = \lim_{y \to +\infty} \left(F^*_Y(y) \right) e^{y\theta^*} + V \sum_{j=1}^{c} R_{-j} e^{-j\theta^*}. \tag{7.58}$$

From this,

$$\lim_{y \to +\infty} F^*_Y(y)e^{y\theta^*} = V \left(1 - \sum_{j=1}^{c} R_{-j} e^{-j\theta^*} \right). \tag{7.59}$$

The calculation (7.53) for V then shows that

$$\lim_{y \to +\infty} F^*_Y(y)e^{y\theta^*} = \frac{\bar{Q}(1 - \sum_{j=1}^{c} R_{-j} e^{-j\theta^*})}{(e^{\theta^*} - 1)(\sum_{k=1}^{d} kQ_k e^{k\theta^*})}. \tag{7.60}$$

This implies

$$\lim_{y \to +\infty} F^*_Y(y)e^{y\theta^*} = e^{-\theta^*} C, \tag{7.61}$$

where

$$C = \frac{\bar{Q}\left(1 - \sum_{j=1}^{c} R_{-j} e^{-j\theta^*}\right)}{(1 - e^{-\theta^*})\left(\sum_{k=1}^{d} kQ_k e^{k\theta^*}\right)}. \qquad (7.62)$$

Recalling that $F_Y^*(y) = \text{Prob}(Y \geq y + 1)$, this equation implies that

$$\text{Prob}(Y \geq y) \sim Ce^{-y\theta^*},$$

so that Y has a geometric-like distribution.

Two special cases deserve attention. First, if the only possible upward step in the walk is $+1$, then $Q_k = 0$ when $k \geq 2$. Equation (7.47) then shows that $Q_1 = e^{-\theta^*}$, so that $\sum_{k=1}^{d} kQ_k e^{k\theta^*} = 1$, and $\bar{Q} = 1 - e^{-\theta^*}$. In this case (7.62) shows that

$$C = 1 - \sum_{j=1}^{c} R_{-j} e^{-j\theta^*}. \qquad (7.63)$$

This is a generalization of equation (7.41). Second, if the only possible downward step in the walk is -1, then $R_{-1} = 1$ and $R_{-j} = 0$ when $j \geq 2$. In this case (7.62) shows that

$$C = \frac{\bar{Q}}{\sum_{k=1}^{d} kQ_k e^{k\theta^*}}. \qquad (7.64)$$

It is also interesting to consider a third, perhaps unexpected, special case. Exactly matching subsequences were considered in Section 5.4, and for these the geometric, rather than the geometric-like, distribution applies. Comparison of equations (1.63) and (1.67) shows that for this distribution, $C = 1$. Any exactly matching subsequence terminates at the first mismatch. In random walk terms, the possible steps sizes can be thought of as $+1$ and $-\infty$. (We could make a more precise analysis by using a step of $-L$, for L large, and then letting $L \to \infty$, but we prefer a more casual approach.) Thus we can put $R_{-\infty} = 1$ and $R_{-j} = 0$ for all finite j. Inserting these values into equation (7.63) we get $C = 1$, as required.

Karlin and Dembo (1992) give several alternative expressions for C. Of the various expressions they give, perhaps the most useful is, in our notation (which differs from theirs),

$$C = \frac{(1 - \sum_j R_{-j} e^{-j\theta^*})^2}{(1 - e^{-\theta^*})AE(Se^{\theta^* S})}, \qquad (7.65)$$

where A is given in equation (7.42) and S is the (random) size of any step in the walk. For the simple random walk of Section 7.2, $E(Se^{\theta^* S}) = q - p$, $A = 1/(q - p)$, and $R_{-1} = 1$. Inserting these values in (7.65) we obtain the value $1 - e^{-\theta^*}$ for C, which is identical to expression in (7.30).

While the formulae for C and A given above are in closed form, they still might not be useful for rapid calculation. Karlin and Dembo (1992) also give some rapidly converging series approximations for both these key parameters. Further simplifying calculations are discussed in Chapter 9.

Problems

7.1. Derive the result (7.37) by solving a cubic equation in e^{θ^*}. (Hint: It is known that one solution of the equation is $e^{\theta^*} = 1$, so that further solutions may be found by solving a quadratic equation.)

7.2. Consider the simple random walk with $h = 0$, $b = 1$ and $a = -L$, where L is positive. Use equation (7.9) to write down the probability that the walk eventually reaches 1 rather than $-L$. For the case $p < q$, show that the limiting value of this probability as $L \to \infty$ is $\frac{p}{q}$.

7.3. *Continuation.* The limiting $(L \to \infty)$ probability found in Problem 7.2 is the probability in an unrestricted random walk which starts at 0 and has $p < q$, that the walk ever reaches $+1$. Show this implies that in the unrestricted case, the probability that the walk ever reaches the value y is $(\frac{p}{q})^y$, for any positive integer y.

7.4. *Continuation.* For the case of the simple random walk with $p < q$, calculate the value of V (defined in equation (7.53)). From this, calculate the right-hand side in the asymptotic expression (7.55). Compare the value found with the value given in Problem 7.3. (Recall that $F_{Y_{\text{unr}}}^*(y)$ is the probability that the unrestricted walk ever reaches a height $y + 1$ or more.)

7.5. Show that, for the simple random walk with $p < q$, the two expressions (7.63) and (7.64) are identical.

8

Statistics (ii): Classical Estimation and Hypothesis Testing

8.1 Introduction

An introduction to classical estimation and hypothesis testing procedures was given in Chapter 3. However, the theoretical aspects of these procedures were not then discussed. In particular, optimality aspects of the two procedures were not addressed. We now give a brief introduction to both classical estimation theory and classical hypothesis testing theory, focusing on optimality aspects of these theories. Extensive treatments are given respectively in Lehmann (1991) and Lehmann (1986).

8.2 Classical Estimation Theory

8.2.1 Criteria for "Good" Estimators

In this section we discuss criteria for "good" estimators of a parameter.

The concepts of estimators and estimates of parameters were introduced in Section 3.3.1. Properties of estimates are found from the properties of the corresponding estimators, and all of the probability theory concerning estimation relates to the probabilistic properties of estimators.

It is not always clear what makes one estimator preferable to another, but there are certain properties of estimators that are often used to choose among various possibilities. It was stated in Section 3.3.1 that one property of an estimator usually taken to be desirable is that it be unbiased, that is, that its mean value be the parameter to be estimated. For example, as shown in equation (2.71), if $X_1, X_2, \ldots, X_n$ are iid, each having a probability distribution with mean μ, then the *estimator* $\bar{X}$ is a random variable also having mean μ, so that $\bar{X}$ is an unbiased estimator of μ.

It is not always possible to find unbiased estimators for a parameter, and even if it is possible, unbiased estimators are not always preferred. This is discussed further below.

A second desirable property is that an estimator be *consistent*. If ξ is the parameter being estimated and $\hat{\xi}_n$ an estimator of ξ derived from n observations, then a consistent estimator is one for which, for any small ϵ, $\mathrm{Prob}(|\hat{\xi}_n - \xi| > \epsilon) \to 0$ as $n \to +\infty$. Application of Chebyshev's inequality and the variance formula (2.71) shows that the sample average of iid random variables is a consistent estimator of the mean of a distribution with finite variance. As the sample size increases, the estimate derived from a consistent estimator is more and more likely to be close to the parameter being estimated.

An estimator can be both unbiased and consistent, one and not the other, or neither (see Problem 8.1), so the concepts of unbiasedness and consistency concern different aspects of the properties of estimators.

A further desirable property of an unbiased estimator is that it have a low variance, ideally lower than that of any other unbiased estimator of that parameter. In some cases such an estimator exists and can be identified, and if so, it might reasonably be thought to be the "best" estimator of that parameter.

For biased estimators a natural desirable property is that it have low mean square error (see (3.14)). In some cases the mean square error of a biased estimator is less than the mean square error, that is, the variance, of an unbiased estimator: See, for example, Problem 8.6. In such cases the biased estimator might be preferred to the unbiased estimator.

Finally, it would also be desirable if an estimator had, at least approximately, a well-known distribution, for example a normal distribution. In this case known theory and results for this distribution can be applied when using this estimator.

Perhaps unexpectedly, there is one estimation procedure that in many practical cases achieves most of, and in some cases all of, these aims, at least in an asymptotic sense for large sample sizes. This is the procedure of *maximum likelihood estimation*.

8.2.2   Maximum Likelihood Estimation

The discrete case

Suppose that $Y_1, Y_2, \ldots, Y_n$ are iid discrete random variables[1] with joint probability distribution $P(Y_1; \xi) \times P(Y_2; \xi) \times \cdots \times P(Y_n; \xi)$. Here ξ is an unknown parameter that is to be estimated, and the notation indicates the dependence of the probability distribution of Y_i on ξ. We denote $(Y_1, Y_2, \ldots, Y_n)$ by $\boldsymbol{Y}$ and define the *likelihood* $L(\xi, \boldsymbol{Y})$ by

$$L(\xi, \boldsymbol{Y}) = P(Y_1; \xi) \times P(Y_2; \xi) \times \cdots \times P(Y_n; \xi). \tag{8.1}$$

While this is identical to the joint probability distribution of $\boldsymbol{Y}$, we now regard it, for a given $\boldsymbol{Y}$, as a function of ξ. The value $\hat{\xi} = \hat{\xi}(Y_1, Y_2, \ldots, Y_n)$ of ξ at which $L(\xi, \boldsymbol{Y})$ reaches a maximum (as a function of ξ) is called the maximum likelihood *estimator* of ξ. This value is a function of $Y_1, Y_2, \ldots, Y_n$. Examples are given below.

This estimator is often found by differentiating $L(\xi, \boldsymbol{Y})$ with respect to ξ and using standard derivative tests for maxima. Care must be used with this procedure, since local maxima may arise, and, as Example 5 below shows, the maximum might be reached at a boundary point.

In those cases where the maximum of $L(\xi, \boldsymbol{Y})$ is found through the differentiation procedure, it is necessary to solve the equation (in ξ)

$$\frac{d}{d\xi} L(\xi, \boldsymbol{Y}) = 0.$$

In practice it is equivalent, and often easier, to solve the equation

$$\frac{d}{d\xi} \log L(\xi, \boldsymbol{Y}) = 0. \tag{8.2}$$

The maximum likelihood *estimate* of ξ is the observed value of the maximum likelihood estimator, once the data have been obtained. That is, this estimate is found by replacing $\boldsymbol{Y} = (Y_1, Y_2, \ldots, Y_n)$ in the above by $\boldsymbol{y} = (y_1, y_2, \ldots, y_n)$ and $L(\xi, \boldsymbol{Y})$ by $L(\xi, \boldsymbol{y})$.

For a given $\boldsymbol{y}$, the value of ξ which maximizes the probability of $\boldsymbol{y}$ is the maximum likelihood estimate of ξ, and in this sense the maximum likelihood procedure is a natural one. We show later that this procedure also has certain optimality properties which justify its use.

---

[1] Maximum likelihood estimators can be defined also for dependent random variables, but we assume the iid case throughout.

Example 1. Suppose that $Y_1, Y_2, \ldots, Y_n$ are iid random variables, each having a Poisson distribution with parameter λ. Then from (2.4),

$$L(\lambda, \boldsymbol{Y}) = \frac{e^{-\lambda}\lambda^{Y_1}}{Y_1!} \times \frac{e^{-\lambda}\lambda^{Y_2}}{Y_2!} \times \cdots \times \frac{e^{-\lambda}\lambda^{Y_n}}{Y_n!}$$
$$= \frac{e^{-n\lambda}\lambda^{\sum Y_i}}{\prod(Y_i!)},$$

all sums and products in this example being over $i = 1, 2, \ldots, n$. Thus

$$\log L(\lambda, \boldsymbol{Y}) = -n\lambda + \left(\sum Y_i\right)\log \lambda - \log\left(\prod(Y_i!)\right)$$

and

$$\frac{d}{d\lambda}\log L(\lambda, \boldsymbol{Y}) = -n + \frac{\sum Y_i}{\lambda}. \tag{8.3}$$

This derivative is zero when $\hat{\lambda} = \bar{Y}$, and it can be checked that this corresponds to a maximum of $L(\lambda, \boldsymbol{Y})$. Thus

$$\text{maximum likelihood estimator of } \lambda = \frac{\sum Y_i}{n} = \bar{Y}. \tag{8.4}$$

From this, the maximum likelihood estimate of λ is $\bar{y}$, the observed average once the data are obtained. Standard properties of the Poisson distribution show that $\bar{Y}$ is an unbiased consistent estimator of λ and has variance λ/n.

Example 2. Consider n independent Bernoulli trials, each of which results in either success (with probability p) or failure (with probability $1 - p$). If $Y_i = 1$ if trial i results in success and $Y_i = 0$ if trial i results in failure, the likelihood (8.1) is, from (1.6),

$$L(p, \boldsymbol{Y}) = \prod_{i=1}^{n} p^{Y_i}(1-p)^{1-Y_i} = p^Y(1-p)^{n-Y}, \tag{8.5}$$

where $Y = \sum_{i=1}^{n} Y_i$ is the total number of successes. Finding the maximum likelihood estimator even in this comparatively simple case is not perhaps as straightforward as might be expected. When $1 \leq Y \leq n-1$ the solution of equation (8.2), with p replacing ξ, is $\hat{p} = Y/n = \bar{Y}$, and this may be shown to be the maximum likelihood estimator of p for these cases. However, the cases $Y = 0$, $Y = n$ must be handled differently, and we illustrate this in the case $Y = n$. Here the likelihood is p^n, and equation (8.2) gives $n/\hat{p} = 0$. There is no real solution of this equation, and this indicates that the maximum is found at a boundary point. Subject to the natural constraint $0 \leq p \leq 1$, the likelihood p^n is maximized, as a function of p, at $p = 1$, and this shows that the maximum likelihood estimator of p when $Y = n$ is $\hat{p} = 1$. This is, however, identical to $\bar{Y}$ when $Y = n$. An analogous remark applies when $Y = 0$; here again the maximum likelihood estimator may be identified with $\bar{Y}$.

The estimator $\bar{Y}$ thus applies in all cases. This estimator has mean p and is thus an unbiased estimator of p. It is also consistent, and has variance $p(1 - p)/n$. The corresponding estimate is y/n, where y is the observed number of successes.

These results generalize naturally to the estimation of the parameters in the multinomial distribution.

The Continuous Case

In the continuous case the likelihood $L(\xi, \boldsymbol{X})$ is defined by

$$L(\xi, \boldsymbol{X}) = f_X(x_1; \xi) f_X(x_2; \xi) \cdots f_X(x_n; \xi), \qquad (8.6)$$

where $f_X(x; \xi)$ is the common density function of the iid continuous random variables $X_1, X_2, \ldots, X_n$, again assumed to depend on some unknown parameter ξ. Maximum likelihood estimators and estimates are found by following a procedure essentially identical to that in the discrete case.

Example 3. Suppose that $X_1, X_2, \ldots, X_n$ are NID(μ, σ^2) random variables and that σ^2 is known. The aim is to find the maximum likelihood estimator of μ. If $(X_1, X_2, \ldots, X_n)$ is denoted by $\boldsymbol{X}$, the likelihood $L(\xi, \boldsymbol{X})$ is

$$L(\xi, \boldsymbol{X}) = \prod_{i=1}^{n} \frac{1}{\sqrt{2\pi}\sigma} e^{-\frac{(X_i - \mu)^2}{2\sigma^2}}. \qquad (8.7)$$

The logarithm of $L(\xi, \boldsymbol{X})$ is

$$\text{constant} - \sum_{i=1}^{n} \frac{(X_i - \mu)^2}{2\sigma^2}, \qquad (8.8)$$

where the constant is independent of μ. Differentiation with respect to μ leads to

$$\text{maximum likelihood estimator of } \mu = \hat{\mu} = \bar{X}. \qquad (8.9)$$

This estimator has mean μ and is thus unbiased. It is also consistent, has variance σ^2/n, and has a normal distribution for all values of n.

Example 4. A more realistic case arises when both the mean and the variance of a normal distribution are to be estimated. There are now two unknown parameters, μ and σ^2, and the estimation procedure follows lines generalizing those described above to many unknown parameters. This procedure leads us to solve the simultaneous equations

$$\frac{\partial \log L(\mu, \sigma^2, \boldsymbol{X})}{\partial \mu} = 0, \quad \frac{\partial \log L(\mu, \sigma^2, \boldsymbol{X})}{\partial (\sigma^2)} = 0,$$

the solutions being $\hat{\mu}$ and $\hat{\sigma}^2$. This gives

$$-\sum_{i=1}^{n} \frac{(X_i - \hat{\mu})}{\hat{\sigma}^2} = 0, \qquad (8.10)$$

$$-\frac{n}{2\hat{\sigma}^2} + \sum_{i=1}^{n} \frac{(X_i - \hat{\mu})^2}{\hat{\sigma}^4} = 0, \qquad (8.11)$$

from which

$$\hat{\mu} = \bar{X}, \quad \hat{\sigma}^2 = \sum_{i=1}^{n} \frac{(X_i - \bar{X})^2}{n}. \qquad (8.12)$$

The former estimator agrees with that in (8.9) and has the same properties as that estimator. The estimator of σ^2 is biased, having mean $(n-1)\sigma^2/n$. It can, nevertheless, be shown to be a consistent estimator of σ^2.

Example 5. Suppose that $X_1, X_2, \ldots, X_n$ are independent random variables, each coming from the uniform distribution

$$f_X(x) = \frac{1}{M}, \quad 0 \le x \le M. \qquad (8.13)$$

The aim is to find the maximum likelihood estimator of M. In this case the likelihood is M^{-n}. The differentiation process associated with the maximization procedure shows that the maximum of the likelihood occurs at a boundary point. Inspection of the likelihood function M^{-n} shows that it is maximized by making M as small as possible, while still being consistent with $X_1, X_2, \ldots, X_n$. Since each X_i must be less than or equal to M, the smallest value $\hat{M}$ that can be chosen for the estimator of M is

$$\hat{M} = X_{\max}. \qquad (8.14)$$

$\hat{M}$ is not an unbiased estimator of M, as is shown (with an appropriate change of notation) by equation (2.94). An unbiased estimator of M is $U = (n+1)X_{\max}/n$, and often some simple adjustment such as this is all that is needed to find an unbiased estimator from a maximum likelihood estimator. Equation (1.116) and the density function for $X_{\max}$ given in equation (2.148) with $i = n$, yields the density function of U as

$$f_U(u) = \frac{n^{n+1}u^{n-1}}{(n+1)^n M^n}, \quad 0 \le u \le \frac{(n+1)M}{n}. \qquad (8.15)$$

Invariance property. A further property of maximum likelihood estimators is that if $\hat{\xi}$ is the maximum likelihood estimator of ξ, and if $g(\xi)$ is a monotonic function of ξ, then the maximum likelihood estimator of $g(\xi)$ is $g(\hat{\xi})$. When $g(\xi)$ is a nonlinear function of ξ, then even if $\hat{\xi}$ happens to be an unbiased estimator of ξ, the maximum likelihood estimator $g(\hat{\xi})$ of $g(\xi)$ is usually biased. However, this bias is normally of order of n^{-1}, where n is the sample size, and as above a simple correction often allows an unbiased estimator of $g(\xi)$ to be found.

Asymptotic properties. Suppose we have an infinite sequence $X_1, X_2, \ldots$ of iid random variables whose distribution depends on a parameter ξ. We consider properties of the maximum likelihood estimator of ξ as n increases. Let $\hat{\xi}_n$ be the maximum likelihood estimator of ξ based on $X_1, X_2, \ldots, X_n$. It can be shown under certain regularity conditions that for large n the maximum likelihood estimator of a parameter is approximately normally distributed, that

$$E(\hat{\xi}_n) \sim \xi, \tag{8.16}$$

and that

$$\mathrm{Var}(\hat{\xi}_n) \sim \frac{-1}{E\left(\frac{d^2}{d\xi^2} \log L\right)}. \tag{8.17}$$

Here L is given by (8.1) for discrete random variables and by (8.6) for continuous random variables.

One reason for the importance of (8.17) is that the right-hand side in this equation is the well-known "Cramér–Rao bound": under certain regularity conditions, no unbiased estimator of the parameter ξ based on an iid sample of size n can have a variance smaller than this. This indicates an asymptotic optimality property of maximum likelihood estimators. For this and other reasons maximum likelihood estimators are used frequently, especially with large samples, and this is the method of estimation of parameters employed often in bioinformatics.

The denominator of the right-hand expression in (8.17) is always negative, so the expression itself is always positive, as is appropriate for a variance. This is seen in the following example of an application of the Cramér–Rao bound.

In Example 1 above, it was shown that, given iid random variables from a Poisson distribution with parameter λ, the maximum likelihood estimator of λ is $\bar{Y}$, and that this estimator is unbiased and has variance $\frac{\lambda}{n}$. From equation (8.3),

$$\frac{d^2}{d\lambda^2} \log L(\lambda, \boldsymbol{Y}) = -\frac{\sum Y_i}{\lambda^2},$$

and thus, since $E(\sum Y_i) = n\lambda$,

$$E\left(\frac{d^2}{d\lambda^2} \log L(\lambda, \boldsymbol{Y})\right) = -\frac{n}{\lambda}.$$

Using this value as the denominator of the right-hand side in (8.17), we find that no unbiased estimator of λ can have a variance less than $\frac{\lambda}{n}$. But we have just seen that this is the variance of the unbiased estimator $\bar{Y}$, so that $\bar{Y}$ is the minimum variance unbiased estimator of λ. The comments in Section 8.2.1 concerning criteria for "good" estimators then suggest that $\bar{Y}$ is a most desirable estimator of λ.

The regularity conditions needed for (8.16) and (8.17) to hold include the requirements that the maximum of the likelihood occurs at a point where the derivative of the likelihood is zero rather than a boundary point, that the range of the random variables not depend on the parameter being estimated, and that the parameter being estimated take values in a continuous interval of real numbers, rather than, for example, only taking integer values. The first two requirements do not hold in Example 5 above. Other examples of this are discussed in Chapter 14. To show why these regularity conditions are needed we provide a sketch of the derivation of (8.17). The proof is given for continuous random variables; the proof for discrete random variables is essentially identical.

Let X be a continuous random variable having density function $f_X(x;\xi)$ and let $X_1, X_2, \ldots, X_n$ be n iid continuous random variables each having the same density function as X. The density function $f_X(x;\xi)$ satisfies

$$\int_R f_X(x;\xi)dx = 1, \tag{8.18}$$

where R is the range of X. Differentiation throughout in equation (8.18) with respect to ξ leads to

$$\int_R \frac{df_X(x;\xi)}{d\xi} dx = 0, \tag{8.19}$$

where it is assumed that $f_X(x;\xi)$ is such that differentiation of the left-hand side in (8.18) can be performed by differentiation under the integral sign. If the range R depends on ξ, this condition usually does not hold. If $\ell(x;\xi)$ is defined by

$$\ell(x;\xi) = \frac{df_X(x;\xi)}{d\xi} \cdot \frac{1}{f_X(x;\xi)}, \tag{8.20}$$

equation (8.19) can be written

$$\int_R \ell(x;\xi)f_X(x;\xi)dx = E\left(\ell(x;\xi)\right) = 0. \tag{8.21}$$

Under the same assumptions as above, a further differentiation with respect to ξ yields

$$\int_R \frac{d\ell(x;\xi)}{d\xi}f_X(x;\xi)dx + \int_R \ell(x;\xi)\frac{df_X(x;\xi)}{d\xi}dx = 0,$$

or, from (8.20),

$$\int_R \frac{d\ell(x;\xi)}{d\xi}f_X(x;\xi)dx + \int_R \left(\ell(x;\xi)\right)^2 f_X(x;\xi)dx = 0.$$

This implies that

$$E\left(\frac{d\ell(X;\xi)}{d\xi}\right) = -E\left((\ell(X;\xi))^2\right).\qquad(8.22)$$

The left-hand side, namely the mean value of $d\ell(X;\xi)/d\xi$, will be denoted by $\mu(\xi)$. An extension of the argument leading to (8.22) shows that

$$E(U) = -E(V),\qquad(8.23)$$

where

$$U = \frac{d^2\log L(\xi,\boldsymbol{X})}{d\xi^2},\quad V = \left(\frac{d\log L(\xi,\boldsymbol{X})}{d\xi}\right)^2,\qquad(8.24)$$

where $L(\xi,\boldsymbol{X})$ is defined in (8.6). Consider now the random variable

$$n^{-1}\frac{d}{d\xi}\left(\sum_{i=1}^{n}\ell(X_i;\xi)\right),$$

which from (8.6) and the definition of $\ell(X_i;\xi)$ is identical to $n^{-1}U$. This average has mean value $\mu(\xi)$, and from Chebyshev's inequality and equation (2.71) has a probability distribution closely concentrated around $\mu(\xi)$ when n is large. We will therefore use the approximations

$$U/n \approx \mu(\xi),\quad V/n \approx -\mu(\xi),\qquad(8.25)$$

the latter approximation following from (8.23).

Denote the maximum likelihood estimator of ξ by $\hat{\xi}$. Then the Taylor series approximation (B.29) gives

$$\left(\frac{d\log L(\xi,\boldsymbol{X})}{d\xi}\right)_{\xi=\hat{\xi}} \approx \frac{d\log L(\xi,\boldsymbol{X})}{d\xi} + (\hat{\xi}-\xi)U.\qquad(8.26)$$

We assume that the maximum likelihood estimator occurs at a point where the derivative of the likelihood is zero, so that the left-hand side in this equation is 0. Thus to a first order of approximation,

$$\hat{\xi}-\xi \approx -\frac{d\log L(\xi,\boldsymbol{X})/d\xi}{U}.\qquad(8.27)$$

Squaring both sides, and using the definition of V in (8.24),

$$(\hat{\xi}-\xi)^2 \approx \frac{V}{U^2}.\qquad(8.28)$$

The expected value of the left-hand side is approximately the variance of $\hat{\xi}$ (and is equal to it if $E(\hat{\xi}) = \xi$). The approximations (8.25) show that the right-hand side is approximately $-\frac{1}{n\mu(\xi)}$. Thus

$$\text{variance of } \hat{\xi} \approx \frac{-1}{n\mu(\xi)}.\qquad(8.29)$$

This is equation (8.17).

The above derivation of the variance of the asymptotic maximum likelihood estimate is a sketch only and can be made more rigorous by paying careful attention to the approximations made. It is given here because it emphasizes the main assumptions made in claiming that equation (8.17) applies. These are (i) that the maximum of the likelihood occurs at a point where the derivative of the likelihood is zero, (ii) that the range of the observations is independent of the parameter being estimated, (iii) that the observations are iid, (iv) that the sample size is large, and (v) since the Taylor series approximation (8.26) is used, that the parameter and its maximum likelihood estimate are both real numbers taking possible values in a continuous interval. In Section 14.7 we discuss the estimation of the topology of a phylogenetic tree by maximum likelihood methods. The theory discussed above shows that one may not automatically assume optimality properties for the maximum likelihood estimator of this topology, since this topology is not a real number taking values in some interval.

The theory above can be generalized to cover the case of several parameters. If the (i, j) element in the *information matrix* I is defined by

$$I_{ij} = - E \left(\frac{d^2}{d\xi_i d\xi_j} \log L \right), \qquad (8.30)$$

then the asymptotic variance of the maximum likelihood estimator of the parameter ξ_i is the (i, i) term in I^{-1}, and the asymptotic covariance of the maximum likelihood estimators of the parameters ξ_i and ξ_j is the (i, j) term in I^{-1}.

8.3   Classical Hypothesis Testing: Simple Fixed Sample Size Tests

8.3.1   The Likelihood Ratio

The five steps of classical hypothesis-testing procedures were outlined in Chapter 3. In this chapter we expand on the procedure of Step 3, the choice of test statistic, and the procedure of Step 4, the step determining when the observed value of the test statistic leads to rejection of the null hypothesis.

We consider first the case where both null and alternative hypotheses are *simple*, that is, that both completely specify the probability distribution of the random variable of interest. While the numerical value α of the Type I error is chosen (in Step 2 of the hypothesis-testing procedure), no specific choice of the numerical value β of the Type II error is made. However, it is desirable to have a procedure that, with the Type I error fixed, minimizes the numerical value β of the Type II error, or equivalently that maximizes the power of the test, that is, the probability $1 - \beta$ of rejecting the null hypothesis when the alternative hypothesis is true.

The methods used to do this are identical for both discrete and continuous random variables, so we describe in detail only the continuous random variable case. Consider iid random variables $X_1, X_2, \ldots, X_n$ each having the same distribution as a random variable X with density function $f_X(x)$. Assume further that the null hypothesis specifies some density function $f_0(x)$ for $f_X(x)$ and that the alternative hypothesis specifies some other density function $f_1(x)$ for $f_X(x)$. The decision about accepting H_0 or H_1 will be based on the observed values $x_1, x_2, \ldots, x_n$ of $X_1, X_2, \ldots, X_n$.

The Neyman–Pearson Lemma

The important Neyman–Pearson lemma states that the most powerful test of H_0 against H_1 (that is, the test that for a given fixed Type I error α maximizes the power $1 - \beta$ of the test) is obtained by using as test statistic the likelihood ratio, or LR, defined as

$$\text{LR} = \frac{f_1(X_1)f_1(X_2)\cdots f_1(X_n)}{f_0(X_1)f_0(X_2)\cdots f_0(X_n)}. \tag{8.31}$$

The null hypothesis is rejected when the observed value of LR is greater than or equal to K, where K is chosen such that

$$\text{Prob(LR} \geq K \text{ when } H_0 \text{ is true)} = \alpha. \tag{8.32}$$

The fact that using LR as test statistic leads to the most powerful test is an important result, and we thus prove it here.

Let A be the region in the joint range of $X_1, X_2, \ldots, X_n$ such that $\text{LR} \geq K$, where K is defined by (8.32). Then

$$\int \cdots \int_A f_0(u_1)f_0(u_2)\cdots f_0(u_n)du_1 du_2 \cdots du_n = \alpha. \tag{8.33}$$

Let B be any other region in the joint range of $X_1, X_2, \ldots, X_n$ such that

$$\int \cdots \int_B f_0(u_1)f_0(u_2)\cdots f_0(u_n)du_1 du_2 \cdots du_n = \alpha. \tag{8.34}$$

Rejection of H_0 when the observed vector $(x_1, x_2, \ldots, x_n)$ is in B thus also leads to a test of the required Type I error. Let A and B overlap in the region C. Define Δ by

$$\Delta = \text{power of the test using } A - \text{power of the test using } B.$$

Then

$$\Delta = \int \cdots \int_A f_1(u_1) \cdots f_1(u_n) du_1 \cdots du_n$$

$$- \int \cdots \int_B f_1(u_1) \cdots f_1(u_n) du_1 \cdots du_n \tag{8.35}$$

$$= \int \cdots \int_{A \backslash C} f_1(u_1) \cdots f_1(u_n) du_1 \cdots du_n$$

$$- \int \cdots \int_{B \backslash C} f_1(u_1) \cdots f_1(u_n) du_1 \cdots du_n, \tag{8.36}$$

where $A \setminus C$ denotes the set of points in A but not in C. Now, in $A \setminus C$,

$$f_1(u_1) \cdots f_1(u_n) \geq K f_0(u_1) \cdots f_0(u_n),$$

while in $B \setminus C$,

$$f_1(u_1) \cdots f_1(u_n) \leq K f_0(u_1) \cdots f_0(u_n).$$

Thus

$$\Delta \geq \int \cdots \int_{A \backslash C} K f_0(u_1) \cdots f_0(u_n) du_1 \cdots du_n$$

$$- \int \cdots \int_{B \backslash C} K f_0(u_1) \cdots f_0(u_n) du_1 \cdots du_n \tag{8.37}$$

$$= \int \cdots \int_A K f_0(u_1) \cdots f_0(u_n) du_1 \cdots du_n$$

$$- \int \cdots \int_B K f_0(u_1) \cdots f_0(u_n) du_1 \cdots du_n \tag{8.38}$$

$$= K\alpha - K\alpha = 0, \tag{8.39}$$

so that the power of the test using A is greater than or equal to the power using B. This completes the proof of the claim that the most powerful test of H_0 against H_1 is provided by using the likelihood ratio as test statistic and rejecting H_0 when the observed value of this likelihood ratio is "too large."

The likelihood ratio procedure has thus not only indicated the appropriate test statistic (Step 3 of the hypothesis testing procedure) but has also indicated which observed values of this test statistic lead to rejection of the null hypothesis (Step 4).

In many cases in practice this procedure reduces to something simple and straightforward, as the following example shows.

Example. Sequence matching. In the sequence-matching example discussed in Section 3.4.1, the simple null hypothesis that the probability of a match is 0.25 was tested against the simple alternative hypothesis that this probability is 0.35. Suppose more generally that under the null hypothesis the probability of a match is p_0 and under the alternative hypothesis this probability is p_1, with $p_1 > p_0$. Then from (8.5) and (8.31) the likelihood ratio LR is

$$\frac{(p_1)^{\sum_{i=1}^{n} Y_i}(1-p_1)^{(n-\sum_{i=1}^{n} Y_i)}}{(p_0)^{\sum_{i=1}^{n} Y_i}(1-p_0)^{(n-\sum_{i=1}^{n} Y_i)}} \tag{8.40}$$

$$= \left(\frac{p_1(1-p_0)}{p_0(1-p_1)}\right)^{\sum_{i=1}^{n} Y_i} \left(\frac{1-p_1}{1-p_0}\right)^{n}. \tag{8.41}$$

The null hypothesis is rejected when the likelihood ratio is sufficiently large, and this happens when the number of matches $Y = \sum_{i=1}^{n} Y_i$ is sufficiently large. That is, it is equivalent, and more convenient, to take Y as the test statistic, and the null hypothesis is rejected if $y \geq K$, where y is the observed value of Y and K is determined by the requirement

$$\text{Prob}(Y \geq K \,|\, p = p_0) = \alpha, \tag{8.42}$$

where α is the Type I error chosen. This justifies the use of Y as test statistic in Section 3.4.1. An example of the calculation of K, together with a discussion of the fact that (8.42) cannot usually be solved exactly for K, is given in Section 3.4.1.

Provided only that $p_1 > p_0$, the calculation of the significance point C is not influenced by the numerical value of p_1. Thus the testing procedure in this example is the most powerful one *whatever* the alternative hypothesis might be, provided only that it defines a value of p exceeding p_0. Thus this testing procedure is *uniformly most powerful* for the (more realistic) composite alternative hypothesis $p = \text{Prob(match)} > p_0$. The theory of composite tests considered below leads to the same conclusion.

8.4 Classical Hypothesis Testing: Composite Fixed Sample Size Tests

8.4.1 Introduction

Tests of a simple null hypothesis against a composite alternative and tests of a composite null hypothesis against a composite alternative occur very often in practice. The tests that we consider are tests about the numerical values of parameters. Whereas a simple hypothesis specifies all the numerical values of all parameters involved in the distribution of the random variable of interest, a composite hypothesis does not completely specify the value of one or more of these parameters. The theory of tests involving composite hypotheses is covered in advanced statistics texts, and we discuss here only those aspects of these tests relevant to the material of this book.

The set of values that the parameters of interest can take is called the parameter space. For example, in the normal distribution there are two parameters, the mean μ and the variance σ^2, and the parameter space is $\{-\infty < \mu < +\infty,\ 0 < \sigma^2 < +\infty\}$. The null hypothesis states that the parameters in the distribution of interest take values in some region ω of the parameter space, while the alternative hypothesis states that they take values in some region Ω of the parameter space. In the normal distribution, for example, the null hypothesis might specify that $\mu = \mu_0$, where μ_0 is a specified numerical value, and make no specification about σ^2, while the alternative hypothesis might make no specification about either μ or σ^2. Here the region ω is the line $\mu = \mu_0$ in the parameter space, while Ω is the entire parameter space.

In practice, ω is usually a subspace of Ω, as in the example above, and we restrict attention to this, the *nested hypotheses*, case. When the null hypothesis is nested within the alternative hypothesis, the aim of the hypothesis-testing procedure is somewhat different from the aim in the unnested case. In the nested case we are interested in whether the more general hypothesis explains the data *significantly* better than the narrower null hypothesis. If the null hypothesis is not rejected, we might decide that the parameter values can reasonably be assumed to be those specified by the null hypothesis.

By analogy with the likelihood ratio test statistic for tests of a simple null hypothesis against a simple alternative hypothesis, the test statistic in the composite hypothesis case is the likelihood ratio λ, defined by

$$\lambda = \frac{L_{\max}(\omega)}{L_{\max}(\Omega)}, \tag{8.43}$$

where $L_{\max}(\omega)$ is the maximum of the likelihood of the observations when the parameters are confined to the region ω of the parameter space defined by the null hypothesis and $L_{\max}(\Omega)$ is the maximum of the likelihood of the observations when the parameters are confined to the region Ω defined by the alternative hypothesis. The numerator on the right-hand side in (8.43) relates to the null hypothesis and the denominator to the alternative hypothesis, in contrast to the ratio (8.31) used for simple hypotheses. This makes no difference to the testing procedure, since the null hypothesis is rejected when λ is sufficiently small (as compared to rejecting the null hypothesis for simple tests when LR is sufficiently large). The concept of being "sufficiently small" is determined by reference to the Type I error chosen for the test. In the example below the choice of a Type I error leads to an explicit determination of how small λ must be for the null hypothesis to be rejected. In more complicated cases approximate procedures are needed, as discussed in Section 8.4.2.

In practice, λ is often a monotonic function of some standard test statistic, and when this is so the test is carried out in terms of that statistic.

Example 1. In this and the following example we discuss the "equal variance" two-sample t-test. In this example we consider the *two-sided* t-test and in the following example the *one-sided* t-test.

Suppose that $X_{11}, X_{12}, \ldots, X_{1m}$ are NID(μ_1, σ^2) and $X_{21}, X_{22}, \ldots, X_{2n}$ are NID(μ_2, σ^2), and that X_{1i} is independent of X_{2j} for all i and j. We wish to test the null hypothesis $\mu_1 = \mu_2$ $(= \mu,$ unspecified) against the alternative hypothesis that leaves the values of both μ_1 and μ_2 unspecified. The value of σ^2 is unknown and is unspecified under both hypotheses. Thus the region ω is $\{\mu_1 = \mu_2, \, 0 < \sigma^2 < +\infty\}$, and the region Ω is

$$\{-\infty < \mu_0 < +\infty, -\infty < \mu_1 < +\infty, 0 < \sigma^2 < +\infty\}.$$

The likelihood of $(X_{11}, X_{12}, \ldots, X_{1m}, X_{21}, X_{22}, \ldots, X_{2n})$ is, from (2.14),

$$\prod_{i=1}^{m} \frac{1}{\sqrt{2\pi}\sigma} e^{-(x_{1i}-\mu_1)^2/(2\sigma^2)} \prod_{i=1}^{n} \frac{1}{\sqrt{2\pi}\sigma} e^{-(x_{2i}-\mu_2)^2/(2\sigma^2)}. \tag{8.44}$$

Under the null hypothesis the two sets of random variables have the same distribution. Writing μ for the mean of this (common) distribution, the maximum of (8.44) occurs when

$$\hat{\mu} = \bar{X} = \frac{X_{11} + X_{12} + \cdots + X_{1m} + X_{21} + X_{22} + \cdots + X_{2n}}{m+n},$$

and the (common) variance σ^2 is

$$\hat{\sigma}_0^2 = \frac{\sum_{i=1}^{m}(X_{1i} - \bar{X})^2 + \sum_{i=1}^{n}(X_{2i} - \bar{X})^2}{m+n}.$$

Inserting $\hat{\mu}$ for both μ_1 and μ_2, and $\hat{\sigma}_0^2$ for σ^2 in (8.44), we get

$$L_{\max}(\omega) = \frac{1}{(2\pi\hat{\sigma}_0^2)^{(m+n)/2}} e^{-(m+n)/2}. \tag{8.45}$$

Under the alternative hypothesis the maximum of (8.44) occurs when $\hat{\mu}_1$ is $\bar{X}_1 = (X_{11} + X_{12} + \cdots + X_{1m})/m$, $\hat{\mu}_2$ is $\bar{X}_2 = (X_{21} + X_{22} + \cdots + X_{2n})/n$, and σ^2 is

$$\hat{\sigma}_1^2 = \frac{\sum_{i=1}^{m}(X_{1i} - \bar{X}_1)^2 + \sum_{i=1}^{n}(X_{2i} - \bar{X}_2)^2}{m+n}.$$

Inserting $\hat{\mu}_1$ for μ_1, $\hat{\mu}_2$ for μ_2 and $\hat{\sigma}_1^2$ for σ^2 in (8.44), we get

$$L_{\max}(\Omega) = \frac{1}{(2\pi\hat{\sigma}_1^2)^{(m+n)/2}} e^{-(m+n)/2}. \tag{8.46}$$

Thus the test statistic λ reduces to

$$\lambda = \frac{L_{\max}(\omega)}{L_{\max}(\Omega)} = \left(\frac{\hat{\sigma}_1^2}{\hat{\sigma}_0^2}\right)^{(n+m)/2} \tag{8.47}$$

The algebraic identity

$$\sum_{i=1}^{m}(X_{1i} - \bar{X})^2 + \sum_{i=1}^{n}(X_{2i} - \bar{X})^2$$

$$= \sum_{i=1}^{m}(X_{1i} - \bar{X}_1)^2 + \sum_{i=1}^{n}(X_{2i} - \bar{X}_2)^2 + \frac{mn}{m+n}(\bar{X}_1 - \bar{X}_2)^2$$

may be used to show that

$$\lambda = \left(\frac{1}{1 + t^2/(m+n-2)}\right)^{(m+n)/2}, \tag{8.48}$$

where t is defined in (3.19). It is equivalent to use t^2 or $|t|$, instead of λ, as test statistic. Small values of λ correspond to large values of $|t|$, so that sufficiently large values of $|t|$ lead to rejection of the null hypothesis. The null hypothesis distribution of the statistic t is known (as the t distribution with $m + n - 2$ degrees of freedom) and significance points are widely available. Thus once a Type I error has been chosen, values of $|t|$ that are sufficiently large to lead to rejection of the null hypothesis can be determined.

Example 2. The theory for one-sided t-tests is slightly different from the theory developed in Example 1 for two-sided tests. Instead of an alternative hypothesis that leaves the values of both μ_1 and μ_2 unspecified, the alternative hypothesis in a one-sided test claims that one mean is greater than or equal to the other, for example that $\mu_1 \geq \mu_2$. In this case the region ω is $\{\mu_1 \geq \mu_2, 0 < \sigma^2 < +\infty\}$. If $\bar{X}_1 \geq \bar{X}_2$, all maximum likelihood estimators are as in Example 1 and the t statistic of that example is recovered as the test statistic. If $\bar{X}_1 < \bar{X}_2$, the maximum likelihood estimators of μ_1 and μ_2 occur at a boundary point of ω where both estimators are equal to the overall average $\bar{X}$. In this case the alternative hypothesis does not explain the observations better than the null hypothesis, the likelihood ratio is 1, and the null hypothesis is not rejected. The outcome of these observations is that the null hypothesis is rejected in favor of the alternative hypothesis $\mu_1 \geq \mu_2$ only for sufficiently large *positive* values of t.

The simplicity of these examples hides several important potential problems. First, the ratio λ might not reduce to a function of a well-known test statistic such as t. More important, there might not even be a unique null hypothesis distribution of the statistic found by the λ ratio procedure. In the case of the above t-test, the null hypothesis distribution of the

t statistic is independent of the unknown variance and also of the mean value unspecified by the null hypothesis. That is, this distribution is independent of the parameters whose values are not prescribed by the null hypothesis. It is fortunate and perhaps unusual that this is the case, and this property does not always hold. If the null hypothesis defines a region in parameter space, there can be no guarantee in general that the distribution of the test statistic is the same for all points in this region.

A case of practical importance arises in the two examples above when the variance of the observations in the first group may not be assumed to be equal to the variance of the observations in the second group. Even when the null hypothesis (that the means of the two groups are equal) is true, various complications arise. First, the estimation of the common mean involves the solution of a cubic equation. This presents no difficulty in practice but leads to the second, and more intrinsic problem, namely that the test statistic found by application of the λ ratio procedure has a probability distribution that depends on the unknown ratio of the variances in the two groups. Thus no unique significance point can be found for the test generalizing the significance points available for the t-tests of Examples 1 and 2.

Various approximate procedures are used in practice for this case. One is to compute the statistic t', defined by

$$t' = \frac{\bar{X}_1 - \bar{X}_2}{\sqrt{A + B}}, \tag{8.49}$$

where

$$A = \frac{S_1^2}{m}, \quad B = \frac{S_2^2}{n},$$

with

$$S_1^2 = \frac{\sum_{i=1}^m (X_{i1} - \bar{X}_1)^2}{m - 1}, \quad S_2^2 = \frac{\sum_{i=1}^n (X_{i2} - \bar{X}_2)^2}{n - 1}.$$

If ν' is defined by

$$\nu' = \frac{(A + B)^2}{\frac{A^2}{m-1} + \frac{B^2}{n-1}},$$

then when the null hypothesis of equal means is true, t' has an approximate t distribution with degrees of freedom given by the largest integer less than or equal to ν' (see Lehmann (1986)). When $m = n$, t' is identical to the t statistic (3.19). However, in this case the number of degrees of freedom appropriate for t' is not equal to the number $2(n - 1)$ appropriate for t in Examples 1 and 2 above: The value of ν' lies in the interval $[n - 1, 2(n - 1)]$, the actual value depending on the ratio of S_1^2 and S_2^2.

8.4.2   The $-2\log\lambda$ Approximation

In this section we continue address the problem that the probability distribution of a test statistic might depend on parameters that are unspecified by the null hypothesis. The approximation discussed above using t' is specific to the t-test. Various approximation procedures have been proposed to address this problem in general. Perhaps the best known is based on the fact that if various regularity assumptions discussed below hold and if the null hypothesis is true, $-2\log\lambda$ (λ as in (8.43)) has an asymptotic chi-square distribution, with degrees of freedom equal to the difference in the numbers of parameters unspecified by null and alternative hypotheses. In this statement the word "asymptotic" means "as the sample size increases indefinitely."

It is important to note that this statement holds only under certain restrictions. Two important requirements are that the parameters involved in the test be real numbers that can take values in some interval and that the maximum likelihood estimator be found from a nonboundary point where the likelihood function is differentiable (and not, for example, a boundary point such as that in Example 5 of Section 8.2.2). A further crucial requirement is that null and alternative hypotheses be nested, as described in Section 8.4.1. The proof of the asymptotic distribution of $-2\log\lambda$ given, for example, in Wilks (1962, pp. 419–421) makes clear the importance of these restrictions. The approximate chi-square null hypothesis distribution of $-2\log\lambda$ has been applied in the literature in several cases of phylogenetic tree construction where these requirements do not hold (see Section 14.9).

The essence of the formal proof may be seen by considering the case of a single unknown parameter θ that takes some given value θ_0 under the null hypothesis and is left unspecified under the alternative hypothesis. If the unconstrained maximum likelihood estimator of θ is $\hat{\theta}$, assumed to be found from a point where the derivative of the likelihood is zero, then the definition (8.43) of λ shows that

$$-2\log\lambda = 2\left(\log L(\hat{\theta};\boldsymbol{X}) - \log L(\theta_0;\boldsymbol{X})\right), \qquad (8.50)$$

where L is the likelihood as defined in Section 8.2.2. We assume that the null hypothesis is true, that is, that $\theta = \theta_0$. When n is large we can assume that $\hat{\theta}$ is close to θ_0, and expansion of $\log L(\theta_0;\boldsymbol{X})$ in a Taylor series approximation as in (B.30), together with the fact that the derivative of the log likelihood is assumed to be zero at the maximum likelihood point, shows that the right-hand side becomes, approximately,

$$-(\hat{\theta} - \theta_0)^2 \frac{d^2}{d\theta^2}\log L, \qquad (8.51)$$

where we may take the derivative to be calculated, to a sufficiently close approximation, at θ_0. We make the further approximation of replacing

$d^2 \log L / d\theta^2$ by its mean value. Then equation (8.17) shows that expression (8.51) is, approximately,

$$\frac{(\hat{\theta} - \theta_0)^2}{\text{variance of } \hat{\theta}}, \tag{8.52}$$

and the asymptotic normality of $\hat{\theta}$ and the discussion below equation (1.70) show that this is asymptotically a chi-square random variable with 1 degree of freedom. Although many details of this proof need to be tidied up, it is enough to demonstrate the various regularity restrictions that are assumed.

We illustrate the importance of these regularity restrictions by considering four examples, the first two of which are taken from the context of the multinomial distribution (2.6). In the second and third examples the regularity conditions do not hold (for two different reasons in the two examples), and the approximate chi-square distribution of $-2 \log \lambda$ does not apply.

Example 1. A frequently occurring test in statistics is that of testing a null hypothesis that specifies the values of the set of parameters $\{p_i\}$ in the multinomial distribution (2.6) as the set of values $\{p_{i0}\}$. The alternative hypothesis makes no restriction on the probabilities $\{p_i\}$, other than the obvious ones that they be nonnegative and sum to 1. This implies that there are $k-1$ free parameters under the alternative hypothesis. In this test the null hypothesis is simple and no parameter estimation is required, and the likelihood is as given in (2.6), with p_i replaced by p_{i0}. The alternative hypothesis is composite, and under this hypothesis the maximum likelihood estimate of p_i is $\hat{p}_i = Y_i/n$, and the likelihood is as given in (2.6) with p_i replaced by $\hat{p}_i$. From this the likelihood ratio λ is given by

$$\lambda = \prod_i \left(\frac{n p_{i0}}{Y_i} \right)^{Y_i}. \tag{8.53}$$

From this, $-2 \log \lambda$ is identical to the statistic defined in equation (3.22). It can be shown (see Problem 8.9) that this statistic and the statistic defined in equation (3.20) are approximately equal when n is large and $Y_i/n \approx p_{i0}$. The asymptotic $-2 \log \lambda$ theory described above then justifies the claim that the statistic (3.20) has an approximate chi-square distribution with $k-1$ degrees of freedom under the null hypothesis. Tables of significance points of this distribution can then be used to assess whether the null hypothesis should be rejected, given an observed value of $-2 \log \lambda$.

Example 2. A more subtle testing procedure associated with the multinomial distribution (2.6) arises in the context of in situ hybridization. In an experimental procedure not described in detail here, and which is here simplified considerably, clones from one species tend to hybridize to homologues in another species. The many random factors involved in the experiment imply that a statistical analysis in needed for data from the in situ hybridization procedure.

As a straightforward example, suppose that chromosome i of species A contains a proportion p_{i0} of the DNA in the genome. If the clone from species B has no homologue in the DNA of species A, hybridization can be taken as random and the probability that the clone hybridizes to chromosome i is taken as p_{i0}, given that hybridization occurs. If some unknown chromosome j in species A contains a homologue of the clone from species B, then given that hybridization occurs, the probability of hybridization to this chromosome is taken to be $1 - \theta + p_{j0}\theta$ and to chromosome i $(i \neq j)$ with probability $p_{i0}\theta$. Here θ is an unknown parameter defining a random hybridization, so that $1 - \theta$ is the probability of a hybridization specifically to the homologue on chromosome j. Given data on the number of hybridizations to the various chromosomes in species A, the aim is to test for a significant excess of hybridizations to one chromosome.

In statistical terms, the null hypothesis (that there is no homologue in species A to the clone from species B) specifies the value p_{i0} for the probability p_i in (2.6). The alternative hypothesis claims that there is some unknown category j such that the probability that any observation falls in category j is $1 - \theta + p_{j0}\theta$, for some unknown parameter θ, while the probability that any observation falls in category i $(i \neq j)$ is $p_{i0}\theta$. Here θ and j are both unknown parameters.

The details of this testing procedure are omitted here: These can be found in Ewens et al. (1992). The $-2\log\lambda$ ratio procedure leads, asymptotically, to

$$-2\log\lambda = (Z_{\max})^2,$$

where

$$Z_{\max} = \max_i \frac{N_i - np_{i0}}{\sqrt{np_{i0}(1 - p_{i0})}}$$

and N_i is the number of observations (out of n) in category i. The probability distribution of the test statistic $Z_{\max}$ is *not* related to a chi-square. The reason for this is that the parameter j is discrete and the continuous real parameter theory underlying the $-2\log\lambda$ approach does not apply.

Example 3. Suppose we wish to test the null hypothesis that the parameter M in the density function of the uniform distribution (8.13) takes some specified null hypothesis value M_0 against an alternative hypothesis that does not specify the value of M. Under the null hypothesis the likelihood of the observations $X_1, X_2, \ldots, X_n$ is M_0^{-n} if each $X_i \leq M_0$ and is 0 otherwise. Under the alternative hypothesis the maximum likelihood estimator of M is $X_{\max}$ and the maximum of the likelihood is $X_{\max}^{-n}$. The likelihood ratio is either 0 (if at least one X_i exceeds M_0) or $(X_{\max}/M_0)^n$ (if all the X_i are less than or equal to M_0). When the null hypothesis is true (implying that all the X_i are less than M_0),

$$-2\log \lambda = 2n \log \frac{M_0}{X_{\max}}.$$

This statistic does not have an asymptotic chi-square distribution as $n \to +\infty$. The reason for this is that the maximum of the likelihood arises at a boundary point.

Examples 2 and 3 illustrate the fact that the above $-2\log \lambda$ theory is applicable only under certain restrictive circumstances. We return to this topic in Section 14.9, when considering tests of hypotheses relating to phylogenetic trees.

Example 4. It was stated in Example 4 of Section 3.4.3 that data often arise in the form of a two-way contingency table such as Table 3.1; examples of this were given in Sections 5.2, 5.3.4, and 6.1. We now show that the test statistic (3.24) that uses the data in Table 3.1 arises from the $-2\log \lambda$ theory, and that often the numerical value of this statistic is close to the numerical value of the chi-square statistic (3.23).

The null hypothesis of no association between row and column classifications implies that the probability p_{jk} that any count falls in the cell in row j, column k is of the form $p_{j.}p_{.k}$, where $p_{j.} = \sum_k p_{jk}, p_{.k} = \sum_j p_{jk}$. This null hypothesis defines some subspace of the parameter space. Under it the likelihood corresponding to the data in Table 3.1 is of the form

$$M \prod_{j=1}^{r} \prod_{k=1}^{c} (p_{j.}p_{.k})^{Y_{jk}},$$

where M is a multinomial constant. This is maximized when

$$\hat{p}_{j.} = \frac{y_{j.}}{y}, \quad \hat{p}_{.k} = \frac{y_{.k}}{y}.$$

From this,

$$L_{\max}(\omega) = M \prod_{j=1}^{r} \prod_{k=1}^{c} \left(\frac{y_{j.}}{y} \frac{y_{.k}}{y} \right)^{Y_{jk}}. \tag{8.54}$$

Under the alternative hypothesis the probability that a count falls in the cell in row j, column k is an arbitrary parameter p_{jk}. The maximum likelihood estimator of p_{jk} is Y_{jk}/y, and this leads to

$$L_{\max}(\Omega) = M \prod_{j=1}^{r} \prod_{k=1}^{c} \left(\frac{Y_{jk}}{y} \right)^{Y_{jk}}. \tag{8.55}$$

From (8.54) and (8.55)

$$-2\log \lambda = 2 \sum_{j=1}^{r} \sum_{k=1}^{c} Y_{jk} \log \left(\frac{Y_{jk}}{E_{jk}} \right), \tag{8.56}$$

where $E_{jk} = y_{j.}y_{.k}/y$. (This notation arises because E_{jk} is often referred to as the expected number of counts in cell (j, k) when the null hypothesis is true.) The right-hand side in (8.56) is the test statistic (3.24).

We now show that the numerical value of this test statistic is often close to that of the chi-square statistic (3.23). Suppose that Y_{jk} is close to E_{jk} and write $Y_{jk} = E_{jk}(1 + \delta_{jk})$. The right-hand side in (8.56) is then

$$2\sum_{j=1}^{r}\sum_{k=1}^{c} E_{jk}(1 + \delta_{jk}) \log(1 + \delta_{jk}).$$

From the logarithmic approximation (B.25) this is approximately

$$2\sum_{j=1}^{r}\sum_{k=1}^{c} E_{jk}(1 + \delta_{jk}) \left(\delta_{jk} - \frac{1}{2}\delta_{jk}^2 \right).$$

The identity $\sum_{j=1}^{r}\sum_{k=1}^{c} E_{jk}\delta_{jk} = 0$ implies that if terms of order δ_{jk}^3 are ignored, this is

$$\sum_{j=1}^{r}\sum_{k=1}^{c} E_{jk}\delta_{jk}^2 = \sum_{j=1}^{r}\sum_{k=1}^{c} \frac{(Y_{jk} - E_{jk})^2}{E_{jk}},$$

and this is the chi-square statistic (3.23).

8.5   Sequential Analysis

8.5.1   The Sequential Probability Ratio Test

Sequential analysis is a statistical hypothesis-testing procedure in which the sample size is not fixed in advance, but depends on the outcomes of the successive observations taken. Several important results in BLAST theory are borrowed from sequential analysis, and there are many parallels between sequential analysis theory and the statistical theory of BLAST, so we consider sequential analysis in some detail. We focus on the case of discrete random variables, since that provides the most appropriate comparison with BLAST.

Hypothesis testing in sequential analysis usually relates to testing for the numerical value of some unknown parameter ξ, so we will discuss it as such a test. Suppose that the null hypothesis specifies a value ξ_0 for this parameter and the alternative specifies a value ξ_1, so that the null hypothesis specifies a probability distribution $P(y; \xi_0)$ for a discrete random variable Y and the alternative hypothesis specifies a probability distribution $P(y; \xi_1)$. Thus both hypotheses are simple. Because the sample size is not chosen in advance, we are free to specify not only the numerical values α of the Type I error but also the numerical value β of the Type II error.

The sequential analysis procedure prescribes that we take observations one by one and calculate, after the nth observation is taken $(n = 1, 2, \dots)$, the discrete analogue of the likelihood ratio (8.31). Sampling stops, and the alternative hypothesis is accepted, whenever this ratio becomes sufficiently large, that is, reaches or exceeds some value B. Similarly, sampling stops, and the null hypothesis is accepted, whenever it becomes sufficiently small, that is, reaches or is less than some value A. Sampling continues while

$$A < \frac{P(y_1; \xi_1)P(y_2; \xi_1) \cdots P(y_n; \xi_1)}{P(y_1; \xi_0)P(y_2; \xi_0) \cdots P(y_n; \xi_0)} < B. \tag{8.57}$$

Our aim is to choose A and B such that the Type I and Type II errors are α and β, respectively. Because of the discrete nature of the process, the bounds A or B may be exceeded in one step. It is a standard result of sequential analysis theory that if this cannot happen, that is, the probability ratio in (8.57) eventually reaches exactly the value A or the value B and not a value less than A or greater than B, the numerical values α and β are achieved by the choice $A = \beta/(1 - \alpha)$ and $B = (1 - \beta)/\alpha$. If, as will almost always be the case in practice, boundary overlap is possible, then this choice of A and B usually leads to Type I and Type II errors close to α and β, respectively.

It is often convenient to take logarithms throughout in (8.57), so that with $A = \beta/(1 - \alpha)$ and $B = (1 - \beta)/\alpha$, sampling continues as long as

$$\log \frac{\beta}{1 - \alpha} < \sum_i \log \frac{P(y_i; \xi_1)}{P(y_i; \xi_0)} < \log \frac{1 - \beta}{\alpha}. \tag{8.58}$$

The null hypothesis is accepted if the lower inequality is eventually broken, and the alternative hypothesis is accepted if the upper inequality is eventually broken.

In Section 1.14.3 we defined the support, or score, $S_{1,0}(y)$ of the observed value y of a random variable Y for the probability distribution claimed by the alternative hypothesis over the probability distribution claimed by the null hypothesis. In the present notation this may be written

$$S_{1,0}(y) = \log \frac{P(y; \xi_1)}{P(y; \xi_0)}. \tag{8.59}$$

The inequalities (8.58) can then be rewritten as

$$\log \frac{\beta}{1 - \alpha} < \sum_i S_{1,0}(y_i) < \log \frac{1 - \beta}{\alpha}, \tag{8.60}$$

the expression $\sum_i S_{1,0}(y_i)$ being the accumulated support provided by the observations for the alternative hypothesis over the null hypothesis.

For continuous random variables the probability $P(y; \xi)$ in (8.57) is replaced by a density function $f(x; \xi)$. We do not discuss the continuous random variable case further, since the theory and procedures are similar to those of the discrete case.

Example. Sequence matching. Suppose that in the sequence-matching test (page 249) the null and alternative hypotheses respectively specify the values $p_0 = 0.25$ and $p_1 = 0.35$ for p, and that both the Type I error and the Type II error are chosen to be 0.01. If Y_i is defined to be 1 when there is a match at site i and is defined to be 0 when there is not, equation (8.58) shows that sampling is continued while

$$\log \frac{1}{99} < \sum_i S_{1,0}(Y_i) < \log 99, \tag{8.61}$$

where

$$S_{1,0}(Y_i) = \log \frac{(0.35)^{Y_i}(0.65)^{(1-Y_i)}}{(0.25)^{Y_i}(0.75)^{(1-Y_i)}} \tag{8.62}$$

is the support offered by Y_i in favor of the alternative hypothesis. In terms of the support concept, the procedure leads to the acceptance of the null hypothesis when the accumulated level of support for the alternative hypothesis reaches $\log(1/99) = -4.595$, so that the accumulated level of support for the null hypothesis reaches 4.595. The alternative hypothesis is accepted when the accumulated level of support for that hypothesis reaches $\log(99) = 4.595$.

Using (8.62), the inequalities (8.61) can be written

$$-9.581 < \sum_i (Y_i - 0.2984) < 9.581. \tag{8.63}$$

The procedure defined by (8.63) is a random walk in which the step size is either 0.7016 (for a match) or -0.2984 (for a mismatch). This illustrates the close link between sequential analysis and random walks.

It is important to note that this sequential test is one of a simple null hypothesis against a simple alternative and that unlike the parallel fixed sample size test discussed in Section 3.4.1, the testing procedure depends on the specifics of the alternative hypothesis. Thus if the alternative hypothesis had been $p = 0.30$ rather than $p = 0.35$, the inequalities (8.63) would be replaced by

$$-18.284 < \sum_i (Y_i - 0.2745) < 18.284. \tag{8.64}$$

This describes a random walk in which the step size is either 0.6255 (for a match) or -0.2745 (for a mismatch). Thus the step sizes of this walk differ from those of the walk described by (8.63), as do the boundaries of the walk.

8.5.2   The Power Curve for a Sequential Test

Consider the sequential test of the null hypothesis that a discrete random variable Y has probability distribution $P(y; \xi_0)$ against the alternative hypothesis that Y has probability distribution $P(y; \xi_1)$, the test having Type I error α and Type II error β. The power $\mathcal{P}(\xi)$ of the test is the probability that the alternative hypothesis is eventually accepted if the distribution of Y is $P(y; \xi)$, that is, if the true value of the parameter is ξ, for any general value ξ.

An approximate formula for $\mathcal{P}(\xi)$, derived by ignoring the possibility of boundary overlap, is (Wilks (1962))

$$\mathcal{P}(\xi) \approx \frac{1 - \left(\frac{\beta}{1-\alpha}\right)^{\theta^*}}{\left(\frac{1-\beta}{\alpha}\right)^{\theta^*} - \left(\frac{\beta}{1-\alpha}\right)^{\theta^*}}, \tag{8.65}$$

where, except in one specific case discussed below, θ^* is the unique nonzero solution for θ of the equation

$$\sum_R P(y; \xi) \left(\frac{P(y; \xi_1)}{P(y; \xi_0)}\right)^{\theta} = 1, \tag{8.66}$$

where R is the range of values of Y. Equivalently, and again except in one specific case, θ^* is the unique nonzero solution for θ in the equation

$$\sum_{y \in R} P(y; \xi) e^{\theta S_{1,0}(y)} = 1, \tag{8.67}$$

where $S_{1,0}(y)$ is defined in equation (8.59). The graph of $\mathcal{P}(\xi)$ as a function of ξ is the power curve of the test.

The value of θ^* depends on the value of ξ. Two cases are of particular interest, namely $\xi = \xi_0$ and $\xi = \xi_1$. Inspection of equation (8.66) shows that when $\xi = \xi_0$ the value of θ^* is 1. Insertion of this value in (8.65) gives $\mathcal{P}(\xi_0) \approx \alpha$, as required. When $\xi = \xi_1$ the value of θ^* is -1, and insertion of this value in (8.65) gives $\mathcal{P}(\xi_1) \approx 1 - \beta$, again as required.

The left-hand side in (8.67) is the moment-generating function of $S_{1,0}(Y)$, assuming that the distribution of Y is $P(y; \xi)$. The argument following (1.82) then shows that if the mean of $S_{1,0}(Y)$ is nonzero, there is a unique nonzero value θ^* solving this equation. When $E(S_{1,0}(Y)) = 0$ there is a double root of equation (8.66) (and of equation (8.67)) at 0, and a special analysis is needed for this value. However, this one isolated value is not usually of any significance, so we ignore it in our future discussion.

The parameter θ^*, first introduced in the literature in the context of sequential analysis, is identical to that discussed in Chapter 7 in describing properties of random walks and also identical to the parameter θ^* that will be used in Chapter 9 to discuss BLAST theory.

The sequence-matching power curve. In the sequence-matching example the parameter of interest is the probability p of a match, and we replace the generic parameter notation ξ used above by the Bernoulli notation p. Suppose that the null hypothesis value of p is 0.25 and the alternative hypothesis value is 0.35. Since the random variable of interest Y can take only two values (+1 for a match, 0 otherwise), the expression (8.66) simplifies to

$$p \left(\frac{0.35}{0.25}\right)^\theta + (1-p) \left(\frac{0.65}{0.75}\right)^\theta = 1. \tag{8.68}$$

This may be solved for θ^* for any chosen value of p and the value found inserted in the right-hand side of equation (8.65). As an example, if $p = 0.3$ (that is, the probability of a match is halfway between the values claimed by the null and alternative hypotheses), the solution of equation (8.68) is found (by numerical methods) to be $\theta^* = -0.32$, and insertion of this value in equation (8.65) shows that the probability that the alternative hypothesis is accepted is about 0.537. Although 0.3 is halfway between the null and alternative hypothesis values 0.25 and 0.35, this probability is not 0.5.

In the discussion above it was in effect assumed that the functional form of the probability distribution of the random variables in the sequential process is known, and that all that is in question is the values of some parameter in that distribution. Thus in equation (8.66) it is assumed that the form of the true probability distribution $P(y; \xi)$ is of the same form as that defined by the null and alternative hypotheses.

This restriction is unnecessary. It might be the case that the true probability distribution is of a different form than that assumed under both null and alternative hypotheses. If the probability that the random variable Y takes the value y is $Q(y)$, the equation defining θ^* is

$$\sum_{y \in R} Q(y) \left(\frac{P(y; \xi_1)}{P(y; \xi_0)}\right)^{\theta^*} = 1. \tag{8.69}$$

Once θ^* is found from this equation, equation (8.65) may be used to find the probability that the alternative hypothesis is accepted.

8.5.3 The Mean Sample Size

The sample size in a sequential test is the (random) number of observations taken until one or the other hypothesis is accepted. An approximation to the mean sample size may be found by ignoring the possibility of boundary overlap. The approximation is found by using equation (7.24), which first arose in the context of sequential analysis.

The sequential procedure continues so long as the inequalities (8.58) hold. If boundary overlap is ignored, the final value of $\sum_i S_{1,0}(y_i)$ is either $\log \beta/(1-\alpha)$ (with probability $1 - \mathcal{P}(\xi)$) or $\log(1-\beta)/\alpha$ (with probability $\mathcal{P}(\xi)$), where $\mathcal{P}(\xi)$ is given by equation (8.65). The mean of the final value of $\sum_i S_{1,0}(Y_i)$ is thus, from equation (1.16),

$$(1 - \mathcal{P}(\xi)) \log \left(\frac{\beta}{1-\alpha} \right) + \mathcal{P}(\xi) \log \left(\frac{1-\beta}{\alpha} \right), \qquad (8.70)$$

and this forms the left-hand side in equation (7.24). The term "$E(S)$" on the right-hand side of equation (7.24) is, in the sequential case,

$$E\left(S_{1,0}(Y_i) \right) = E \left(\log \frac{P(Y_i; \xi_1)}{P(Y_i; \xi_0)} \right) = \sum_{y \in R} P(y; \xi) \log \frac{P(y; \xi_1)}{P(y; \xi_0)}. \qquad (8.71)$$

Assuming this mean to be nonzero, it follows from equation (7.24) that the mean sample size is

$$\frac{(1 - \mathcal{P}(\xi)) \log(\frac{\beta}{1-\alpha}) + \mathcal{P}(\xi) \log(\frac{1-\beta}{\alpha})}{\sum_{y \in R} P(y; \xi) \log \frac{P(y; \xi_1)}{P(y; \xi_0)}}. \qquad (8.72)$$

Both numerator and denominator in this expression depend on the value of $\mathcal{P}(\xi)$, and hence on the parameter θ^*, so that the mean sample size itself also depends on the value of this parameter.

The mean of $S_{1,0}(Y_i)$ is zero for only one value of the parameter ξ, so we do not pursue the mean sample size formula for this case.

A generalization of (8.72), namely

$$\frac{(1 - \mathcal{P}(\xi)) \log(\frac{\beta}{1-\alpha}) + \mathcal{P}(\xi) \log(\frac{1-\beta}{\alpha})}{\sum_{y \in R} Q(y) \log \frac{P(y; \xi_1)}{P(y; \xi_0)}}, \qquad (8.73)$$

applies in the case where the probability distribution $Q(y)$ of Y is of a different form from that assumed in null and alternative hypotheses.

The sequence-matching mean sample size. In the sequence-matching example where the null hypothesis value of p is 0.25 and the alternative hypothesis value of p is 0.35, and the Type I and Type II errors are both 0.01, the expression (8.72) reduces to

$$\frac{9.190 \mathcal{P}(p) - 4.595}{p \log \frac{7}{5} + (1-p) \log \frac{13}{15}}. \qquad (8.74)$$

Thus when the null hypothesis is true, so that $p = 0.25$, the probability $\mathcal{P}(0.25)$ that the alternative hypothesis is accepted is 0.01, and the mean sample size is 194. When $p = 0.30$, the probability $\mathcal{P}(0.30)$ that the alternative hypothesis is accepted is 0.537 (calculated above), and the mean sample size is found from (8.74) to be 441. When the alternative hypothesis is true, $p = 0.35$, $\mathcal{P}(0.35) = 0.99$, and the mean sample size is 182.

One of the motivations for the development of sequential methods was to find a procedure that led to a decision to accept some hypothesis with a smaller mean sample size than that needed for a fixed sample size test with the same Type I and Type II errors. It is thus interesting to note that in the above example, the sample size of this fixed-size test is about 450.

It sometimes causes concern that there is a value of p for which the denominator in the expression (8.74) is zero. However, this is the value of p for which the double root of equation (8.66) arises and for which a different analysis is needed. Since this refers to one isolated value of p, we do not pursue the analysis of this case.

The mean sample size curve, as with the power curve, depends on the specific choice made under the alternative hypothesis for the probability p of a match at any site.

8.5.4 The Effect of Boundary Overlap

The sequential analysis theory described above assumes that there is no boundary overlap. In practice, there will almost always be boundary overlap, and in this case the *actual* Type I and Type II errors of the test differ from the nominal values α and β that were used to define the boundaries in the testing procedure. Random walk theory can be used to assess how significant the effects of boundary overlap are.

To illustrate this, consider the sequential test of the null hypothesis that the Bernoulli parameter p is 0.45 against the alternative hypothesis that it is 0.55, with nominal Type I and Type II errors of 5%. Sampling continues while

$$-14.67 < \sum_i (2y_i - 1) < 14.67,$$

where $y_i = 1$ if trial i results in success and $y_i = 0$ if trial i results in failure. Defining $u_i = 2y_i - 1$, this means that sampling continues while

$$-14.67 < \sum_i u_i < 14.67,$$

where $u_i = 1$ if trial i results in success and $u_i = -1$ if trial i results in failure. Thus $\sum_i u_i$ performs a simple random walk. Since this sum can take only integer values, the walk will continue in practice until $\sum_i u_i$ reaches either -15 or 15. Thus there must be boundary overlap. Equation (7.10) can then be used to show that if $p = 0.45$ (that is, the null hypothesis is true), the actual Type I error is 4.7% rather then the nominal 5%. The actual Type II error is also 4.7%.

It can be shown that in any sequential procedure the sum of the actual Type I and Type II errors is less than the sum $\alpha + \beta$ of the nominal values, and usually (as in the above example) both are less than their respective nominal values.

The boundary overlap problem can, however, be more important than is illustrated in this example. BLAST theory recognizes this and deals with it in a novel and interesting way, as described in Chapter 9.

Problems

8.1. Show that the estimator of μ given in (8.12) is both unbiased and consistent and that the estimator of σ^2 is biased but consistent. Write down a biased estimator of μ that is not consistent and, by considering X_1 only, find an unbiased estimator of μ that is not consistent.

8.2. Show that the maximum likelihood estimators of the probabilities for the various outcomes in a multinomial distribution are the observed proportions of observations in the respective outcomes.

8.3. Replacing ξ by p in the Cramér–Rao bound (8.17), find a lower bound for the variance of any unbiased estimator of the Bernoulli parameter p, using the likelihood L defined in (8.5). Find an unbiased estimator of p whose variance reaches this lower bound.

8.4. Write down the likelihood L defined by n iid random variables X_1, $X_2, \ldots, X_n$ from the exponential distribution (1.59). By putting $1/\lambda = \xi$ in this likelihood, find the Cramér–Rao bound (8.17) for the variance of any unbiased estimator of $1/\lambda$. Find an unbiased estimator of $1/\lambda$ whose variance reaches this lower bound.

8.5. Use the Cramér–Rao bound (8.17) to find a lower bound for the variance of any unbiased estimator of the parameter λ in the Poisson distribution (1.15). Find an unbiased estimator of λ whose variance achieves this bound.

8.6. Suppose that X_1 and X_2 are independent random variables, each having the exponential distribution (1.59). An unbiased estimator of the mean $1/\lambda$ of this distribution is $\bar{X} = (X_1 + X_2)/2$, and the variance (which is also the mean square error) of this estimator is $1/(2\lambda^2)$. The aim of this and the next two questions is to investigate estimators with a smaller mean square error than that of $\bar{X}$.

(i) Use equations (2.55) and (B.43) to show that $E(\sqrt{X_iX_j}) = \pi/4\lambda$.

(ii) Show from (i) that the bias of $\sqrt{X_iX_j}$ as an estimator of $1/\lambda$ is $(\pi - 4)/4\lambda$.

(iii) The variance of $\sqrt{X_i X_j}$ is, from equation (1.21), $E(X_i X_j) - \left(\pi^2/16\lambda^2\right)$. Use a result of Problem 1.15 together with equation (2.55) to show that this variance is $(16 - \pi^2)/16\pi^2$.

(iv) Use the results of (ii) and (iii) together with equation (3.15) to show that the mean square error of $\sqrt{X_i X_j}$ as an estimator of $1/\lambda$ is $(4 - \pi)/2\lambda^2$. Show from this that the mean square error of $\sqrt{X_i X_j}$ is less than that of $\bar{X}$.

8.7. *Continuation.* A second estimator of $1/\lambda$ is $k\bar{X}$, for some constant k. Find the range of values of k for which the mean square error of this estimator is less than that of $\bar{X}$.

8.8. *Continuation.* Suppose that $X_1, \ldots, X_n$ are independent random variables, each having the exponential distribution (1.59). The variance (and mean square error) of $\bar{X}$ as an estimator of $1/\lambda$ is now $1/n\lambda^2$. Compare this mean square error with that of the estimator $\sqrt{X_1 \cdots X_n}$ for the cases $n = 3$, $n = 4$, using the fact that $\Gamma(1.25) = 0.9064025$, $\Gamma(\frac{4}{3}) = 0.8929795$, $\Gamma(\frac{5}{3}) = 0.9027453$.

8.9. Suppose that the null hypothesis of Example 1 of Section 8.4.2 is true. Then one can write $y_i/n = p_i + h_i$, where h_i is a small deviation from zero, typically of order n^{-1}. The λ ratio (8.53) can then be written as

$$\lambda = \prod_i \left(\frac{p_i}{p_i + h_i}\right)^{y_i}. \tag{8.75}$$

This implies that

$$-2\log\lambda = 2\sum_{i=1}^{k} y_i \log\left(1 + \frac{h_i}{p_i}\right). \tag{8.76}$$

(i) Approximate the logarithmic term in (8.76) by an expression calculated from the approximation (B.25).

(ii) Write y_i as $np_i + nh_i$ in the expression obtained in (i).

(iii) Ignoring terms of order h_i^3, show that the resulting expression reduces to the X^2 statistic (3.21).

8.10. Suppose that the null hypothesis being tested in Example 2 of Section 3.4.1 is true. Use the fact that the mean of Y_j is np_j and the variance of Y_j is $np_j(1 - p_j)$ to show that the mean value of the statistic (3.20) is exactly $k - 1$ (the mean of a chi-square random variable with $k - 1$ degrees of freedom; see Problem 1.19).

(More difficult). Find the variance of the statistic (3.21) when the null hypothesis is true, and compare this with the chi-square distribution value $2(k-1)$.

8.11. Let $X_1, X_2, \ldots, X_n$ be independent continuous random variables, each having density function $f(x) = e^{-(x-\theta)}$, $\theta \leq x < +\infty$. Let $X_{(1)}, X_{(2)}, \ldots, X_{(n)}$ be the corresponding order statistics. Use (2.143) to find the density functions of $X_{(1)}$ and $X_{(1)}$, and from this show that

$$E\left(X_{(1)}\right) = \theta + \frac{1}{n}, \ E\left(X_{(2)}\right) = \theta + \frac{2n-1}{n(n-1)}.$$

Thus show that $X_{(1)} - (n-1)(X_{(2)} - X_{(1)})/n$ is an unbiased estimator of θ.

8.12. Develop the sequential likelihood ratio test of the null hypothesis that the mean of a Poisson distribution is λ_0 against the alternative hypothesis that the mean of the distribution is λ_1, with Type I error α and Type II error β. Here λ_0 and λ_1 are given constants with $\lambda_1 > \lambda_0$.

8.13. *Continuation.* If the true mean of the Poisson distribution is λ, show that θ^* is the nonzero solution of the equation

$$\theta^*(\lambda_0 - \lambda_1) - \lambda + \lambda \left(\frac{\lambda_1}{\lambda_0}\right)^{\theta^*} = 0. \tag{8.77}$$

If $\lambda_0 = 4$, $\lambda_1 = 5$, solve equation (8.77) for λ for the values $\theta^* = -1.2$, -1.0, -0.8, -0.6, -0.4, -0.2, $+0.2$, $+0.4$, $+0.6$, $+0.8$, $+1.0$, $+1.2$.

8.14. *Continuation.* Use the results of Problem 8.13 to sketch the power curve in the case $\alpha = \beta = 0.05$.

8.15. *Continuation.* Use the results of Problem 8.14 to sketch the mean sample size curve.

8.16. For the two respective cases $\xi = \xi_0$ and $\xi = \xi_1$, discuss the denominator in equation (8.72) as a relative entropy.

9
BLAST

9.1 Introduction

BLAST is the most frequently used method for assessing which DNA or protein sequences in a large database have significant similarity to a given query sequence. Many of the results derived in previous chapters, especially those relating to the renewal theorem, random walks, and sequential analysis, were discussed because they are needed in the statistical theory associated with the BLAST procedure. In this chapter we describe how they are used for this purpose. For concreteness the discussion is in terms of protein (amino acid) sequences; the analysis for DNA sequences is similar to, but simpler than, that for protein sequences.

Various sophisticated versions of BLAST, some of which are mentioned at the end of this chapter, are now available and used widely. The theory for these is far more complicated than that which we discuss. Our aim here is to give an indication of the nature of BLAST theory by considering a simple version of BLAST, leading to a readily understood statistical analysis. In particular, the version of BLAST discussed in this chapter is an older one and does not allow for the alignment gaps accommodated in more recent versions such as Washington University and NCBI versions 2.0. We will describe Washington University's version 1.4, which was used to generate the examples of Section 9.5.

9.2   The Comparison of Two Aligned Sequences

9.2.1   Introduction

We start by considering as given an ungapped global alignment of two protein sequences, both of length N, as shown, for example, in (7.2). This is done mainly as a preliminary step to the generalizations in the following sections. In particular, the generalization to finding the best among all local alignments of two sequences will be considered in Section 9.3. The further generalization to database searches will be considered in Section 9.4.

The null hypothesis to be tested is that for each aligned pair of amino acids, the two amino acids were generated by independent mechanisms, so that if amino acid j occurs at any given position in the first sequence with probability p_j and amino acid k occurs at any given position in the second sequence with probability p'_k, the probability that they occur together in a given aligned pair is

$$\text{null hypothesis probability of the pair } (j,k) = p_j p'_k. \qquad (9.1)$$

If the alternative hypothesis is also specified, we write

$$\text{alternative hypothesis probability of the ordered pair } (j,k) = q(j,k).$$
$$(9.2)$$

Substitution matrices were discussed in Section 6.5 and are discussed at length below. They are central to the theory and implementation of BLAST. In principle, the elements in such a matrix can be chosen without reference to any theory. In practice, when the alternative hypothesis probabilities $q(j,k)$ are specified, there is a close theoretical relation between the $q(j,k)$ and the elements in the substitution matrix. This relation derives from the relation between the hypotheses in a hypothesis-testing procedure and the test statistic used to choose between these hypotheses, as introduced in Section 8.3.1. A discussion of this relationship in the context of BLAST is given in Section 9.2.4. For the moment we assume that the elements in the substitution matrix are given, either through this theory or otherwise.

9.2.2   Ladder Points and Excursions

We number the positions in the alignment from left to right as positions $1, 2, \ldots, N$. A score $S(j,k)$ is allocated to each position where the aligned amino acid pair (j,k) is observed, where $S(j,k)$ is the (j,k) element in the substitution matrix chosen. It is required in the theory, and is assumed throughout, that at least one element in the substitution matrix be positive and that the null hypothesis mean score $\sum_{j,k} p_j p'_k S(j,k)$ be negative. In order to apply the theory of Chapter 7 we also assume throughout that the greatest common divisor of the scores is 1.

An accumulated score at position i is calculated as the sum of the scores for the various amino acid comparisons at positions $1, 2, \ldots, i$. As i increases, this accumulated score undergoes a random walk, described in (7.3) for the protein sequence comparison (7.2). When the null hypothesis is true, the walk has negative drift and will go through a succession of increasingly negative ladder points (recall that a ladder point is a new low in the walk). Because the substitution matrix will usually include elements, or scores, whose values are -2 or less, the accumulated score at any ladder point will not necessarily be one less than the accumulated score at the preceding ladder point. This implies that, in random walk terms, boundary overlap can occur. The analysis of this overlap is needed for subsequent calculations.

Let $Y_1, Y_2, \ldots$ be the respective maximum heights of the excursions of this walk after leaving one ladder point and before arriving at the next, and let $Y_{\max}$ be the maximum of these maxima. $Y_{\max}$ is in effect the test statistic used in BLAST, so it is necessary to find its null hypothesis distribution.

$Y_1, Y_2, \ldots$ are independent, and ignoring end effects for now, can be taken as being identically distributed. The asymptotic probability distribution of any Y_i was shown in Chapter 7 to be the geometric-like distribution (1.67). The values of C and θ^* in this distribution depend on the substitution matrix used and the amino acid frequencies $\{p_j\}$ and $\{p_k'\}$. The probability distribution of $Y_{\max}$ also depends on n, the mean number of ladder points in the walk. We now discuss the computation of C, θ^*, and n.

9.2.3 Parameter Calculations

The formulae required for the BLAST random walk were derived in Chapter 7. The expression for C is given in equation (7.62), and requires only notational amendments for application to BLAST. The step size is identified with a score $S(j, k)$, and the null hypothesis probability of taking a step of any size is found from the two sets of frequencies $\{p_j\}$ and $\{p_j'\}$.

The computation of θ^* also follows the random walk principles laid down in Chapter 7. As noted below equation (7.43), θ^* is found from an equation involving the mgf of the step size in this random walk. When the null hypothesis is true, this equation is

$$\sum_{j,k} p_j p_k' e^{\theta^* S(j,k)} = 1. \tag{9.3}$$

The convention in BLAST printouts is to denote θ^* by λ, so we change notation in the remainder of this chapter and use this convention. Since the mean score is negative, the proof of Theorem 1.1 (page 35) shows that λ is the unique positive solution of the equation

$$\sum_{j,k} p_j p_k' e^{\lambda S(j,k)} = 1. \tag{9.4}$$

This change of notation was foreshadowed in the choice of the notation for the geometric-like distribution. The calculation of λ from equation (9.4) will usually require numerical methods: See Section B.15.

The calculation of the null hypothesis probability distribution of $Y_{\max}$ depends not only on C and λ but also on the mean number n of ladder points in the walk. This mean number in turn depends on the mean distance A between ladder points. A general formula for A is given in equation (7.42) and is readily converted to the situation discussed here. However, the arguments leading to this formula do not necessarily provide an efficient general formula for finding the constants R_{-j} in equation (7.42), and we now describe two alternative approaches.

The first alternative approach uses a decomposition of paths. Consider as a simple example a walk in which the possible steps are $+1$ and -2, with respective probabilities p and $q = 1 - p$. Any ladder point reached in the walk is at a distance 1 or 2 below the previous one. The respective probabilities of these two cases are denoted by R_{-1} and $R_{-2} = 1 - R_{-1}$, as in Chapter 7.

The probability that -2 is a ladder point is the probability that the walk goes immediately to -2, together with the probability of the event that the walk first goes to $+1$, and then starting from $+1$, reaches 0 as the first point reached below $+1$ and then -2 as the first ladder point below 0. This implies that

$$R_{-2} = q + p(1 - R_{-2})R_{-2}.\tag{9.5}$$

The positive solution of this equation is

$$R_{-2} = \frac{-q + \sqrt{4pq + q^2}}{2p}.\tag{9.6}$$

From this the value of R_{-1} follows as $1 - R_{-2}$, and then the value of A follows from equation (7.42).

For general substitution matrices this method might not be effective. In such a case, Karlin and Altschul (1990) provide rapidly converging series expansions that give accurate values of A using only a few terms in the series. We assume from now on that a value of A, arrived at by one method or another, is in hand.

Since the two sequences compared are each of length N, and the mean distance between ladder points is A, the mean number n of ladder points is equal for all practical purposes to N/A. While various approximations are involved with this calculation, the intuitive interpretation is clear: If, for example the length N is 1000 and the mean distance A between successive ladder points is 50, one expects about 20 ladder points in the walk involved with the comparison of the two sequences, and this is the value given by the expression N/A.

9.2.4   The Choice of a Score

So far, we have taken the score $S(j, k)$ as given, and have not discussed what might be a reasonable choice, on statistical grounds, for this score. In applications of BLAST this score, whether found by a BLOSUM or a PAM matrix, is a log likelihood ratio (as discussed briefly in Section 6.5), and we now indicate why this is appropriate.

The random walk described in Section 9.2.2 is determined by the sum of the scores $S(j, k)$ at each position during the walk. In sequential analysis one also considers the sum of scores. In sequential analysis the score used is a likelihood ratio, arrived at through statistical optimality methods. Specifically, if the random variable Y whose probability distribution is being assessed is discrete, this is the "score" statistic $S_{1,0}(y)$, defined in equation (8.59) as the log likelihood ratio

$$S_{1,0}(y) = \log \frac{P(y; \xi_1)}{P(y; \xi_0)}.$$

Based on the comparison of the BLAST and the sequential analysis procedures, it can be argued that a suitable score to use in BLAST should also be the logarithm of a likelihood ratio. Under this argument, if the amino acid pair (j, k) is observed at any position, and if $p_j p'_k$ and $q(j, k)$ are, respectively, the null and the alternative hypothesis probabilities of this pair, the (discrete random variable) score $S(j, k)$ becomes

$$S(j, k) = \log \frac{q(j, k)}{p_j p'_k}. \tag{9.7}$$

Any score proportional to $S(j, k)$ is also reasonable.

The second argument favoring the choice (9.7) for the score associated with the pair (j, k) is more subtle (Karlin and Altschul (1990)). This argument also leads to the choice of a specific proportionality constant. Suppose some arbitrary substitution matrix is chosen, with (j, k) element $S(j, k)$. Now let $q(j, k)$ be defined implicitly by

$$S(j, k) = \lambda^{-1} \log \frac{q(j, k)}{p_j p'_k}, \tag{9.8}$$

where λ is defined in equation (9.4), and thus explicitly by

$$q(j, k) = p_j p'_k e^{\lambda S(j,k)}. \tag{9.9}$$

The right-hand side is the typical term on the left hand-side in equation (9.4). Therefore $\sum_{j,k} q(j, k) = 1$. Thus the $q(j, k)$ (which are all positive) form a probability distribution. This is not an arbitrary distribution. Karlin and Altschul (1990) and Karlin (1994) show that in practice, when the null hypothesis is true, the frequency with which the observation (j, k) arises in high-scoring excursions, where the score used is as given in equation (9.8), is asymptotically equal to $q(j,k)$. They then argue that a scoring scheme is "optimal" if the frequency of the observation (j,k) in high-scoring excursions is asymptotically equal to the "target" frequency $q(j,k)$, the frequency arising if the alternative hypothesis is true, i.e., the frequency in the most biologically relevant alignments of conserved regions. This, then, argues for the use of $S(j,k)$ as defined in equation (9.8) as the score statistic.

These arguments lead us to adopt the following procedure. Suppose that the alternative hypothesis specifies a well-defined probability $q(j,k)$ for the amino acid pair (j,k), while the null hypothesis specifies a probability $p_j p'_k$ for this pair. Then we define the score $S(j,k)$ associated with this pair as that given by equation (9.8).

These arguments do not yet specify how to determine the most appropriate $q(j,k)$'s. There are various possibilities for this. One frequently adopted choice is that deriving from the evolutionary arguments that lead to the PAMn matrix construction described in Section 6.5.3. In the notation of Section 6.5.3,

$$q(j,k) = p_j m_{jk}^{(n)}, \tag{9.10}$$

so that

$$S(j,k) = \log \frac{m_{jk}^{(n)}}{p'_k}. \tag{9.11}$$

The values of the $q(j,k)$'s for the simple symmetric model of Section 6.5.3 are given in equation (6.34) for one specific value of n. The derivation of these values emphasizes that $q(j,k)$ is a function of n, and that some extrinsic choice of a reasonable value of n must be made to use PAMn matrices in BLAST methods. We discuss aspects of this choice in Section 9.6.

The choice of $S(j,k)$ as the logarithm of a likelihood ratio can be related to the concepts of relative entropy and support discussed in Section 1.14.2. Specifically, the score defined by equation (9.8) is proportional to the support given by the observation (j,k) in favor of the alternative hypothesis over the null hypothesis. Equation (1.114) shows that when the alternative hypothesis is true, the mean H of this support is

$$H = \sum_{j,k} q(j,k) \log \frac{q(j,k)}{p_j p'_k}, \tag{9.12}$$

and this is the relative entropy defined in equation (1.109). Equation (9.8) shows that this relative entropy can be written as

$$H = \sum_{j,k} q(j,k) \lambda S(j,k) = \lambda E\left(S(j,k)\right), \tag{9.13}$$

the expected value being taken assuming that the alternative hypothesis is true.

From the discussion following (9.9), if the score $S(j, k)$ for the pair (j, k) is defined as in (9.8), the mean score in high-scoring segments is asymptotically $\sum_{j,k} q(j, k)S(j, k)$, and from (9.13) this is

$$\lambda^{-1}H. \tag{9.14}$$

This asymptotic result is used in BLAST calculations (see Section 9.3.3).

Simulations, however, show that the convergence to this asymptotic value is very slow. For the symmetric PAMn substitution matrix discussed in Section 6.5.4 with $n = 259.0675$, and with equal amino acid frequencies, the asymptotic value $\lambda^{-1}H$ of the mean step size in high-scoring segments, found from computation of λ and H from (9.4) and (9.12), respectively, is 0.446. This is identical to the value given in equation (6.35), found from Markov chain considerations. For this example, Table 9.1 shows simulation estimates of this mean for various values of N, the length of the alignment. The slow rate of convergence to the asymptotic value 0.446 is clear. This

N	500	5,000	50,000	500,000	5,000,000	limiting value
mean step	1.021	.712	.608	.560	.533	.446

TABLE 9.1. Simulation values for the mean step size in maximally-scoring segments, as a function of N. Simulations performed with 10,000 to 1,000,000 repetitions.

observation will be relevant to the edge correction formula discussed in Section 9.3.3.

The value of the relative entropy H appears on BLAST printouts. However, the calculation used for these printouts is slightly different from that implied by (9.13). The value of $q(j, k)$ used to compute the score $S(j, k)$ may well be unknown, so that while the values of λ and $S(j, k)$ are known, direct computation of H as defined in (9.13) is not possible.

The BLAST printout value of H uses an indirect approach. With the values of λ, the $S(j, k)$, and the p_j and p'_k in hand, $q(j, k)$ is calculated by using equation (9.9). The printout value of H is now calculated as in (9.13), using the values of $q(j, k)$ so calculated.

9.2.5 Bounds for the P-Value

Using values of $n = N/A$ (derived in Section 9.2.3), of C (given by equation (7.62)), and of λ (given by equation (9.4)), asymptotic bounds for the null hypothesis distribution of $Y_{\max}$ may be found by substituting these values in the inequalities (2.134). It is convenient to replace y in these inequalities by $x + \lambda^{-1} \log N$, and to introduce a new parameter K defined by

$$K = \frac{C}{A} e^{-\lambda}. \tag{9.15}$$

This leads to

$$e^{-Ke^{-\lambda(x-1)}} \leq \mathrm{Prob}(Y_{\max} - \lambda^{-1} \log N \leq x) \leq e^{-Ke^{-\lambda x}}, \tag{9.16}$$

or equivalently

$$1 - e^{-Ke^{-\lambda x}} \leq \mathrm{Prob}(Y_{\max} > \lambda^{-1} \log N + x) \leq 1 - e^{-Ke^{-\lambda(x-1)}}. \tag{9.17}$$

Allowing for notational changes, these are identical to the inequalities (1.13) in Karlin and Dembo (1992). This leads immediately to bounds for the P-value associated with any observed value $y_{\max}$ of $Y_{\max}$, as well as to a conservative P-value found from (2.141) and an approximate conservative P-value found from (2.142).

These calculations ignore edge effects. If N is very large, these effects are negligible. However, short sequences are often used in practice, and then edge effects become important. These are discussed in Section 9.3.3.

The calculation of the bounds in (9.17) requires calculation of both λ and K. The computation of λ via equation (9.4) is comparatively straightforward. The calculation of K from the right-hand side in equation (9.15) would require the calculation of C, A, and λ. However, an exact calculation of K is straightforward in at least two cases, the first where the largest of the $S(j, k)$ is $+1$, and the second where the smallest $S(j, k)$ is -1. Using the notation S for the (random) step in the walk, the two respective formulae for K are

$$K = \left(e^{-\lambda} - e^{-2\lambda} \right) E\left(Se^{\lambda S} \right) \tag{9.18}$$

and

$$K = \left(e^{-\lambda} - e^{-2\lambda} \right) \frac{(E(S))^2}{E\left(Se^{\lambda S} \right)}, \tag{9.19}$$

the expectations being taken assuming that the null hypothesis (9.1) is true.

Equation (9.17) allows us to approximate the P-value of $Y_{\max}$ efficiently, and to a reasonable degree of accuracy. It is often difficult to calculate P-values even for relatively simple random variables, so it is remarkable that P-values can be closely approximated for the random variable $Y_{\max}$, and even more remarkable that this can be done with such efficiency.

9.2.6   The Normalized and the Bit Scores

Karlin and Altschul (1993) call the expression

$$\lambda Y_{\max} - \log(NK) \tag{9.20}$$

a "normalized score," denoted here by S'. In terms of this score, the inequalities (9.16) can be written as

$$e^{-e^{\lambda}e^{-s}} \le \mathrm{Prob}(S' \le s) \le e^{-e^{-s}}. \tag{9.21}$$

A conservative P-value corresponding to any observed value of S' can be calculated from the lower inequality. From the upper inequality we obtain

$$\mathrm{Prob}(S' \ge s) \approx 1 - e^{-e^{-s}}. \tag{9.22}$$

When s is large, (9.22) may be approximated by

$$\mathrm{Prob}(S' \ge s) \approx e^{-s}. \tag{9.23}$$

The P-value corresponding to an observed value $s' = \lambda y_{\max} - \log(NK)$ of S' is, from (9.22),

$$P\text{-value} \approx 1 - e^{-e^{-s'}}. \tag{9.24}$$

The close similarity between (9.22) and equation (2.123) is of course no coincidence. The approximation (9.22) and the fact that the mean value of the distribution described in (2.123) is γ (Euler's constant) show that the mean of S' is approximately γ. From (9.20) and the linearity property of a mean (see Section 1.4), the approximate null hypothesis mean of $Y_{\max}$ is then $\lambda^{-1}(\log(KN) + \gamma)$. The value of γ is usually much smaller than KN, and in BLAST calculations this mean is approximated by

$$\lambda^{-1}(\log(KN)). \tag{9.25}$$

BLAST printouts or published papers record a score similar to the normalized score S', namely the "bit" score. In more recent printouts this is defined by

$$\text{bit score} = \frac{\lambda Y_{\max} - \log K}{\log 2}. \tag{9.26}$$

Previous printouts recorded a bit score defined by

$$\text{bit score} = \frac{\lambda Y_{\max}}{\log 2}. \tag{9.27}$$

Both the normalized score and the bit score have an invariance property. The values of both scores, and hence the respective probability distributions of both scores, do not change if all entries in the substitution matrix used are multiplied by the same constant, say G. This can be seen from the fact that such a multiplication changes the value of $Y_{\max}$ by a multiplicative factor of G, but at the same time (see equation (9.4)) changes the value of λ by a multiplicative factor $1/G$. Thus the quantity $\lambda Y_{\max}$, the central feature of both scores, remains unchanged, and consequently the normalized score and the bit score remain unchanged also.

A much stronger result than this is true. Whereas the value of $Y_{\max}$ has no absolute interpretation if the substitution matrix from which it is calculated is not specified, the normalized score S' and the bit score do have such an interpretation. In the case of S' this is made clear by approximations such as (9.22): Here the right-hand side is free of any parameters, so that the (approximate) distribution of S' is known whatever the details of the substitution matrix. If N is given, the same can be said for the bit score (9.26).

9.2.7 The Number of High-Scoring Excursions

In this section we define and discuss the quantity E', whose calculation leads ultimately to the quantity "Expect" found on BLAST printouts. Throughout the discussion we ignore edge effects: These are discussed in detail in Section 9.3.3.

Consider excursions from a ladder point in the random walk described by the comparison of the two sequences. We have seen that under the null hypothesis, for each such excursion, the maximum height Y has a geometric-like distribution whose parameters can be calculated. Denoting as above the maximum heights of the excursions from the various ladder points by $Y_1, Y_2, \ldots$, the relation (1.67) shows that the probability that any Y_i takes a value v or larger is approximately $Ce^{-\lambda v}$, where C and λ are those appropriate to the walk in question. Since to a close approximation the number of excursions can be taken to be N/A, as discussed in Section 9.2.3, the mean number of excursions reaching a height v or more is approximately $\frac{NC}{A}e^{-\lambda v}$. This mean ultimately forms the basis of the BLAST P-value calculation. In the standard BLAST calculations discussed below, this is replaced by the approximating value

$$NKe^{-\lambda v}, \tag{9.28}$$

where K is given by (9.15). All calculations below assume this value.

Since $Y_1, Y_2, \ldots$ are iid random variables, the number of excursions having a height v or more has a binomial distribution with mean given by (9.28). The theory developed in Section 4.2 shows that when v is large, the number of excursions reaching a height greater than or equal to v, that is, the number of high-scoring segment pairs (HSPs) with a score v or more, has, using the Poisson approximation to the binomial, a Poisson distribution with mean given in (9.28). (A more sophisticated analysis, based on equations such as (2.77) that allow for the fact that the number of ladder points is a random variable, arrives at the same conclusion.) Thus, to test for significance, the actual number of such excursions achieving a score exceeding v can be compared with the tail probability of this Poisson distribution.

The expected value of the number of excursions corresponding to the observed maximal score $y_{\max}$ is found by replacing the arbitrary number v

in equation (9.28) by y_{max}. This expected value is denoted by E', so that

$$E' = NKe^{-\lambda y_{max}}. \tag{9.29}$$

The relation between E' and the normalized score S' defined in (9.20) is

$$S' = -\log E', \tag{9.30}$$

and the relation between E' and the P-value approximation is found from (9.24) as

$$P\text{-value} \approx 1 - e^{-E'}, \quad E' = -\log(1 - P\text{-value}). \tag{9.31}$$

It follows from the approximation (B.21) that the approximate P-value is very close to E' when E' is small.

Similar calculations may be made for any high-scoring excursion.

9.2.8   The Karlin–Altschul Sum Statistic

Focusing on the value of Y_{max} loses the information provided by the heights of the second-largest, third-largest, etc., excursions in the random walk. In this section we discuss a statistic that uses information from these other excursions.

Consider the r largest excursion heights, that is, the r largest Y_i values, assuming that there are at least r ladder points. It is convenient to use a notation that is different from the notation for order statistics used in Chapter 2, and assume that $Y_1(= Y_{max}) \geq Y_2 \geq \cdots \geq Y_r$. By analogy with the definition in equation (9.20) we can compute r normalized scores $S'_1, S'_2, \ldots, S'_r$ from $Y_1, Y_2, \ldots, Y_r$, where

$$S'_i = \lambda Y_i - \log(NK). \tag{9.32}$$

Note that $S'_1 = S'$ as defined in equation (9.20).

Karlin and Altschul (1993) show that to a close approximation, the null hypothesis joint density function $f_{\boldsymbol{S}}(s_1, \ldots, s_r)$ of $\boldsymbol{S} = (S'_1, \ldots, S'_r)$ is

$$f_{\boldsymbol{S}}(s_1, \ldots, s_r) = \exp\left(-e^{-s_r} - \sum_{k=1}^{r} s_k\right). \tag{9.33}$$

We can use any reasonable function of $S'_1, S'_2, \ldots, S'_r$ as test statistic. Transformation methods such as those introduced in Chapter 2 can then be used to find the distribution of this test statistic, and this in turn allows the computation of a P-value, and also an E, or Expect, value, corresponding to any observed value of this statistic.

The specific statistic suggested by Karlin and Altschul (1993) is the sum $T_r = S'_1 + \cdots + S'_r$ of the normalized scores. This is called the Karlin–Altschul sum statistic. Using transformation methods such as those described in Section 2.15, Karlin and Altschul use the joint density function

(9.33) to calculate the null hypothesis density function $f(t)$ of T_r. The resulting expression is

$$f_{T_r}(t) = \frac{e^{-t}}{r!(r-2)!} \int_0^{+\infty} y^{(r-2)} \exp(-e^{(y-t)/r}) dy. \qquad (9.34)$$

As an exercise in transformation theory we confirm this calculation for the case $r = 2$ in Section D. When t is sufficiently large, this density function can be used to find the approximate expression

$$\text{Prob}(T_r \geq t) \approx \frac{e^{-t} t^{r-1}}{r!(r-1)!}. \qquad (9.35)$$

In the case $r = 1$, this is the approximation given in equation (9.23). The approximation (9.35) is sufficiently accurate when $t > r(r+1)$, and popular implementations of BLAST use (9.35) when this inequality holds.

If t is the observed value of T_r, the right-hand side in (9.35) then provides the approximate P-value corresponding to this observed value. This is used as a component of the eventual BLAST printout P-value.

Karlin and Altschul (1993) provide an example (see their Table 1) in which the observed values of the highest two normalized scores are $s_1' = 4.4$ and $s_2' = 2.5$. Using the value $r = 1$ in the approximation (9.35), the P-value corresponding to the highest normalized score 4.4 is $e^{-4.4} = 0.012$. Using the value $r = 2$, the P-value corresponding to the sum 6.9 of the highest two normalized scores is calculated from (9.35) as $\frac{6.9}{2} e^{-6.9} = 0.0035$, and these calculations confirm those given by Karlin and Altschul. For further aspects of these calculations, and of the calculations in their Table 2, see Problems 9.9 and 9.10.

A further aspect of the use of a test statistic based on T_r is that of consistent ordering. We say that r HSPs, HSP1, HSP2, ..., HSPr, between two sequences are *consistently ordered* if whenever the midpoint in the first sequence in HSPi comes before the midpoint in the first sequence in HSPj, then the same is true for the midpoints of the second sequence. More generally, one might require that the sequences in the different HSPs not overlap, or overlap no more than some fixed proportion (in popular implementations of BLAST, the default value of this proportion is 0.125). When consistent ordering is required, the P-value calculations must be amended. In the case where overlaps are unrestricted, this requirement cuts down the search space by a factor of $r!$, the number of orderings of the r HSPs. This implies that the P-value calculated from (9.35) should be divided by $r!$. A simple approximation (Karlin and Altschul (1993)) is that P-value calculations are amended by replacing t, the observed value of T_r, by $t + \log(r!)$ in the right-hand side of (9.35), or by a corresponding amendment to the calculations using (9.34). The popular implementations of BLAST use this approach, and furthermore allow the degree to which the HSPs overlap to be restricted. Restricting overlaps should require a further adjustment of the P-value. This is not apparently done by the popular implementations. However, an adjustment is made to the "edge correction" factor discussed below, which may or may not account for this (see Section 9.3.3).

A further complication introduced by the use of the sum statistic in BLAST is that of multiple testing. In practice, the value of r is not fixed in advance and is allowed to vary. Thus the problem of multiple testing, discussed in Section 3.8, arises. We delay discussion of the way in which this problem is handled in BLAST calculations until Section 9.3.4.

In BLAST printouts the notation r is replaced by N. We have used r here because N is used to denote a sequence length. Further, r is the notation used in the fundamental paper of Karlin and Altschul (1993).

9.3   The Comparison of Two Unaligned Sequences

9.3.1   Introduction

The theory of Section 9.2 considered calculations relevant to a fixed un-gapped alignment in the comparison of two sequences each of length N. In this section we consider a more general question. We are given two sequences of lengths N_1 and N_2, but we are not given any specific alignment between them. The goal is to find the significance of high-scoring segment pairs between all possible (ungapped) local alignments. The highest-scoring pair is called the maximal-scoring segment pair (MSP).

9.3.2   Theoretical and Empirical Background

BLAST considers all ungapped alignments determined by all possible relative positions of the two sequences. For each relative position, the alignment is extended as far as possible in either direction, giving a total of $N_1 + N_2 - 1$ ungapped alignments. Figure 9.1 shows the first five alignments between two sequences of length 11 and 9 respectively.

Each such alignment yields a random walk similar to that considered in Section 9.2.2 arises, giving a collection of random walks. There are $N_1 N_2$ amino acid comparisons that can be made as the two sequences take all possible positions relative to each other.

The theory for this case is far more complicated than that outlined in Section 9.2, where only one alignment occurs. Among other matters the question of the dependence of the walks arising in different alignments must be addressed. The key papers developing the theory are Dembo et al. (1994a, 1994b). The theory is too advanced for this book, and here we simply reproduce the relevant results, the most important of which is that, to a sufficient approximation, many of the conclusions of Section 9.2 carry over to the present case, with N replaced by $N_1 N_2$.

```
sequence 1    . . . . . . . . . . .
sequence 2                        . . . . . . . . . .

sequence 1    . . . . . . . . . .
sequence 2                        . . . . . . . . . .

sequence 1    . . . . . . . . . .
sequence 2                    . . . . . . . . . .

sequence 1    . . . . . . . . . .
sequence 2                    . . . . . . . . .

sequence 1    . . . . . . . . . .
sequence 2                . . . . . . . . .
```

FIGURE 9.1.

However, there are several qualifications to make about this statement. First, several conditions (given by Dembo et al. (1994a, 1994b)) need to be satisfied before the theory of Section 9.2 can be used. Second, some of the theoretical results proved apply only in the limit as both N_1 and N_2 become large. Thus the theory might not hold in the case of interest in practice, where both sequences might be of length only a few hundred or less. Thus many simulations have been carried out to assess the extent to which the theory of Section 9.2 carries through to cases of practical interest; see, for example, Altschul and Gish (1996) and Pearson (1998). A broad conclusion reached from these simulations is that the theory of Section 9.2 does carry over to a reasonable approximation if N is replaced by the product $N_1 N_2$, or by a more refined function allowing for edge effects. Thus with much but not complete theoretical and empirical support, and remembering that cases can arise that are not covered by the theory of Section 9.2, we now use that theory for the comparison of two sequences, replacing N by $N_1 N_2$ for the moment, and by a more refined expression in Section 9.3.3.

We consider first the random variable Y_{max}, the maximum score achieved in the random walk comparing the sequences, using all possible ungapped local alignments between the two. This score corresponds to the MSP. Any MSP or HSP starts at a ladder point in the BLAST random walk and finishes the first time that the maximum upward excursion from this ladder point is reached. Under the heuristic adopted, Y_{max} is the maximum of a number of geometric-like random variables, whose distribution depends on the parameters λ, C, and n. The calculations for λ and C follow as in Section 9.2.3. The mean number n of ladder points in this random walk corresponding to the collection of all alignments of the two sequences is approximated by

$$\frac{N_1 N_2}{A},$$

(9.36)

where A is the mean distance between ladder points. This value is used for n throughout the following theory. The discussion at the end of the Section 9.2.3 applies equally well to explain this formula. The theory discussed in Section 9.2.3 can now be used, with the value of n given by equation (9.36), of C given by equation (7.62), and of λ given by equation (9.4).

The key formulae in Section 9.2.3 are now taken over to the present case with these parameter values. Thus assuming that the null hypothesis (9.1) is true,

$$1 - e^{-Ke^{-\lambda x}} \leq \mathrm{Prob}(Y_{\max} > \lambda^{-1} \log(N_1 N_2) + x) \leq 1 - e^{-Ke^{-\lambda(x-1)}}, \quad (9.37)$$

and if the normalized score S' is redefined as

$$S' = \lambda Y_{\max} - \log(N_1 N_2 K), \quad (9.38)$$

the inequality (9.21) and the approximations (9.22) and (9.23) continue to hold. As a result, the right-hand side in the latter approximation also has the interpretation of an approximate P-value corresponding to the observed value s of S' as defined in (9.38).

Similarly, the approximate null hypothesis mean of $Y_{\max}$ is

$$\lambda^{-1} \left(\log(N_1 N_2 K) + \gamma \right). \quad (9.39)$$

9.3.3   Edge Effects

The calculations of the preceding sections do not allow for edge effects, an important factor in the comparison of two comparatively short sequences. In this section we discuss the adjustments to the previous calculations that are used in BLAST calculations to allow for edge effects.

A high-scoring random walk excursion induced by the comparison of the two sequences might be cut short at the end of a sequence match, so that the height of high-scoring excursions, and the number of such excursions, will tend to be less than that predicted by the above theory. Whereas much of BLAST theory concerns two long sequences for which edge effects are of less importance, in practice BLAST considers databases made up of a large number of often short sequences, for which edge effects are important. Thus BLAST calculations allow for edge effects, and do this by subtracting from both N_1 and N_2 a factor depending on the mean length of any high-scoring excursion. The justification for this is largely empirical (Altschul and Gish (1996)).

Equation (9.14) shows that the mean value of the step in a high-scoring excursion asymptotically approaches the value $\lambda^{-1} H$. Given that the height achieved by a high-scoring excursion is denoted by y, equation (7.24) suggests that the mean length $E(L|y)$ of this excursion, conditional on y, is given by

$$E(L|y) = \frac{\lambda y}{H}. \quad (9.40)$$

BLAST theory then replaces N_1 and N_2 in the calculations given above respectively by $N_1' = N_1 - E(L)$, $N_2' = N_2 - E(L)$. Specifically, the normalized score (9.20) is replaced by

$$\lambda Y_{\max} - \log(N_1'N_2'K), \tag{9.41}$$

with

$$N_1' = N_1 - \frac{\lambda Y_{\max}}{H}, \quad N_2' = N_2 - \frac{\lambda Y_{\max}}{H}. \tag{9.42}$$

The expression (9.28) for the expected number of excursions scoring v or higher is correspondingly replaced by

$$N_1'N_2'Ke^{-\lambda v}, \tag{9.43}$$

with $N_1' = N_1 - \lambda v/H$, $N_2' = N_2 - \lambda v/H$. Similarly, the calculation of E' given in (9.29) is replaced by

$$E' = N_1'N_2'Ke^{-\lambda y_{\max}}. \tag{9.44}$$

The use of edge corrections using (9.42) assumes that the asymptotic formula (9.14) for the mean step size in a high-scoring excursion is appropriate. The simulations discussed in Section 9.2.4 show that this might not be the case, at least when N_1 and N_2 are both of order 10^2. Table 9.2 shows empirical MSP mean lengths (from simulations with 10,000 to 1,000,000 replications) and the values calculated from (9.40) for the simulation leading to the data of Table 9.1. Clearly the values calculated from (9.40) are inaccurate for anything other than very large values of N. Thus while the calculated values approach the empirical values as N increases (in line with the convergence of the mean step sizes to the asymptotic value in Table 9.1), the use of the edge correction implied by (9.42) might in practice lead to P-value estimates less than the correct values, that is, to anticonservative tests, for anything other than very large values of N. The use of the observed value of the length of the MSP appears to give more accurate results (Altschul and Gish (1996)).

N	500	5,000	50,000	500,000	5,000,000
Empirical mean length	43.2	106.8	181.1	258.0	335.9
Calculated mean length	98.7	168.4	237.4	301.8	373.9

TABLE 9.2. Empirical values for the mean length of the MSP and the value found from (9.40) and empirical values of $y_{\max}$. Simulations performed with 10,000 to 1,000,000 repetitions.

In the popular implementations of BLAST the edge effect correction factor for the Karlin–Altschul sum statistic T_r is calculated as follows.

First, a raw edge effect correction is calculated as $\lambda(Y_1 + Y_2 + \cdots + Y_r)/H$, generalizing the term $\lambda Y_{max}/H$ given in (9.42). When consistent ordering is required and overlaps are restricted by a factor f, this is then multiplied by a factor $1 - (r+1)f/r$, where f is an "overlap adjustment factor" that can be chosen by the investigator. The default value of f is 0.125, implying that overlaps between segments of up to 12.5% are allowed. The use of f is illustrated by an example in Section 9.5.2. To this the value $r - 1$ is added, leading eventually to an edge correction value $E(L)$, defined by

$$E(L) = \frac{\lambda}{H}(Y_1 + Y_2 + \cdots + Y_r)\left(1 - \frac{r+1}{r}f\right) + r - 1. \qquad (9.45)$$

While this formula is used in BLAST, there appears to be no publication justifying its validity. It could be tested empirically in the spirit of Altschul et al. (1996). The values of N_1 and N_2 in the normalized score formula (9.32) are then replaced, respectively, by

$$N_1' = N_1 - E(L), \ N_2' = N_2 - E(L). \qquad (9.46)$$

The normalized scores in (9.32) are now redefined as

$$S_i' = \lambda Y_i - \log(N_1' N_2' K), \qquad (9.47)$$

and with this new definition the sum statistic T_r is redefined as

$$T_r = S_1' + S_2' + \cdots + S_r'. \qquad (9.48)$$

The problems discussed above concerning the accuracy of the approximation (9.40) leading to the expressions for N_1' and N_2', and hence of calculations derived from S_i' and T_r, apply here also.

If the r HSPs are required to be consistently ordered, a term $\log r!$ is added to T_r (as discussed in Section 9.2.8), and if the sum so calculated is t, the P-value is then calculated as in (9.35).

9.3.4 Multiple Testing

There is no obvious choice for the value of r when the sum statistic is used in the test procedure. It is natural to consider all $r = 1, 2, 3, \ldots$, and choose the set of HSPs with lowest sum statistic P-value as the most significant, regardless of the value of r, and this is what is done in BLAST calculations. However, this procedure implies that a sequence of tests, one for each r, rather than a single test, is performed, so that the issue of multiple testing, discussed in Section 3.8, arises. Green (unpublished results) has found through simulations that ignoring the multiple testing issue leads to a significant overestimate of BLAST P-values, so that an amendment to formal P-value calculations is indeed necessary.

Unfortunately, there is no rigorous theory available to deal with this issue, and in practice it is handled in an ad hoc manner. For example, in the Washington University versions of BLAST, the P-value is adjusted when $r > 1$ by dividing the formal P-value by a factor $(1 - \pi)\pi^{r-1}$. The parameter π has default value .5, but its value can be chosen by the user. The default value 0.5 is used in the example in Section 9.5.

When $r = 1$ the procedure is slightly different. The factor $(1 - \pi)\pi^{r-1}$ in this case is $1 - \pi$, and this implies that the value of E' given in (9.29) is divided by $1 - \pi$ to find the amended expected value E. The BLAST default value 0.5 of π implies that $E = 2E'$, so that E is calculated to be

$$E = 2N_1' N_2' K e^{-\lambda_{\max}}. \tag{9.49}$$

The P-value corresponding to this is then found, using the analogy with (9.31), to be

$$P\text{-value} \approx 1 - e^{-E}. \tag{9.50}$$

The P-value and Expect calculations used in BLAST embody the amendments discussed above to the theoretical values (given in Section 9.2). These amendments relate to edge effect corrections, multiple testing corrections and, in the case of the sum statistic, the consistent ordering and overlap corrections. Some details of these amendments appear not to be mentioned in BLAST documentation in the popular implementations of BLAST, and only become clear by careful reading of the code.

9.4   The Comparison of a Query Sequence Against a Database

We now consider the case that is most relevant in practice. In this case we have a single "query" sequence, and we wish to search an entire database of many sequences for those with significant similarity to the query sequence. To do this, first a (heuristic) search algorithm is employed to find the high-scoring HSPs, or sets of HSPs. The P-values and Expect values of these HSPs are then approximated. These approximations are discussed in this section.

Whereas query sequence amino acid frequencies are taken from the query sequence at hand, database frequencies often are taken from some (different) published set of estimated amino acid frequencies, for example those in Robinson and Robinson (1991). These might be different again from those used to create the substitution matrix.

In approximating database P-values and Expect values, the size of the entire database must be taken into account. This raises another multiple testing problem in addition to that discussed above. What is done in practice is first to use the results of the last section to compare the query sequence to each individual database sequence, to obtain P-values for individual sequence comparisons. Then the individual sequence P-values are adjusted to account for the size of the database. If all the sequences in the database were the same size, then we could just multiply the Expect values by the number of sequences, using the linearity property of means (see (2.66)). As an approximation to this, what is done in practice is to multiply by D/N_2, where D is the total length of the database, i.e., the sum of the lengths of all of the database sequences, and N_2 is the length of the database sequence that aligns with the query to give the HSP (or HSPs) in question. These Expect values are then converted to P-values. The details are as follows.

We consider first the case of single HSPs ($r = 1$). Because of its linearity properties, the most useful quantity for database searches is the quantity E, defined in (9.49), and its generalizations for other HSPs. Suppose that in the database sequence of interest there is some HSP with score v. The Poisson distribution is then used to approximate the probability that in the match between query sequence and database sequence at least one HSP scores v or more. This probability is

$$1 - e^{-E}. \tag{9.51}$$

Since the entire database is D/N_2 times longer than the database sequence of interest, the mean number of HSPs scoring v or more in the entire database, namely the BLAST printout quantity Expect, is given by

$$\text{Expect} = \frac{(1 - e^{-E})D}{N_2}. \tag{9.52}$$

From this value of Expect an approximate P-value is calculated from (9.31) as

$$P\text{-value} \approx 1 - e^{-\text{Expect}}. \tag{9.53}$$

This is the BLAST printout P-value for the case $r = 1$.

For the case $r > 1$, sum statistics for various database sequences are calculated as described in Section 9.2.8, and P-values are calculated either from (9.34) or (9.35), using all the amendments discussed above. From each such P-value a total database value of Expect is calculated using a formula generalizing that derived from (9.52), namely

$$\text{Expect} = \frac{(P\text{-value})D}{N_2}, \tag{9.54}$$

where N_2 is the length of the database sequence from which the sum is found. From this a P-value is calculated as in (9.53).

Finally, all single (i.e., $r = 1$) HSPs or summed ($r > 1$) HSPs with sufficiently low values of Expect (or, equivalently, sufficiently low P-values) are listed, and eventually printed out in increasing order of their Expect values. The value of r, given as N in the printout, is also listed.

9.5   Printouts

In this section we relate the above theory to an actual BLAST printout.

BLAST printouts give the values of λ, calculated from (9.4), of K, calculated from (9.15) amended appropriately for sequence comparisons, and H, found from the procedure described at the end of Section 9.2.4. They also list the statistics "Score" or "High Score," which in the case $r = 1$ are the values of the maximal scores $y_{\max}$ or other high-scoring HSPs. In the case of the sum statistic T_r (with $r > 1$), the score of the highest-scoring component in the sum is listed. Also listed are the "bit scores" associated with these, together with "Expect" values and P-values calculated as described in Section 9.4. We repeat that variants of these calculations are possible for different versions of BLAST and that sometimes more sophisticated calculations, taking into account factors not discussed above, are used.

9.5.1   Example

A partial printout of Example 3 from the Washington University BLAST 1.4 program follows:[1]

```
BLASTP 1.4.10MP-WashU [29-Apr-96] [Build 22:25:52 May 19 1996]


Query= gi|557844|sp|P40582|YIV8_YEAST HYPOTHETICAL 26.8 KD PROTEIN IN HYR1
                     3'REGION.
        (234 letters)

Database:  SWISS-PROT Release 34.0
           59,021 sequences; 21,210,388 total letters.


-------------------------------------------------------------------

                                                    Smallest
                                                      Sum
                                               High  Probability
Sequences producing High-scoring Segment Pairs: Score P(N)      N

sp|P46429|GTS2_MANSE GLUTATHIONE S-TRANSFERASE 2 (EC 2....  53  0.010   3
sp|P46420|GTH4_MAIZE GLUTATHIONE S-TRANSFERASE IV (EC 2... 70  0.14    1
sp|P41043|GTS2_DROME GLUTATHIONE S-TRANSFERASE 2 (EC 2.... 54  0.19    2
sp|P34345|YK67_CAEEL HYPOTHETICAL 28.5 KD PROTEIN C29E4... 50  0.42    2
sp|Q04522|GTH_SILCU  GLUTATHIONE S-TRANSFERASE (EC 2.5.... 62  0.87    1


-------------------------------------------------------------------


>sp|P46429|GTS2_MANSE GLUTATHIONE S-TRANSFERASE 2 (EC 2.5.1.18) (CLASS-SIG).
          Length = 203
```

---

[1] http://sapiens.wustl.edu/blast/blast/example3-14.html

```
 Score = 53 (24.4 bits), Expect = 0.010, Sum P(3) = 0.010
 Identities = 10/19 (52%), Positives = 15/19 (78%)

Query:   167 ISKNNGYLVDGKLSGADIL 185
             I+KNNG+L  G+L+ AD +
Sbjct:   136 ITKNNGFLALGRLTWADFV 154

 Score = 46 (21.2 bits), Expect = 0.010, Sum P(3) = 0.010
 Identities = 8/21 (38%), Positives = 13/21 (61%)

Query:    45 PELKKIHPLGRSPLLEVQDRE 65
             PE K    P G+ P+LE+ ++
Sbjct:    39 PEFKPNTPFGQMPVLEIDGKK 59

 Score = 36 (16.6 bits), Expect = 0.010, Sum P(3) = 0.010
 Identities = 8/26 (30%), Positives = 12/26 (46%)

Query:   202 EDYPAISKWLKTITSEESYAASKEKA 227
             E YP  K ++T+ S    A  + A
Sbjct:   173 EQYPIFKKPIETVLSNPKLKAYLDSA 198


>sp|P46420|GTH4_MAIZE GLUTATHIONE S-TRANSFERASE IV (EC 2.5.1.18) (GST-IV)
             (GST-27) (CLASS PHI).
             Length = 222

 Score = 70 (32.3 bits), Expect = 0.15, P = 0.14
 Identities = 17/56 (30%), Positives = 27/56 (48%)

Query:    18 RLLWLLDHLNLEYEIVPYKRDANFRAPPELKKIHPLGRSPLLEVQDRETGKKKILA 73
             R L  L+  ++YE+VP  R      PE   +P G+ P+LE D   + + +A
Sbjct:    18 RALLALEEAGVDYELVPMSRQDGDHRRPEHLARNPFGKVPVLEDGDLTLFESRAIA 73


>sp|Q04522|GTH_SILCU GLUTATHIONE S-TRANSFERASE (EC 2.5.1.18) (CLASS-PHI).
             Length = 216

 Score = 62 (28.6 bits), Expect = 2.1, P = 0.87
 Identities = 15/43 (34%), Positives = 21/43 (48%)

Query:    18 RLLWLLDHLNLEYEIVPYKRDANFRAPPELKKIHPLGRSPLLE 60
             R+L  L  +LE+E VP   A    P  ++P G+ P LE
Sbjct:    15 RVLVALYEKHLEFEFVPIDMGAGGHKQPSYLALNPFGQVPALE 57

-----------------------------------------------------------------------

Matrix name    Lambda    K       H
----------------------------------------
 BLOSUM62      0.320   0.137   0.401
```

We first verify the calculations leading to the Maize Glutathione match sequence value of 0.15 for Expect. For this case, the printout above shows

that
$$N_1 = 234, \quad N_2 = 222, \quad y_{\max} = 70.$$

Equation (9.42), in conjunction with the printout values of λ and H, gives
$$N_1' = 234 - \frac{0.32(70)}{0.401} = 178, \quad N_2' = 222 - \frac{0.32(70)}{0.401} = 166.$$

Inserting these values and the printout value of K in (9.44), we get
$$E' \approx (178)(166)(0.137)e^{-0.32(70)} \approx 7.6(10)^{-7}. \tag{9.55}$$

Multiplying by the multiple testing factor 2 gives $E \approx 15.2(10)^{-7}$. Inserting this value in (9.52), we get
$$\text{Expect} \approx \left(1 - e^{-15.2(10)^{-7}}\right) \frac{21{,}210{,}388}{222} \approx 0.15, \tag{9.56}$$

in agreement with the value 0.15 for Expect found in the printout.

Given this value, equation (9.53) gives an approximate P-value of 0.14, in agreement with the printout calculation. Further, equation (9.27) gives a value $0.320(70)/\log 2 \approx 32.3$ for the bit score, in agreement with the printout value.

A similar set of calculations gives, to a close approximation, the value 2.1 for Expect in the Silcu Glutathione match.

We finally consider the Manse Glutathione match, for which $r = 3$, and describe the calculations leading to the printout value 0.010 for Expect. As noted above, this value of Expect is found using a series of amendment calculations, starting with the edge correction. The expression (9.45), together with data in the printout and the default value 0.125 for f, leads to an edge correction of
$$\frac{0.32}{0.401}(53 + 46 + 36)\left(1 - \frac{4}{3}(0.125)\right) + 2 = 91.78.$$

Thus from (9.46), $N_1' = 142.2835$ and $N_2' = 111.2835$. Using these values in (9.32), the amended observed value of T_3 is computed as
$$0.32(53 + 46 + 36) - 3\log\left((0.137)(142.22)(111.22)\right) = 20.16.$$

The consistent ordering requirement holds, so we add $\log 3! = 1.79$ to this to get the value 21.95. The P-value corresponding to this is found from (9.35) to be $1.181(10)^{-8}$. Multiplying by the multiple testing factor $2^3 = 8$ yields a value of $9.448(10)^{-8}$. The value of E is essentially identical to this.

Finally, the total database Expect value is found by multiplying this by $21{,}210{,}388/203$, and this gives the value 0.010 found in the printout.

It might be a matter of concern that various somewhat arbitrary constants enter into the above calculations. This concern is reinforced by the fact that the distribution function of maximum statistics changes very sharply, as demonstrated in Table 2.5. As a result, calculated P-values are quite sensitive to the somewhat arbitrary numerical values of these constants. In practice, this concern is not important, since users of BLAST printouts seldom view a P-value even as small as 10^{-5} as interesting, and use the numerical P-values together with significant biological judgment.

9.5.2 A More Complicated Example

The way in which some BLAST outputs are formatted can be confusing. The partial output from a BLAST search against Swissprot is given below,[2] in which only the set of HSPs between the query and one database sequence are shown. There are 12 HSPs in total; however, since consistent ordering is required, the smallest sum P-value comes from a set of 8 HSPs.

```
Query=  gi|604369|sp|P40692|MLH1_HUMAN MUTL PROTEIN HOMOLOG 1 (DNA MISMATCH
        (756 letters)


                                                        Smallest
                                                          Sum
                                             High    Probability
Sequences producing High-scoring Segment Pairs:    Score   P(N)       N

sp|P38920|MLH1_YEAST MUTL PROTEIN HOMOLOG 1 (DNA MIS...   675   1.7e-138   8

>sp|P38920|MLH1_YEAST MUTL PROTEIN HOMOLOG 1 (DNA MISMATCH REPAIR PROTEIN.)
          Length = 769

 Score = 675 (309.6 bits), Expect = 1.7e-138, Sum P(8) = 1.7e-138
 Identities = 127/222 (57%), Positives = 170/222 (76%)

Query:   8 IRRLDETVVNRIAAGEVIQRPANAIKEMIENCLDAKSTSIQVIVKEGGLKLIQIQDNGTG 67
           I+ LD +VVN+IAAGE+I  P NA+KEM+EN +DA +T I ++VKEGG+K++QI DNG+G
Sbjct:   5 IKALDASVVNKIAAGEIIISPVNALKEMMENSIDANATMIDILVKEGGIKVLQITDNGSG 64

Query:  68 IRKEDLDIVCERFTTSKLQSFEDLASISTYGFRGEALASISHVAHVTITTKTADGKCAYR 127
           I K DL I+CERFTTSKLQ FEDL+ I TYGFRGEALASISHVA VT+TTK + +CA+R
Sbjct:  65 INKADLPILCERFTTSKLQKFEDLSQIQTYGFRGEALASISHVARVTVTTKVKEDRCAWR 124

Query: 128 ASYSDGKLKAPPKPCAGNQGTQITVEDLFYNIATRRKALKNPSEEYGKILEVVGRYSVHN 187
           SY++GK+   PKP AG  GT I VEDLF+NI +R +AL++ ++EY KIL+VVGRY++H+
Sbjct: 125 VSYAEGKMLESPKPVAGKDGTTILVEDLFFNIPSRLRALRSHNDEYSKILDVVGRYAIHS 184

Query: 188 AGISFSVKKQGETVADVRTLPNASTVDNIRSIFGNAVSRELI 229
           I FS KK G++  +  P+ + D IR++F +V+ LI
Sbjct: 185 KDIGFSCKKFGDSNYSLSVKPSYTVQDRIRTVFNKSVASNLI 226

 Score = 215 (100.6 bits), Expect = 1.7e-138, Sum P(8) = 1.7e-138
 Identities = 39/85 (45%), Positives = 58/85 (68%)
```

---

[2]http://blast.wustl.edu/blast/example2-14.html

```
Query: 259 LLFINHRLVESTSLRKAIETVYAAYLPKNTHPFLYLSLEISPQNVDVNVHPTKHEVHFLH 318
            + FIN+RLV    LR+A+ +VY+ YLPK   PF+YL + I P   VDVNVHPTK EV FL
Sbjct: 259 IFFINNRLVTCDLLRRALNSVYSNYLPKGNRPFIYLGIVIDPAAVDVNVHPTKREVRFLS 318

Query: 319 EESILERVQQHIESKLLGSNSSRMY 343
            ++ I+E++  + ++L   ++SR +
Sbjct: 319 QDEIIEKIANQLHAELSAIDTSRTF 343


 Score = 136 (64.7 bits), Expect = 1.7e-138, Sum P(8) = 1.7e-138
 Identities = 40/121 (33%), Positives = 58/121 (47%)


Query: 636 LIGLPLLIDNYVPPLEGLPIFILRLATEVNWDEEKECFESLSKECAMFYSIRKQYISEES 695
            L  LPLL+   Y+P L  LP FI RL  EV+W++E+EC + + +E A+ Y      + S
Sbjct: 649 LKSLPLLLKGYIPSLVKLPFFIYRLGKEVDWEDEQECLDGILREIALLYIPDMVPKVDTS 708

Query: 696 TLSGQQSEVPGSIPNSWKWTVEHIVYKALRSHILPPKHFTEDGNILQLANLPDLYKVFERC 756
               S + E   I    +                  +++++ANLPDLYKVFERC
Sbjct: 709 DASLSEDEKAQFINRKEHISSLLEHVLFPCIKRRFLAPRHILKDVVEIANLPDLYKVFERC 769


 Score = 93 (45.2 bits), Expect = 1.7e-138, Sum P(8) = 1.7e-138
 Identities = 21/52 (40%), Positives = 29/52 (55%)


Query: 539 ALAQHQTKLYLLNTTKLSEELFYQILIYDFANFGVLRLSEPAPLFDLAMLAL 590
            A  QH  KL+L++   +  ELFYQI + DFANFG + L      D+ +  L
Sbjct: 549 AAIQHDLKLFLIDYGSVCYELFYQIGLTDFANFGKINLQSTNVSDDIVLYNL 600


 Score = 76 (37.4 bits), Expect = 1.7e-138, Sum P(8) = 1.7e-138
 Identities = 17/49 (34%), Positives = 30/49 (61%)


Query: 501 INLTSVLSLQEEINEQGHEVLREMLHNHSFVGCVNPQWALAQHQTKLYL 549
            +NLTS+ L+E++++  H  L ++   N ++VG V+ +  LA  Q  L L
Sbjct: 509 VNLTSIKKLREKVDDSIHRELTDIFANLNYVGVVDEERRLAAIQHDLKL 557


 Score = 42 (22.0 bits), Expect = 1.7e-138, Sum P(8) = 1.7e-138
 Identities = 8/26 (30%), Positives = 16/26 (61%)


Query: 609 EYIVEFLKKKAEMLADYFSLEIDEEG 634
            E I+ +   + ML +Y+S+E+  +G
Sbjct: 614 EKIISKIWDMSSMLNEYYSIELVNDG 639


 Score = 41 (21.5 bits), Expect = 1.7e-138, Sum P(8) = 1.7e-138
 Identities = 9/33 (27%), Positives = 20/33 (60%)


Query: 365 SLTSSSTSGSSDKVYAHQMVRTDSREQKLDAFL 397
            S T++++    K   +++VR D+ + K+ +FL
Sbjct: 381 SYTTANSQLRKAKRQENKLVRIDASQAKITSFL 413


 Score = 39 (20.6 bits), Expect = 1.5e-21, Sum P(5) = 1.5e-21
 Identities = 9/27 (33%), Positives = 14/27 (51%)


Query: 411 IVTEDKTDISSGRARQQDEEMLELPAP 437
            + T+ K D  + R    + +MLE P P
```

```
Sbjct: 112 VTTKVKEDRCAWRVSYAEGKMLESPKP 138


 Score = 37 (19.7 bits), Expect = 1.7e-132, Sum P(7) = 1.7e-132
 Identities = 7/22 (31%), Positives = 13/22 (59%)


Query: 503 LTSVLSLQEEINEQGHEVLREM 524
           +TS LS ++ N +G   R++
Sbjct: 409 ITSFLSSSQQFNFEGSSTKRQL 430


 Score = 36 (19.3 bits), Expect = 4.2e-46, Sum P(7) = 4.2e-46
 Identities = 9/40 (22%), Positives = 20/40 (50%)


Query:  14 TVVNRIAAGEVIQRPANAIKEMIENCLDAKSTSIQVIVKE 53
           TV N+  A  +I   + ++++   +D K  ++  I K+
Sbjct: 215 TVFNKSVASNLITFHISKVEDLNLESVDGKVCNLNFISKK 254


 Score = 34 (18.4 bits), Expect = 1.7e-138, Sum P(8) = 1.7e-138
 Identities = 7/20 (35%), Positives = 12/20 (60%)


Query: 242 MNGYISNANYSVKKCIFLLF 261
           ++G + N N+  KK I  +F
Sbjct: 241 VDGKVCNLNFISKKSISPIF 260


 Score = 34 (18.4 bits), Expect = 9.1e-106, Sum P(5) = 9.1e-106
 Identities = 6/23 (26%), Positives = 14/23 (60%)


Query: 209 NASTVDNIRSIFGNAVSRELIEI 231
           N +++  +R    +++ REL +I
Sbjct: 510 NLTSIKKLREKVDDSIHRELTDI 532
```

Information about the 12 HSPs is summarized in Table 9.3. The HSPs are numbered 1 to 12 as they occur above. The P-values indicated for any HSP are calculated from some Karlin–Altschul sum statistic associated with the HSP. Thus these P-values do not apply to the HSP itself, but rather to the HSP in conjunction with other HSPs with which it forms a consistent set. When more than one consistent set contains an HSP, the P-value reported for any HSP is the smallest one. Consistent sets have not been given on the standard printout, so that determining which HSPs form which consistent sets has been left to the user. An option, however, has recently been implemented in the Washington University version of BLAST 2.0 that will allow the output of consistent sets.

The eight HSPs forming the most significant set are 1, 11, 2, 7, 5, 4, 6, 3, listed in their consistent order. Notice that HSP 5 overlaps HSP 4, as shown by the HSP spans. This overlap is allowed under the default option, since there is no overlap after removing the right 12.5% of residues from HSP 5 and the left 12.5% of residues from HSP 4. The first seven HSPs are consistent; the eighth is not consistent with the previous seven, and it forms the consistent set containing HSPs 8, 5, 4, 6, 3, listed in their consistent order. HSP 9 is part of the set 1, 11, 2, 9, 4, 6, 3. HSP 10 is part of the set 10, 2, 7, 5, 4, 6, 3, and HSP 12 is part of the set 1, 12,

HSP	1	2	3	4	5	6
N	8	8	8	8	8	8
P-value	1.7e-138	1.7e-138	1.7e-138	1.7e-138	1.7e-138	1.7e-138
query span	8-229	259-343	636-756	539-590	501-549	609-634
target span	5-226	259-343	649-769	549-600	509-557	614-639

HSP	7	8	9	10	11	12
N	8	5	7	7	8	5
P-value	1.7e-138	1.5e-21	1.7e-132	4.2e-46	1.7e-138	9.1e-106
query span	365-397	411-437	503-524	14-53	242-261	209-231
target span	381-413	112-138	409-430	215-254	241-260	510-532

TABLE 9.3.

4, 6, 3. It might not always be so easy to find consistent sets, especially when there are hundreds of HSPs and very long HSPs. Furthermore, there may be ambiguities in that a given HSP may report an $N(= r)$ of 5, yet be consistent with two different sets of 4 HSPs. In this case BLAST reports the set with lower P-value. However, it might not be clear from the printout which set this is, and it might be necessary to calculate the significance values to find it.

9.6   Minimum Significance Lengths

9.6.1   A Correct Choice of n

When sequences are distantly related, the similarities between them might be subtle. Thus we will not be able to detect significant similarity unless a long alignment is available. On the other hand, if sequences are very similar, then a relatively short alignment is sufficient to detect significant similarity. In this section we discuss how this issue can be put on a more rigorous foundation.

If the similarity is subtle, each aligned pair will tell us less, in terms of information, than each aligned pair in more similar sequences. This will lead us to the concept of information content per position in an alignment. The theory to be developed relates to a fixed ungapped alignment of length N.

The PAMn substitution matrix has been discussed extensively above. In this section we take for granted the evolutionary model underlying these matrices. Our analysis follows that of Altschul (1991). In particular, we assume for convenience, with Altschul, that the amino acid frequencies in the two sequences compared are the same. However, in some other respects our analysis differs from his.

The analysis of Section 9.2.4 shows that an investigator using a PAMn substitution matrix in a BLAST procedure is in effect testing the alternative hypothesis that n is the correct value to use in the evolutionary process leading to the two protein sequences compared against the null hypothesis that the appropriate value of n is $+\infty$. In this section we assume that the alternative hypothesis is correct (that is, that the correct value of n has been chosen), and in effect explore aspects of the power of the testing procedure by finding the mean length of protein sequence needed before the alternative hypothesis is accepted. In the following section we explore the effects of an incorrect choice of n.

Suppose that, in formal statistical terms, we decide to adopt a testing procedure with Type I error α. Equation (9.23) shows that the value s of the normalized score statistic S' needed to meet this P-value requirement is approximately given by $s = -\log \alpha$. From equation (9.20) the corresponding value $y_{\max}$ of $Y_{\max}$ is

$$y_{\max} = \lambda^{-1} \log \left(\frac{NK}{\alpha} \right). \tag{9.57}$$

When the alternative hypothesis is true, the mean score for the one amino acid comparison at any position is, from (9.8),

$$\sum_{j,k} q(j,k) S(j,k) = \lambda^{-1} \sum_{j,k} q(j,k) \log \frac{q(j,k)}{p_j p_k}. \tag{9.58}$$

Equation (7.24) shows that if the mean final position in a random walk is F and the mean step size is G, the mean number of steps needed to reach the final position is F/G. This then suggests that the mean sequence length needed in the maximally scoring local alignment in order to obtain significance with Type I error α is the ratio of the expressions in (9.57) and (9.58), namely

$$\frac{\log \left(\frac{NK}{\alpha} \right)}{\sum_{j,k} q(j,k) \log \frac{q(j,k)}{p_j p_k}}. \tag{9.59}$$

Altschul (1991) calls this the "minimum significance length." The expression (9.59) does not change if we change the base of both logarithms. The choice of the base 2 for these logarithms has an "intuitive appeal" (Altschul (1991)), since then various components in the resulting expression can be interpreted in terms of bits of information, as discussed in Section B.10. We thus make this choice in the following discussion, and write the ratio (9.59) as

$$\frac{\log_2 \left(\frac{NK}{\alpha} \right)}{\sum_{j,k} q(j,k) \log_2 \left(\frac{q(j,k)}{p_j p_k} \right)}. \tag{9.60}$$

We consider first the denominator in (9.60). This can be thought of as the mean of the relative support, in terms of bits, provided by one observation for the alternative hypothesis against the null hypothesis, given that the alternative hypothesis is true.

It follows that the numerator in (9.60) can be thought of as the mean total number of bits of information needed to claim that the two sequences are similar. The value of K is known from experience to be typically about 0.1, and α is typically 0.05 or 0.01. Thus the value of the numerator is largely determined by the length N, and to a close approximation is $\log_2 N$. Given the value $N = 1,000$, for example, this approximate numerator expression shows that about 9.97 bits of information are needed in order to claim significant similarity between the two sequences.

Our main interest, however, is not in the numerator or the denominator of (9.60), but in the ratio of the two, that is, the minimum significant length. When n is large, $q(j,k)$ is close to $p_j p_k$; the mean information per aligned pair given in the denominator is small and the minimum significant length is large. This is as expected: If null and alternative hypotheses specify quite similar probabilities for any aligned pair, many observations will in general be needed to decide between the two hypotheses. On the other hand, if n is small, the mean relative support for the alternative hypothesis provided by each aligned pair is large, and the minimum significant length is small. The limiting $(n \to 0)$ values $q(j,j) = p_j$, $q(j,k) = 0$ for $j \neq k$, together with the convention that $0 \log 0 = 0$ (see Section B.7), show that as $n \to 0$, the denominator in (9.60), that is, the mean support from each position in favor of the alternative hypothesis, approaches $- \sum_j p_j \log_2 p_j$. If all amino acids are equally frequent, this mean support is $\log_2 20 = 4.32$, and we can think of this as 4.32 bits of information. In practice, the actual frequencies of the observed amino acids imply that a more appropriate value is about 4.17. Thus the minimum significant length is $(\log_2 N)/4.17$. If $N = 1,000$, this is about 2.39.

When $N = 1,000$ and $n = 250$, corresponding to a PAM250 substitution matrix, the probabilities $q(j,k)$ are such that each amino acid pair provides a mean of only 0.36 bits of information, and a minimum significance length of about $\log(1000)/0.36 = 9.97/0.36 = 28$ is required on average to accept the alternative hypothesis.

9.6.2   An Incorrect Choice of n

The above calculations all assume that the correct value for n has been chosen, and thus the correct alternative hypothesis probabilities $q(j,k)$ were used. Suppose now that an incorrect value n was chosen, the correct value being m. Then the score would have been calculated using a PAMn matrix but should have calculated using a PAMm matrix. What does this imply?

Suppose that with the correct choice m the probability of the ordered pair (j, k) is $r(j, k)$. The mean score is then

$$\lambda^{-1} \sum_{j,k} r(j, k) \log \frac{q(j, k)}{p_j p_k}.$$ (9.61)

Clearly, $r(j, k) = q(j, k)$ when $n = m$, and equation (1.110) then shows that the mean score is positive. More generally, the mean score is positive if n and m are close. However, as $m \to +\infty$, $r(j, k) \to p_j p_k$, and for this value of $r(j, k)$ the mean score is negative. Thus for any choice of n there will be values of m sufficiently large compared to n so that the mean score is negative. This matter is discussed further below.

In cases where the mean score (9.61) is positive, the minimal significance length is

$$\frac{\log \left(\frac{NK}{\alpha}\right)}{\sum_{j,k} r(j, k) \log \frac{q(j,k)}{p_j p_k}}.$$ (9.62)

This minimal length depends on $q(j, k)$, that is, on the choice of n. This choice of n may well involve substantial extrinsic guesswork, and it is thus important to assess the implications of an incorrect choice. Altschul (1991) gives examples of the effect on the minimal significance length of using scores derived from one PAM matrix when another is appropriate.

The fact that the mean (9.61) can be negative requires some discussion. Negative means arise when m is sufficiently large compared to n, that is, when the two species being compared diverged a long time in the past relative to the time assumed by the PAM matrix used in the analysis. In this case the data are better explained by assuming no similarity between the two sequences than by assuming a close similarity between the two sequences. The more negative this mean, the more likely it is that the null hypothesis will be accepted, and in the limit $m \to +\infty$, when $r(j, k) = p_j p_k$, the probability of rejecting the null hypothesis is equal to the chosen Type I error.

As an example of this effect, if in the simple symmetric model of Section 6.5.4 the value $n = 100$ is chosen, the mean score (9.61) is negative when m is 193 or more.

These observations indicate the perils of deciding on too small a value of n. Whereas a correctly chosen small value of n leads to shorter minimal significant lengths, as discussed above, an incorrectly small choice may lead to the possibility that a real similarity between the two sequences will not be picked up. The practice sometimes adopted of using a variety of substitution matrices to overcome this problem must be viewed with some caution, particularly in the light of the multiple testing problem discussed in Section 3.8.

9.7   BLAST: A Parametric or a Nonparametric Test?

In parametric tests the test statistic is found from likelihood ratio arguments, as discussed in Chapter 8. By contrast, the test statistic in a nonparametric test is often found on reasonable but nevertheless arbitrary grounds.

Many of the calculations and arguments used in the immediately preceding sections derive from the derivation of the score $S(j, k)$ in a substitution matrix from likelihood ratio arguments: See, for example, equations (9.7) and (9.8). In this sense the BLAST testing theory can be thought of as a parametric procedure deriving from the likelihood ratio theory in Section 8.3.1.

The assumptions made in this theory are, however, subject to debate. For example, Benner et al. (1994) claim that the time homogeneity assumption implicit in the calculations cannot be sustained, claiming, for example, that the genetic code influenced substitutions earlier in time and various chemical properties influenced substitutions more recently. Thus comparisons of distantly related species can be problematic. Even in the comparison of more closely related species, it is not clear that a uniform set of rules governs substitutions.

Even if this and similar claims are true, the statistical aspects of the BLAST procedure are still valid, in the sense that the P-value calculations are still correct. The P-value calculations take the scores in the substitution matrix as given, so that even if these scores were chosen in any more or less reasonable way, rather than from theoretical deductions using Markov chain and likelihood ratio theory, no problems arise with the correctness of the P-value calculations. In this sense the BLAST testing process can be thought of as a nonparametric procedure, where the choice of test statistic does not derive from a likelihood ratio or any other optimality argument but is chosen instead on commonsense grounds, as was, for example, the nonparametric Mann–Whitney test statistic discussed in Section 3.5. On the other hand, if the various assumptions implicit in finding a substitution matrix from likelihood ratio arguments are not correct, some of the theory in the preceding sections, particularly that associated with the optimal choice of n for a PAMn matrix, needs amendment.

9.8   Relation to Sequential Analysis

There are many similarities between the BLAST calculations given in the chapter and sequential analysis calculations discussed in Section 8.5. First and foremost, the central BLAST parameter λ $(= \theta^*)$ was first introduced into probability theory in the context of sequential analysis, being used in that theory to calculate power curves (see equations (8.65) and (8.67)), as well as mean sample size (see equation (8.72)). Second, both sequential analysis and BLAST theory center on running sums of iid random variables, and further, the random variables in both cases are either likelihood ratios or multiples of likelihood ratios.

It is therefore interesting to compare further the calculations deriving from (9.59) with the analogous calculation for a sequential test of hypothesis. If the alternative hypothesis is true, the mean step size in the sequential procedure defined by (8.58) is

$$\sum_y p(y; \xi_1) \log \left(\frac{p(y; \xi_1)}{p(y; \xi_0)} \right).$$

From (8.58), the accumulated sum in the sequential procedure necessary to reject the null hypothesis is $\log((1 - \beta)/\alpha)$, where α and β are the Type I and Type II errors, respectively. If these errors are both small, as is normally the case, this is close to $\log(1/\alpha)$. If we argue as in the derivation of the ratio (9.59) above, the mean number of observations needed to reject the null hypothesis when the alternative hypothesis is true, in a test with Type I error α, would be the ratio

$$\frac{\log(1/\alpha)}{\sum_y p(y; \xi_1) \log \left(\frac{p(y; \xi_1)}{p(y; \xi_0)} \right)}. \tag{9.63}$$

If we identify the observation y in a sequential test with the pair (j, k) in a sequence comparison, the denominators in the two expressions (9.59) and (9.63) are identical. The comparison between the two expressions thus concerns only their respective numerators. The numerator in (9.59) can be written as $\log(1/\alpha) + \log(N_1 N_2 K)$. The difference between the two numerators is, then, the additive factor $\log(N_1 N_2 K)$. This factor arises because in the BLAST procedure the test statistic is essentially the maximum of $N_1 N_2 / A$ iid geometric-like random variables, and the mean of such a maximum, like the mean of the maximum of n iid geometric random variables given in equation (2.119), is approximately $\log(N_1 N_2 K)$, as shown in (9.25). This comparison shows how much more stringent a test based on a maximal test statistic must be compared to one based, in the sequential procedure, on the typical value of a statistic. Once allowance for this difference is made, the similarity between the two procedures becomes apparent.

A second connection between sequential analysis and BLAST testing derives from the comparison of the denominator in the sequential analysis expression (8.73) and the denominator in the BLAST expression (9.62). In the sequential analysis case the form of the correct probability distribution $Q(y)$ of Y differs from that assumed under the null and alternative hypotheses. In the BLAST case the parallel comment might be, for example, that the elements in the substitution matrix were calculated from the evolutionary process leading to some PAM matrix, whereas some quite different evolutionary model might be appropriate.

A further connection between the BLAST and the sequential analysis testing procedures is that in both cases the step size in the testing procedure depends implicitly on some alternative hypothesis. In this respect both procedures differ from the fixed sample size test of Section 3.4.1 for the parameter p in a binomial distribution, where the testing procedure is independent of the alternative hypothesis value of p (so long as it exceeds the null hypothesis value).

Despite these connections between BLAST and the sequential testing procedure, the two procedures are rather different, and in some respects the BLAST procedure is more like the fixed sample size test. For example, the sample size is in effect fixed in advance and the test does not rely on achieving some specified Type II error.

9.9 Further Developments of BLAST and Further Theory

As stated at the beginning of this chapter, the BLAST model described in this book is the simplest one possible, and in practice versions of BLAST are used that are more sophisticated than that described here, and further extensions continue to appear. An important generalization allows gaps in the sequence alignments (Altschul et al. (1997)). Some of the parameter calculations for these do not follow from the explicit theory described above, but follow rather from simulations and approximations derived from statistical regression theory. A full generalization of the theory to cover such cases is one central area of current research, and the papers by Mott and Tribe (1999) and Siegmund and Yakir (2000) should be consulted for a discussion of this development. A second generalization is to PSI (position specific iterated) BLAST (Altschul et al. (1997)), again having very complex theory. Another generalization is to the case of Markov-dependent sequences, the theory for which is developed by Karlin and Dembo (1992). However, the theory for these generalizations is well beyond that appropriate for an introductory book. Our aim here is to give some idea, by considering a comparatively simple case only, of the flavor of the statistical and probability theory involved with BLAST searches, and to point to areas of current and further research.

Problems

9.1. Consider the calculation that led to equation (9.6). Use the path decomposition method to do the analogous calculation for the probability u

that the generalized random walk under consideration reaches -1 as its first ladder point. Check that $u + v = 1$.

9.2. For the simple random walk of Section 7.2 the value of θ^* is given in (7.7), the value of C is $1 - e^{-\theta^*}$, and the value of A is $(q-p)^{-1}$. From this, the value of K, calculated from equation (9.15), is $(q-p)(e^{-\theta^*} - e^{-2\theta^*})$. Making the change of notation $\theta^* = \lambda$, check that both equations (9.18) and (9.19) give this value.

9.3. For the symmetric PAM matrix discussed in Section 6.5.4, the case $c = 1$ corresponding to the value $n = 259.0675$ leads to a mean step size, when the alternative hypothesis is true, of 0.446 (see equation (6.35)). BLAST theory shows that this value should also be given by the expression $\lambda^{-1}H$ (see (9.14)). Use the values for $q(j,k)$ and $q(j,j)$ given in (6.34), the values $p_j = p'_k = 0.05$ in the expression (9.12) to compute H, and equation (9.4) to compute λ (in the case $S(j,j) = 12$, $S(j,k) = -1$ $(j \neq k)$), to verify this.

9.4. Suppose that the PAM model of the "simple symmetric" example of Section 6.5.4, for which in particular $p_j = p'_k = 0.05$, leads to a substitution matrix in which $S(j,j) = 10$ $(j = 1, 2, \ldots, 20)$ and $S(j,k) = -1$ $(j \neq k)$.

(i) Use equation (9.4) to find the associated value of λ. (This will require numerical methods.)

(ii) From the result of (i), use equation (9.19) to find K.

(iii) Use equation (9.9) to find the (common) values of $q(j,j)$ $(j = 1, 2, \ldots, 20)$ and the (common) values of $q(j,k)$ $(j \neq k)$.

(iv) From the results of (ii) and (iii), find the relative entropy H defined in (9.12).

(v) Use equations (6.28) and (6.29) to find the value of n implied by the values of $q(j,j)$ and $q(j,k)$ found in (iii) above.

(vi) Use the value of n found in (v) in the expression (6.31) to confirm the ratio -10 for $S(j,j)/S(j,k)$.

9.5. Repeat Problem 9.4 with the value 10 for $S(j,j)$ replaced by (i) 6, 8, 12, and 14, with $S(j,k) = -1$ $(j \neq k)$ and $p_j = p'_j = 0.05$ (as in Problem 9.4). Compare your values of λ, K, and H with those on BLAST printouts.

9.6. Repeat Problem 9.4 with $S(j,j)$ replaced by 20 and $S(j,k)$ replaced by -2. Comment on the similarities and differences between your calculations and those of Problem 9.4.

9.7. Suppose that in the simple symmetric example of Section 6.5.4 the value $n = 50$ is chosen to calculate the simple PAM substitution matrix. Find the values of the true value m for this model for which the mean score (9.61) is negative.

9.8. Suppose that only two amino acids "X" and "Y" exist, occurring with respective frequencies 0.6 and 0.4. Suppose that a PAM matrix is used in a sequence alignment and that the match probabilities corresponding to this matrix are $q(X, X) = 0.46$, $q(X, Y) = 0.28$, $q(Y, Y) = 0.26$. Compute the mean score (9.61) in the cases (i) $r(X, X) = 0.38$, $r(X, Y) = 0.44$, $r(Y, Y) = 0.18$, (ii) $r(X, X) = 0.40$, $r(X, Y) = 0.40$, $r(Y, Y) = 0.20$, (iii) $r(X, X) = 0.42$, $r(X, Y) = 0.36$, $r(Y, Y) = 0.22$, (iv) $r(X, X) = 0.44$, $r(X, Y) = 0.32$, $r(Y, Y) = 0.24$. Comment on your answers.

9.9. This problem refers to Table 1 of Karlin and Altschul (1993). Given the values of λ, K, and $N(= N_1 N_2)$ referred to in their paper, confirm that the three normalized scores listed can be derived from the three corresponding scores listed, using equation (9.20).

9.10. This problem refers to Table 2 of Karlin and Altschul (1993). Given the values of "Score," λ, K, and $N(= N_1 N_2)$ referred to in their paper, confirm their calculations for the various normalized scores and the various segment and sum P-values.

10
Stochastic Processes (iii): Markov Chains

10.1 Introduction

Introductory aspects of the theory of Markov chains were discussed in Chapter 4. In the present chapter further details of the theory of Markov chains will be discussed, first for Markov chains with no absorbing states and second for Markov chains with absorbing states. Our analysis of finite Markov chain theory is often oversimplified, since an examination of some of the subtleties involved in the full theory is not appropriate for bioinformatics. A recent and more complete exposition of the theory can be found in Norris (1997).

The two distinguishing Markov characteristics were introduced in Chapter 4, as were the concepts of the states of a Markov chain, transition probabilities between these states, and the concept of a transition matrix displaying these transition probabilities. It was shown that the random walk of Chapter 7 is a special example of a Markov chain. The statistical analysis associated with BLAST can also in principle be approached through Markov chain theory, as discussed below in Example 2 of Section 10.6.2. However, the special features of BLAST make the analyses of Chapter 7 and Chapter 9 more straightforward than a Markov chain analysis.

10.2 Markov Chains with No Absorbing States

10.2.1 Introduction

Finite, aperiodic irreducible Markov chains were introduced and discussed in Chapter 4. It was shown in that chapter that perhaps the most important feature of such a chain is its stationary distribution. Several properties of this distribution were introduced but no formal proof was given of them. Here we outline these proofs, derived here under the assumption, made throughout, that the Markov chain of interest is finite, aperiodic, and irreducible.

10.2.2 Convergence to the Stationary Distribution

In Chapter 4 it was shown that the n-step transition matrix $P^{(n)}$ of a Markov chain is the nth power of the single-step transition matrix, that is, that $P^{(n)} = P^n$. It was also claimed that as $n \to +\infty$, $P^{(n)}$ approaches a matrix all of whose rows are identical and all of which display the stationary distribution of the Markov chain. Computer programs such as Maple or Mathematica have subroutines for finding powers of matrices, so that these can be used to provide a straightforward computational method for finding n-step transition probabilities, for any value of n, and for finding the stationary distribution to any desired level of approximation.

However, it is appropriate to consider a mathematical, rather than a numerical, approach to finding $P^{(n)}$. Matrix theory shows why the high powers of the transition matrix approach the "stationary distribution" matrix (4.24) and, further, indicates the rate at which these high powers converge. The relevant linear algebra theory is outlined in Section B.19, and we now discuss its application to Markov chains.

Suppose that P is the transition matrix of a finite aperiodic irreducible Markov chain. It can be shown that this matrix has one eigenvalue λ_1 equal to 1 and that all other eigenvalues have absolute value less than 1.

We choose right and left eigenvectors r_1 and ℓ'_1 corresponding to the eigenvalue 1 by the requirements $\ell'_1 \mathbf{1} = 1$ and $\ell'_1 r_1 = 1$, where $\mathbf{1}$ is a column vector all of whose elements are 1. The eigenvector ℓ'_1 satisfies the equation

$$\ell'_1 = \ell'_1 P. \tag{10.1}$$

This, however, is the same equation (4.22) as that satisfied by the stationary distribution φ. It can be shown that there is a unique solution of this equation that also satisfies the normalization requirement $\ell'_1 \mathbf{1} = 1$, so that $\ell'_1 = \varphi'$.

Since the n-step transition probabilities are given by the elements in the matrix P^n, it follows from the identity of ℓ'_1 and φ' and the spectral expansion (B.48) that as n increases, the n-step transition matrix $P^{(n)}$

approaches the matrix

$$1\varphi', \tag{10.2}$$

where φ' is the stationary distribution of the Markov chain with transition matrix P. This is a matrix all of whose rows are equal to φ'. This completes the outline of the proof of the claim made concerning expression (4.24).

It is also important to consider the rate of convergence to the stationary distribution. The spectral expansion of P^n in (B.48) shows that this rate depends on the magnitude of the nonunit eigenvalues of P, and in particular on that of the largest absolute nonunit eigenvalue(s).

10.2.3   Stationary Distributions: A Numerical Example

The eigenvalues of the transition matrix P given in (4.25) are

$$\lambda_1 = 1.0000, \quad \lambda_2 = 0.5618, \quad \lambda_3 = 0.4000, \quad \lambda_4 = 0.3382.$$

Sets of left and right eigenvectors satisfying $\ell_j' r_j = 1$ $(j = 1, 2, 3, 4)$ are

$$(0.2414, 0.3851, 0.2069, 0.1667), \quad (-.3487, 0.5643, -.2155, 0.0000),$$
$$(0.0000, .6667, 0.0000, -0.6667), \quad (0.3800, 0.2348, -0.6148, 0.0000),$$

and

$$\begin{bmatrix} 1 \\ 1 \\ 1 \\ 1 \end{bmatrix}, \begin{bmatrix} -1.3088 \\ .7662 \\ -.5162 \\ .7662 \end{bmatrix}, \begin{bmatrix} .25 \\ .25 \\ .25 \\ -1.25 \end{bmatrix}, \begin{bmatrix} .7953 \\ .0679 \\ -1.1090 \\ 0.0679 \end{bmatrix}.$$

The matrix P^n is, then,

$$\begin{bmatrix} .2414 & .3851 & .2069 & .1667 \\ .2414 & .3851 & .2069 & .1667 \\ .2414 & .3851 & .2069 & .1667 \\ .2414 & .3851 & .2069 & .1667 \end{bmatrix} + (.5618)^n \begin{bmatrix} .4564 & -.7385 & .2820 & 0 \\ -.2672 & .4324 & -.1651 & 0 \\ .1800 & -.2913 & .1112 & 0 \\ -.2672 & .4324 & -.1651 & 0 \end{bmatrix}$$

$$+ (.4000)^n \begin{bmatrix} 0 & .1667 & 0 & -.1667 \\ 0 & .1667 & 0 & -.1667 \\ 0 & .1667 & 0 & -.1667 \\ 0 & -.8333 & 0 & .8333 \end{bmatrix} + (.3382)^n \begin{bmatrix} .3022 & .1867 & -.4890 & 0 \\ .0258 & .0159 & -.0417 & 0 \\ -.4214 & -.2604 & .6818 & 0 \\ .0258 & .0159 & -.0417 & 0 \end{bmatrix}.$$

For the values $n = 2, 4, 8, 16$ the matrices found from this expansion agree with those given in the expressions (4.27)–(4.30). However, the spectral expansion above indicates clearly the geometric rate at which the stationary distribution is reached.

10.2.4 Reversibility and Detailed Balance

The use of a substitution matrix for BLAST calculations, as well as the evolutionary considerations to be discussed in Chapter 13, indicate the importance of examining the relationship between a Markov chain running forward in time (the direction in which evolution has actually proceeded) and the corresponding "time-reversed" chain running backward in time (the direction relevant to many evolutionary inferences made on the basis of contemporary data). This issue is particularly relevant in comparing two related contemporary sequences, where one sequence is reached from the other by going backward in time to an assumed common ancestor and then forward in time to the other sequence. We now consider aspects of this relation for a finite, irreducible, and aperiodic Markov chain with stationary distribution φ.

Suppose that at time 0 the initial state of the Markov chain with transition matrix P is chosen at random in accordance with the stationary distribution φ. The discussion surrounding equation (4.20) shows that this distribution then applies at all future times. During the course of t transitions the chain will move through a succession of states, which we denote by $S(0)$, $S(1)$, ..., $S(t)$. Define $S^*(i)$ by $S^*(i) = S(t - i)$. We show below that the probability structure determining the properties of the reversed sequence of states $S^*(0)$, $S^*(1)$, ..., $S^*(t)$ is also that of a finite aperiodic irreducible Markov chain, whose typical element p_{ij}^* is found from the typical element p_{ij} of P by the equation

$$p_{ij}^* = \frac{\varphi_j p_{ji}}{\varphi_i}. \tag{10.3}$$

Furthermore, the stationary distribution of the reversed chain is also φ.

These claims are proved as follows. The fact that the reversed process has transition probabilities given by (10.3) follows from the conditional probability formula (1.95) when A_i is the event "$S^*(u) = E_i$" and A_j is the event "$S^*(u+1) = E_j$." The fact that $\sum_j p_{ij}^* = 1$ follows immediately from the stationary distribution property $\sum_j \varphi_j p_{ji} = \varphi_i$: See equation (4.21). The fact that the stationary distribution of the time-reversed process is φ follows from the equations

$$\sum_i \varphi_i p_{ij}^* = \sum_i \varphi_j p_{ji} = \varphi_j.$$

The equality of the extreme left- and right-hand expressions is the defining property of a stationary distribution: See equation (4.20). Irreducibility and aperiodicity are equally quickly demonstrated.

Suppose that a probability distribution $\boldsymbol{\lambda} = \{\lambda_j\}$ can be found such that

$$\lambda_i p_{ij} = \lambda_j p_{ji} \tag{10.4}$$

for all (i, j) pairs. Summation of both sides of equation (10.4) over all possible values of i yields

$$\sum_i \lambda_i p_{ij} = \sum_i \lambda_j p_{ji} = \lambda_j.$$

But this is the defining equation of the stationary distribution. This observation sometimes provides the most convenient way of finding the stationary distribution of an irreducible aperiodic Markov chain: If a distribution λ can be found for which equation (10.4) holds for all (i, j) pairs, then this is the stationary distribution φ of the Markov chain, which then satisfies

$$\varphi_i p_{ij} = \varphi_j p_{ji} \quad \text{for all } (i, j). \tag{10.5}$$

These are the so-called *detailed balance* equations. From equation (10.3), the detailed balance equations hold if and only if $p_{ij}^* = p_{ij}$, in which case the properties of the reversed-time Markov chain are equivalent to those of the forward-time chain. An observer watching the successive states in the reversed chain would have no way of telling whether he/she was watching the reversed or the original chain. Examples of reversible and irreversible Markov chains used in evolutionary models are given in Section 13.1.

A version of the reversibility criterion in a form more convenient for evolutionary processes was given by Tavaré (1986): see the expression (13.21). An arbitrary Markov chain is unlikely to be reversible; for example, the Markov chain with transition matrix (4.19) is not reversible, since the terms in this matrix and in the stationary distribution (4.26) do not satisfy the detailed balance equations.

The reversibility property in often assumed implicitly in comparing sequences from two different contemporary species, since to get from one species to another one must travel backward in time to a common ancestor and then forward in time to the other species. The calculations leading to PAM matrices, discussed in Section 6.5.3, consider only Markov chains going forward in time. It is thus of interest to show that the transition matrix M_1, defined in Section 6.5.3 and used to compute various PAMn substitution matrices, is reversible for all practical purposes. This is done as follows.

M_1 is a particular case of the matrix P, defined through equations (6.20) and (6.21), so it is sufficient to prove that P is reversible. It is reasonable to assume that the quantity $\sum_m A_{jm}$ appearing in the denominator of (6.19) is, for all practical purposes, proportional to the stationary probability φ_j of amino acid j. This implies that $A_{jk} = b\varphi_j a_{jk}$ for some constant b. Equation (6.20) then implies that $A_{jk} = d\varphi_j m_{jk}$ for some constant d. But the matrix $\{A_{jk}\}$ is by construction symmetric, so that $A_{jk} = A_{kj}$ for all (j, k). From this,

$$d\varphi_j p_{jk} = d\varphi_k p_{kj}.$$

Cancellation of the constant d shows that this is the reversibility requirement (10.5).

10.3   Higher-Order Markov Dependence

10.3.1   Testing for Higher-Order Markov Dependence

The test of Markov dependence (more exactly, first-order Markov dependence) of successive nucleotides in a DNA sequence was discussed in Section 5.2. This was presented as a chi-square test, and the calculations of Example 4 of Section 8.4.2 show that this is an approximation to a $-2 \log \lambda$ test. In this section we extend the analysis to tests of higher-order Markov dependence (defined in the following paragraph) via the $-2 \log \lambda$ procedure, following the discussion by Tavaré and Giddings (1989). For concreteness we present the discussion in terms of DNA sequences.

The probability structure of the nucleotides in a DNA sequence is described by a Markov chain of order $k \geq 1$ if the probability that any nucleotide occurs at a given site depends on the nucleotides at the preceding k sites. The transition probabilities for the case $k = 3$, for example, are of the form

$$p_{h:mnr} = \text{Prob}(Y_{j+3} = a_h \,|\, Y_{j+2} = a_m, Y_{j+1} = a_n, Y_j = a_r),$$

where Y_q denotes the nucleotide at site q, and the a_k are specified nucleotide types. This notation shows that for general k there are 3×4^k such transition probabilities.

Suppose that the null hypothesis is that the Markov chain is of order $k - 1$, and the alternative that it is of order k, for some value of k. (The test discussed in Section 5.2 is for the case $k = 1$.) The likelihood of the data under both hypotheses may be written down in terms of arbitrary parameters of the form $p_{h:mnr}$ under both null and alternative hypotheses. The likelihoods under null and alternative hypotheses are then maximized with respect to these parameters and the $-2 \log \lambda$ statistic calculated.

The number of arbitrary parameters under null and alternative hypotheses are, respectively, $3 \times 4^{k-1}$ and 3×4^k, so that the $-2 \log \lambda$ test has $9 \times 4^{k-1}$ degrees of freedom. Thus once k exceeds 2, extensive data would be required to ensure that the asymptotic chi-square properties of $-2 \log \lambda$ apply. Further details are given by Tavaré and Giddings (1989) and Reinert et al. (2000). In general, it has been found that even DNA in nonfunctional intergenic regions tend to have high-order dependence.

10.3.2   Testing for a Uniform Stationary Distribution

If the hypothesis of Markov dependence is accepted, a further test of interest is whether the stationary distribution of the Markov chain is uniform. We will discuss this in the case of first-order dependence; the discussion for higher-order dependence is similar.

In the first-order case, equation (4.20) shows that a necessary and sufficient condition that the stationary distribution be uniform is that the transition probabilities in each column of the transition matrix sum to 1. This implies that, for example, the elements in the fourth row of the transition matrix are determined by the elements in the first three rows.

The probability of any observed DNA sequence can be calculated under the null hypothesis that the stationary distribution is uniform. Since under this hypothesis the elements in each row and each column of the transition matrix must sum to 1, there are 9 free parameters. The probability of the observed sequence can be maximized with respect to these parameters. Under the alternative hypothesis the only constraint is that the elements in any row must sum to 1 and therefore there are 12 free parameters. The probability of the observed sequence can then be maximized with respect to these parameters. From these two maximum likelihoods a $-2 \log \lambda$ statistic may be calculated. Under the null hypothesis this statistic has an asymptotic chi-square distribution with 3 degrees of freedom. This then provides a testing procedure for uniformity of the stationary distribution.

10.4 Patterns in Sequences with First-Order Markov Dependence

If the test of independence in the previous section suggests that there is Markov dependence in a DNA sequence, as indeed is usually the case in reality, it is of interest to extend results obtained under the independence assumption to the dependent case. In this section we extend the calculations concerning patterns in Section 5.7.3, and in particular the generating functions (5.55) and (5.58), to the case of first-order dependence. Details of this generalization are given in Robin and Daudin (1999) and Reinert et al. (2000), and we only discuss aspects of them here. The discussion is in terms of nucleotide sequences, with attention focused on a word of arbitrary length k.

We assume that some 4×4 transition matrix P of the Markov chain is given or postulated, describing the transition probabilities from one nucleotide to the next. It is notationally convenient in this section to write the typical term in this matrix as $p(i,j)$. It is assumed that this Markov chain admits a stationary distribution and that for all practical purposes stationarity has been reached.

We write the k-letter word of interest as $(m_1, m_2, \ldots, m_k)$ and suppose that this word has just appeared or, more exactly, concluded, in the sequence. We call this "appearance A." We now ask for the probability $q(y)$ that the word *first* reappears y sites after appearance A. The argument that led to equation (5.32) in the independent case shows that this is the probability that it does appear y sites after appearance A, less the probability that it first appears at some position j sites after appearance A ($j < y$) and then also appears $y - j$ sites after this.

If $y < k$, it may be impossible for the word to appear y sites after appearance A, unless it can overlap with itself appropriately. For example, the word *gaga* can overlap if $y = 2$, but not if $y = 1$ or 3. If w_j is defined as in Section 5.7.2, it is possible for the word to reappear y sites after appearance A if $w_{k-y} = 1$, and if it is not possible, $w_{k-y} = 0$. If $w_{k-y} = 1$, then since the letter m_k occurred in the final site corresponding to appearance A, the probability that the word appears y sites after appearance A is

$$p(m_k, m_{k-y+1})p(m_{k-y+1}, m_{k-y+2}) \cdots p(m_{k-1}, m_k).$$

We denote this probability, which is the probability that starting with m_k the next y letters to appear are the last $k - y$ letters of the word of interest, by $Q(y)$. Thus the probability that the word occurs y sites after appearance A is $w_{k-y}Q(y)$.

The probability that the word *first* occurs j sites after appearance A and also $y - j$ sites after this is the sum over j ($j = 1, 2, \ldots, y - 1$) of the probability $q(j)$ that the word of interest first appears j sites after appearance A, multiplied by the probability, given that the word appears j sites after appearance A, that it appears $y - j$ sites after this. These arguments lead to

$$q(y) = w_{k-y}Q(y) - \sum_{j=1}^{y-1} q(j)w_{k-y+j}Q(y - j). \qquad (10.6)$$

The case $y \geq k$ can be handled similarly. These equations lead to the generating function for the probability distribution of the distance until the word next appears after appearance A, which generalizes the "independence" generating function given in equation (5.55). The mean and variance of the distance in question can then be found directly from this generating function.

Similar, and indeed simpler, calculations can be made for the probability distribution and generating function of the distance to the first occurrence of the word of interest after the origin. These may be found in Robin and Daudin (1999).

10.5   Markov Chain Monte Carlo

Markov chains can be used for a variety of calculation and optimization purposes in bioinformatics to which they are not initially clearly related. The methods used are described as "Markov Chain Monte Carlo" methods. We discuss two of these in this section, describing first the Hastings–Metropolis algorithm on which they are based.

10.5.1   The Hastings–Metropolis Algorithm

The aim of the Hastings–Metropolis algorithm is to construct an aperiodic irreducible Markov chain having some prescribed stationary distribution $\varphi' = (\varphi_1, \varphi_2, \ldots, \varphi_s)'$, where $\varphi_j > 0$, $j = 1, \ldots, s$.

We choose a set of constants $\{q_{ij}\}$ such that $q_{ij} > 0$ for all (i,j) and $\sum_j q_{ij} = 1$ for all i. We then define a_{ij} by

$$a_{ij} = \min\left(1, \frac{\varphi_j q_{ji}}{\varphi_i q_{ij}}\right) \tag{10.7}$$

and p_{ij} by

$$p_{ij} = q_{ij} a_{ij}, \quad i \neq j, \tag{10.8}$$

and $p_{ii} = 1 - \sum_{j \neq i} p_{ij}$. Since $q_{ij} > 0$ for all (i,j), this construction shows that $p_{ij} > 0$ for all (i,j) including the case $i = j$ (see Problem 10.7), so that the Markov chain defined by the $\{p_{ij}\}$ is aperiodic and irreducible.

Theorem 10.1. The stationary distribution of the Markov chain defined by (10.8) is φ'.

This theorem is checked by showing that the detailed balance requirements (10.5) hold. Suppose without loss of generality that $\varphi_j q_{ji}/\varphi_i q_{ij} < 1$. Then $\varphi_i q_{ij}/\varphi_j q_{ji} > 1$ and

$$a_{ij} = \frac{\varphi_j q_{ji}}{\varphi_i q_{ij}}, \quad p_{ij} = \frac{\varphi_j q_{ji}}{\varphi_i}, \quad a_{ji} = 1, \quad p_{ji} = q_{ji}. \tag{10.9}$$

From these results the detailed balance requirement $\varphi_i p_{ij} = \varphi_j p_{ji}$ follows. The special case where $\varphi_j q_{ji}/\varphi_i q_{ij} = 1$ is easily handled separately.

The above proof can be extended to the case where some q_{ij} are zero, so long as $q_{ji} > 0$ whenever $q_{ij} > 0$ and the Markov chain defined by the resultant p_{ij} is aperiodic and irreducible.

Different Markov chains may be constructed by different choices of the q_{ij}, all having the desired stationary distribution φ', and for any given application one choice will often be more useful than another.

10.5.2   Gibbs Sampling

Let Y_i, $i = 1, 2, \ldots, k$, be discrete finite random variables, $\mathbf{Y}$ the random vector $(Y_1, Y_2, \ldots, Y_k)'$, and $P_{\mathbf{Y}}(\mathbf{y})$ the distribution of $\mathbf{Y}$. Assume furthermore that $P_{\mathbf{Y}}(\mathbf{y}) > 0$ for all $\mathbf{y}$.

We will define a Markov chain whose states are the possible values of $\mathbf{Y}$. Enumerate the vectors in some order as vectors $1, 2, \ldots, s$ and identify vector j with the jth state in a Markov chain whose transition probabilities we now define. If vectors i and j differ in more than one component, put $p_{ij} = 0$. If they differ in at most one component, suppose, to be concrete, that they differ in their first component (if they differ at all). Write vector i as $(y_1, y_2, \ldots, y_k)$ and vector j as $(y_1^*, y_2, \ldots, y_k)$. Then we define

$$p_{ij} = \text{Prob}(Y_1 = y_1^* \mid Y_2 = y_2, Y_3 = y_3, \ldots, Y_k = y_k) \qquad (10.10)$$

$$= \frac{\text{Prob}(Y_1 = y_1^*, Y_2 = y_2, Y_3 = y_3, \ldots, Y_k = y_k)}{\text{Prob}(Y_2 = y_2, Y_3 = y_3, \ldots, Y_k = y_k)}, \qquad (10.11)$$

where the probabilities in both numerator and denominator are calculated using $P_Y(y)$.

We claim that this Markov chain is irreducible and aperiodic, and furthermore has stationary distribution $P_Y(y)$.

Aperiodicity follows from the fact that $p_{ii} > 0$, and irreducibility follows from the fact that each state in the Markov chain can be reached, after a finite number of steps, from every other state. The fact that the stationary distribution is $P_Y(y)$ is proved as follows.

Define $q_{ij} = p_{ij}$. Then from (10.7) it follows that if the denominator in (10.11) is denoted by Q,

$$a_{ij} = \min\left(1, \frac{P_Y(y_1^*, y_2, \ldots, y_k)P_Y(y_1, y_2, \ldots, y_k)/Q}{P_Y(y_1, y_2, \ldots, y_k)P_Y(y_1^*, y_2, \ldots, y_k)/Q}\right) = 1.$$

This implies that $p_{ij} = q_{ij}a_{ij}$, and from Theorem 10.1, $P_Y(y)$ is the stationary distribution of the Markov chain defined by the $\{p_{ij}\}$.

Example. In this example we show that the segment alignment procedure of Section 6.6 is essentially a Gibbs sampling procedure. The alignment procedure was described in Section 6.6 as a Markov chain process with S states, each state corresponding to an array of amino acids having N rows and W columns. We define $c_{ij}(s)$ as the number of times that amino acid j occurs in column i in the array corresponding to state s in this Markov chain, and put

$$q_{ij}^*(s) = \frac{c_{ij}(s) + b_j}{N + B}, \quad q_{ij}(s) = \frac{c_{ij}(s)}{N},$$

where the b_j are pseudocounts, $B = \sum b_j$, as defined in Section 6.6.

Define p_j is the background frequency of amino acid j. The relative entropy between $\{q_{ij}^*(s)\}$ and $\{p_j\}$ is

$$\sum_{i=1}^{W} \sum_{j=1}^{20} q_{ij}^*(s) \log\left(\frac{q_{ij}^*(s)}{p_j}\right). \qquad (10.12)$$

States for which this relative entropy is high are those corresponding to good alignments. Our aim is therefore to find states for which this relative entropy is high.

We associate with state s in this Markov chain the probability λ_s, defined by

$$\lambda_s = \text{const} \prod_{i=1}^{W} \prod_{j=1}^{20} \left(\frac{q_{ij}^*(s)}{p_j} \right)^{c_{ij}(s)}, \qquad (10.13)$$

where the constant is chosen so that $\sum_{s=1}^{S} \lambda_s = 1$.

We will say that states s and u are neighbors if either $s = u$ if the arrays corresponding to these two states differ only in the entries in one row. This implies that in the step-by-step procedure described in Section 6.6 it is possible to move in one step from state s to state u and also from state u to state s.

If states s and u are neighbors, for any position i the respective values of the counts $c_{ij}(s)$ and $c_{ij}(u)$ either will be identical for all 20 values of j or will be identical for 18 values of j and differ by $+1$ and -1 for the remaining two values of j. The values $+1$ and -1 arise when there are different amino acids in position i in the row where states s and u differ.

For any state a, define a_r as the reduced array obtained by removing row r from state a. Then if states s and u differ in row r, $q_{ij}(s_r) = q_{ij}(u_r)$. For each (i,j) pair, the values of $q_{ij}^*(s)$ and $q_{ij}^*(u)$ are very close to this common value $q_{ij}(s_r)$, and from now on we make the approximation that both are equal to $q_{ij}(s_r)$.

Making this approximation, the ratio λ_s/λ_u becomes, in the notation adopted in equation (6.38),

$$\frac{\lambda_s}{\lambda_u} = \frac{q_{1,s(1)} q_{2,s(2)} \cdots q_{W,s(W)}}{q_{1,u(1)} q_{2,u(2)} \cdots q_{W,u(W)}} \cdot \frac{p_{u(1)} p_{u(2)} \cdots p_{u(W)}}{p_{s(1)} p_{s(2)} \cdots p_{s(W)}}. \qquad (10.14)$$

Equations (6.38) and (10.14) jointly imply that with the approximation made above,

$$\frac{\lambda_s}{\lambda_u} = \frac{p_{us}}{p_{su}},$$

so that

$$\lambda_s p_{su} = \lambda_u p_{us} \qquad (10.15)$$

for all neighboring states. Equation (10.15) is also true for states that are not neighbors, since in this case both sides of the equation are 0. But this equation is identical in form to equation (10.4) and the detailed balance equation (10.5). Thus the step procedure described in Section 6.6 is, for all practical purposes, that of a Gibbs sampling process, and λ_s is the stationary probability of state s in the procedure.

States with high stationary probability are visited comparatively frequently in the procedure and thus may be recognized. Since $\log \lambda_s$ is approximately a linear function of the relative entropy (10.12), we have achieved the aim of ensuring that states such that $\{q_{ij}^*(s)\}$ and $\{p_j\}$ have high relative entropy are visited comparatively frequently and can thus be identified.

This is the reason for the choice of the transition probabilities introduced in Section 6.6.

10.5.3 Simulated Annealing

The goal of the simulated annealing procedure is to find, at least approximately, the minimum of some positive function defined on an extremely large number s of "states" $E_1, \ldots, E_s$, and to find those states for which this function is minimized (or approximately minimized). Write the value of this function for state E_j as $f(j)$.

To illustrate some aspects of the simulated annealing procedure, it is useful to consider the traveling salesman problem, which is equivalent in complexity to many problems that arise in bioinformatics. In this classical problem, a salesman wishes to find the minimal traveling path linking n cities, returning finally to his city of origin. The number of possible paths is $O(n!)$, which is extremely large when n exceeds 20 or 30. In this case there is no hope of listing the total distance along each path and thus finding the minimum distance by exhaustive search. We may call any such path a "state" E_j, and define $f(j)$ as the total distance traveled along the path. Although the number of paths is very large, once a path E_j is given, the distance $f(j)$ along that path is easily computed, and "similar" paths should have similar total distances.

The concept of a collection of states and a function $f(j)$ corresponding to state E_j may be generalized. Let T be a fixed positive parameter whose value is as yet unspecified. The aim is to construct a Markov chain with states $E_j, j = 1, 2, \ldots, s$, such that the stationary distribution probability $\varphi(j; T)$ of the state E_j is

$$\varphi(j; T) = C \cdot \exp(-f(j)/T). \qquad (10.16)$$

Here the constant C is chosen to ensure that the sum of the probabilities in the stationary distribution is 1.

The reason for this construction is the following. Suppose that for some state E_j, $f(j)$ is small. If a Hastings–Metropolis procedure can be used to build a Markov chain with stationary distribution $\{\varphi(j; T)\}$, then at stationarity the chain should visit state E_j comparatively often, since from (10.16) the stationary distribution probability for this state is comparatively large. Thus starting in any state and following the procedure for a large number of iterations, states with low values of $f(\cdot)$ should become recognizable.

We construct a "neighborhood" of each state, the neighborhood of state E_j being chosen as a set of states in some sense "close" to E_j, and to which the variable in the Markov chain can move, in one step, from E_j. Moves in one step to states outside the neighborhood are not allowed. We make the following four requirements about these neighborhoods.

(a) If E_j is in the neighborhood of E_k, then E_k is in the neighborhood of E_j.

(b) The number of states in the neighborhood of any given state is independent of that state. This number is denoted by N.

(c) The neighborhoods are linked in the sense that the Markov chain can eventually move, after a finite number of steps, from E_j to E_m, for all (j, m).

(d) Given that the Markov chain is in E_j, it may next move only to a state in the neighborhood of E_j.

Suppose that in the traveling salesman problem the cities to be visited are A, B, C, D, E. Any choice of a path through these cities by the salesman is equivalent to an ordering of these cities, for example ADEBC. If a neighboring path is defined by switching the order of two adjoining cities, for example in the above case to ADBEC, then requirements (a) and (b) above are satisfied.

Requirements (a)–(d) can be relaxed with care, but for simplicity we impose them as stated. Subject to these requirements, the choice of N and of the states in the neighborhood of any given state are in principle arbitrary, but in practice the usefulness of the simulated annealing algorithm depends to some extent on a wise choice of these.

Suppose that the variable in the Markov chain is currently in state E_i. The Hastings–Metropolis quantity q_{ij} is chosen in the simplest possible way, namely

$$q_{ij} = \begin{cases} N^{-1} & \text{if } E_j \text{ is in the neighborhood of } E_i, \\ q_{ij} = 0 & \text{otherwise.} \end{cases}$$

For each (i, j) for which $q_{ij} = N^{-1}$, this choice implies that the Hastings–Metropolis quantity a_{ij} is given by

$$a_{ij} = \min\left(1, \frac{\varphi(j; T)N}{\varphi(i; T)N}\right) = \min\left(1, e^{(f(i) - f(j))/T}\right). \tag{10.17}$$

Consequently, the desired Markov chain has transition probabilities p_{ij} as follows: When $i \neq j$,

$$p_{ij} = \begin{cases} N^{-1}\frac{\varphi(j;T)}{\varphi(i;T)} & \text{if } E_j \text{ is a neighbor of } E_i \text{ and } \varphi(i; T) > \varphi(j; T), \\ N^{-1} & \text{if } E_j \text{ is a neighbor of } E_i \text{ and } \varphi(i; T) < \varphi(j; T), \\ 0 & \text{if } E_j \text{ is not a neighbor of } E_i, \end{cases}$$

and

$$p_{ii} = 1 - \sum_{j \neq i} p_{ij}.$$

In the above procedure a large value of T implies that all states in the neighborhood of the present state of the Markov chain are chosen with approximately equal probability, and also that the stationary distribution of the Markov chain tends to be uniform. A small value of T implies that different states in the neighborhood of E_i tend to have rather different stationary distribution probabilities. However, too small a choice of T might lead to the procedure tending to stay near local maxima and not move to other higher maxima. Part of the art of the process is in choosing a value of T that allows comparatively rapid movement from one neighborhood to another (large T) but at the same time picks out states within neighborhoods with comparatively large stationary probabilities (small T).

Example. Waterman (1995) describes an application of the simulated annealing process to the so-called double digest problem. Restriction endonucleases were described briefly in Section 5.8.3. In the double digest procedure a length of DNA is subjected to two restriction endonucleases, both separately and together. When the DNA is subjected to the first restriction endonuclease it will be cut wherever the recognition sequence for that restriction endonuclease occurs, and a parallel comment applies for the second restriction endonuclease. When both are applied together the DNA is cut whenever either recognition sequence occurs. The data thus consist of three sets of sequence lengths, the sum of the lengths in each set being the length of the original DNA sequence, and the aim is to reconstruct, from these lengths, the original set of locations of the two recognition sequences. This problem is "NP-complete," implying in particular that no polynomial-time algorithm is known (or, conjecturally, can ever be known) for its solution. Thus heuristic methods are required, and the simulated annealing method is one such approach.

Suppose that the application of both restriction endonucleases together cuts the DNA into s segments of lengths $c(1), c(2), \ldots, c(s)$, numbered so that $c(1) \leq c(2) \leq \cdots \leq c(s)$. Suppose also that the first restriction endonuclease cuts the DNA into n segments, which we write $A_1, A_2, \ldots, A_n$, and the second cuts the DNA into m segments, which we write $B_1, B_2, \ldots, B_m$. Without mixing the A_i's with the B_i's, these segments may be placed jointly into $n!m!$ different orderings. Each ordering defines a sequence of points on the DNA as the cut points of one of the two restriction endonucleases, and for any ordering these j cut points will divide the DNA into s segments, some of whose lengths can be 0. These lengths are now numbered $d_j(1), d_j(2), \ldots, d_j(s)$ in such a way that $d_j(1) \leq d_j(2) \leq \cdots \leq d_j(s)$.

If all measurements are error-free, at least one of these $n!m!$ orderings will produce the lengths $c(1), c(2), \ldots, c(s)$. To find these orderings, or more generally if measurement errors are possible to find those orderings that most closely achieve this, it is reasonable to attempt to find the values of

j that minimize a function $f(j)$ of the form

$$f(j) = \sum_{u=1}^{s} \frac{(d_j(u) - c(u))^2}{c(u)}.$$

An interesting choice of the neighborhood of any ordering for the simulated annealing procedure is the generalization of that given above for the traveling salesman. Consider any of the $n!m!$ orderings of the segments produced by the two restriction endonucleases, for example (with $n = 5, m = 4$) $A_5A_2A_1A_4A_3$ and $B_3B_4B_1B_2$. The neighborhood of this ordering can be taken as any ordering for which at most one switch between adjoining segments in each ordering is made, for example $A_5A_1A_2A_4A_3$ and $B_3B_4B_2B_1$. Waterman (1995) discusses more efficient choices than this in both the double digest and the traveling salesman problems.

10.6   Markov Chains with Absorbing States

10.6.1   Theory

We now turn to Markov chains where there are absorbing states. These were introduced and discussed briefly in Section 4.6. In this section we discuss two questions asked for these Markov chains, namely, "If there are two or more absorbing states, what is the probability that a specified absorbing state is the one eventually entered?" and "What is the mean time until an absorbing state is eventually entered?"

Let $Y(1), Y(2), \ldots$ be a sequence of random variables such that

$$\text{Prob}(Y(t + 1) = j \mid Y(t) = i) = p_{ij}, \quad i, j = 1, 2, \ldots, s.$$

Thus the successive values of $Y(\cdot)$ have probabilistic behavior governed by a Markov chain with s states $E_1, E_2, \ldots, E_s$.

We first consider the case with two absorbing states, which are assumed to be E_1 and E_s, so that $p_{1j} = 0$ for $j \neq 1$ and $p_{sj} = 0$ for $j \neq s$. For $i = 2, 3, \ldots, s - 1$, let w_i be the probability that eventually $Y(t) = s$, given that $Y(0) = i$. Then a generalization of the argument that led to equation (7.4) shows that the w_i are the solution of the simultaneous equations

$$w_i = \sum_{j=2}^{s-1} p_{ij}w_j + p_{is}, \quad i = 1, 2, 3, \ldots, s. \tag{10.18}$$

A generalization of the argument that led to (7.11) shows that if t_i is the mean number of transitions in the Markov chain until either $Y(t) = 1$ or $Y(t) = s$, given that $Y(0) = i$, then the t_i are the solution of the simultaneous equations

$$t_i = \sum_{j=2}^{s-1} p_{ij} t_j + 1, \quad i = 2, 3, \ldots, s-1. \tag{10.19}$$

These equations are subject to the boundary conditions $t_1 = t_s = 0$.

We next consider the case where there is only one absorbing state, which we take to be E_s. In this case the probability of eventual absorption in E_s is 1, and is thus not of interest. On the other hand, the probability that absorption in E_s takes place at or before some designated number of steps in the chain might well be of interest: See Example 2 in Section 10.6.2. The mean time until the absorbing state E_s is eventually entered is found by replacing equations (10.19) by

$$t_i = \sum_{j=1}^{s-1} p_{ij} t_j + 1, \quad i = 1, 2, \ldots, s-1, \tag{10.20}$$

with boundary condition $t_s = 0$.

10.6.2   Examples

Example 1. Word recurrence lengths. The theory of Markov chains with absorbing states can be used to find the mean distance between successive occurrences of a word consisting of a short DNA sequence, as discussed in the "nonoverlapping" analysis of Sections 5.8 and 5.9. The Markov chain approach is due to Karlin and Brendel (1996), and we initially follow their analysis.

We assume that the nucleotides $a, g, c,$ and t occur with respective frequencies $p_a, p_g, p_c,$ and p_t and that successive nucleotides are independent. Suppose first that the word of interest is atg. A calculation in Section 5.9.2 shows that when overlaps are not counted, the mean distance between successive occurrences of this word is $1/(p_a p_t p_g)$. This result can be found using the theory of absorbing Markov chains as follows.

We consider all words consising of three nucleotides and divide these into four states: atg, xya, xat, and "*other.*" Here x and y are arbitrary nucleotides, possibly identical, and "*other*" consists of all words of length three other than those specifically listed. The DNA sequence is scanned from left to right, starting immediately following an occurrence of atg, and our aim is to find the mean number of nucleotide sites until the next occurrence. This is done by making atg an absorbing state and seeking the mean number of sites visited until it is entered, assuming that the initial state is "*other.*" The Markov chain transition matrix is then

$$
\begin{array}{cccc}
 & atg & xya & xat & other \\
\begin{array}{c} atg \\ xya \\ xat \\ other \end{array} &
\left[\begin{array}{cccc}
1 & 0 & 0 & 0 \\
0 & p_a & p_t & p_c + p_g \\
p_g & p_a & 0 & p_c + p_t \\
0 & p_a & 0 & 1 - p_a
\end{array}\right]
\end{array}. \tag{10.21}
$$

Let t_2, t_3, and t_4 be the mean number of sites until atg is first observed, given respectively that the current state is xya, xat, and "*other.*" Our aim is to find t_4. This Markov chain has only one absorbing state, so equations (10.20) show that t_2, t_3, and t_4 satisfy the simultaneous equations

$$
\begin{aligned}
t_2 &= p_a t_2 + p_t t_3 + (p_c + p_g)t_4 + 1, \\
t_3 &= p_a t_2 + (p_c + p_t)t_4 + 1, \\
t_4 &= p_a t_2 + (1 - p_a)t_4 + 1.
\end{aligned}
$$

Solving these equations gives $t_4 = 1/(p_a p_t p_g)$, as found in Section 5.9.2.

We next consider the word ata. A calculation in Section 5.9.2 shows that the mean number of nucleotides before this sequence first occurs is $(1 + p_a p_t)/p_a^2 p_t$. To find this result using Markov chain theory it is necessary to consider the four states ata, $\overline{at}a$, xat, and "*other.*" Here x is an arbitrary nucleotide, and $\overline{at}a$ is any word of length three finishing with a and not starting with at. The transition matrix for this Markov chain, with ata being regarded as an absorbing state, is

$$
\begin{array}{cccc}
 & ata & \overline{at}a & xat & other \\
\begin{array}{c} ata \\ \overline{at}a \\ xat \\ other \end{array} &
\left[\begin{array}{cccc}
1 & 0 & 0 & 0 \\
0 & p_a & p_t & p_c + p_g \\
p_a & 0 & 0 & 1 - p_a \\
0 & p_a & 0 & 1 - p_a
\end{array}\right]
\end{array}. \tag{10.22}
$$

If t_2, t_3, and t_4 are the mean numbers of sites until ata is first observed, given that the current state is $\overline{at}a$, xat, and "*other,*" respectively, equation (10.20) shows that

$$
\begin{aligned}
t_2 &= p_a t_2 + p_t t_3 + (p_c + p_g)t_4 + 1, \\
t_3 &= (1 - p_a)t_4 + 1, \\
t_4 &= p_a t_2 + (1 - p_a)t_4 + 1.
\end{aligned}
$$

These equations yield $t_4 = (1 + p_a p_t)/(p_a^2 p_t)$, again as found in Section 5.9.2.

This approach generalizes naturally to the case where there is a Markov dependence between successive nucleotides, but the size of the transition matrix that is then needed implies that in practice the approach of Section

5.8.2, using generating functions, is probably to be preferred.

Example 2. Markov chains and the alignment of two sequences. The BLAST theory relating to two aligned sequences of equal length discussed in Section 9.2 relies on asymptotic results for geometric-like random variables. In this section we discuss an exact as opposed to an asymptotic analysis, following the work of Daudin and Mercier (2000). To simplify the discussion we consider the analysis in terms of DNA sequences. An essentially identical analysis applies for protein sequences.

The two sequences are scanned from left to right, and a score $S(i, k)$, determined from some score matrix, is allocated at any site if the two respective nucleotides at that site are nucleotides i and k. The accumulated score after the first j sites have been scanned is denoted by S_j. As in Chapter 9, the test statistic of interest is $Y_{\max}$, which is redefined here by

$$Y_{\max} = \max_{a,b}(S_b - S_a),$$

where a runs over all ladder points and $b \geq a$.

The calculations in Chapter 9 provide an approximate asymptotic null hypothesis probability distribution for $Y_{\max}$, and this allows a readily computed P-value approximation for the eventual observed value $y_{\max}$ of $Y_{\max}$. By contrast, the Daudin and Mercier analysis provides, at least in principle although not necessarily easily in practice, an exact P-value calculation for the observed value $y_{\max}$ of $Y_{\max}$.

It is assumed, as in Chapter 9, that the mean score at any site is negative. The random walk defined by the accumulated score S_j, $j = 1, 2, 3, \ldots$, proceeds through a sequence of increasingly negative ladder points, as described in Chapter 9. We define $S^*(j)$ as the accumulated score at the last ladder point before site j. (The point $(0, 0)$ is classified as a ladder point.) We chose some fixed positive constant s. From the accumulated score we define a new sequence $\{U_j\}$, the height of the current excursion up to site j, given by

$$U_0 = 0, \quad U_j = \max(0, S_j - S^*(j)), \tag{10.23}$$

but subject to the further requirement that if $U_j \geq s$ for some j, then U_j is immediately set to the value s and stays there thereafter, that is, that $U_j = U_{j+1} = \cdots = s$.

The probabilistic behavior of the successive values of the U_j is thus governed by a Markov chain in which the possible values of U_j are $0, 1, \ldots, s$, with the value $U_j = s$ corresponding to an absorbing state. The transition probability matrix of this Markov chain, which we denote by P, is found from the probability distribution of the steps $S(i, k)$.

Let the length of each of the aligned sequences be N. The event that $Y_{\max} \geq s$ is the event that this Markov chain enters the state $U_j = s$ at or before the Nth step in the chain. This probability can be calculated from the Nth power of P, and this allows the calculation of an exact P-value associated with an observed value s of $Y_{\max}$.

The matrix P could be quite large, and N could be extremely large. Thus the calculation of P^N could involve substantial computing, but would still be feasible for values of s and N arising in practice. However, the above calculations refer to a fixed alignment. The general BLAST procedure of comparing two *unaligned* sequences requires that this calculation be repeated a large number of times, which might be impractical with current computing power.

The spectral expansion (B.48) has interesting implications concerning the form of the P-value. The matrix P has one eigenvalue λ_1 equal to 1. In any case, arising in practice there will be a unique nonunit eigenvalue λ_2 less than 1 in absolute value but larger in absolute value than the remaining eigenvalues. If the contribution of the remaining eigenvalues to the spectral expansion (B.48) is small, as will be the case when N is large, this expansion shows that, to a first order of approximation, the P-value corresponding to the observed maximum s is of the form

$$P\text{-value} \approx 1 - A\lambda_2^N, \tag{10.24}$$

where A is some constant depending on the left and right eigenvectors of P corresponding to λ_2. This expression is of the same general mathematical form as that found from the procedures of Chapter 9, as we would expect, but the two expressions are derived from two different procedures for approximating the P-value.

10.7   Continuous-Time Markov Chains

10.7.1   Definitions

Markov processes can be in either discrete or continuous time, and in either discrete or continuous space; that is to say, there are four types of Markov processes. The Markov chains considered in Chapter 4 and so far in this chapter are in discrete time and discrete space. In this section we consider a random variable Y that takes values in some discrete space but whose values can change in continuous time. The value of Y at time t is denoted by $Y(t)$.

The Markov chain "memoryless" and the "time homogeneity" assumptions are defined for the continuous-time process as follows. The *memoryless* assumption is that given that $Y = i$ at any time u, the probability that $Y = j$ at any future time $u + t$ does not depend on the values before time u. The *time homogeneity* assumption is that the conditional probability

$$\text{Prob}(Y(u + t) = j \mid Y(u) = i) \tag{10.25}$$

is independent of u, so we write it as $P_{ij}(t)$.

We focus on the case where the possible values of $Y(t)$ are $1, 2, \ldots, s$ and where it is assumed that the transition probability equations (10.25) take the form

$$P_{ij}(h) = q_{ij}h + o(h), \quad j \neq i, \tag{10.26}$$
$$P_{ii}(h) = 1 - q_i h + o(h), \tag{10.27}$$

as $h \to 0$, with q_i defined by

$$q_i = \sum_{j \neq i} q_{ij}. \tag{10.28}$$

We will call the q_{ij} the *instantaneous transition rates* of the process. If q_i is independent of i, the expression (10.27) becomes identical in form to (4.1). The justification for the choice of the mathematical forms on the right-hand sides of (10.26) and (10.27) is the same as that given in Chapter 4 leading to the expression (4.1).

In Section 13.3.1 we consider evolutionary models in which equations of the form (10.26) and (10.27) are assumed. The justification for this is, in effect, that a change at a position in a population from i to j in a small time interval of length $2h$ is approximately twice the probability that this happens in a time interval of length h. Therefore, a generalization of the argument leading to the expression (4.1) leads to equations (10.26) and (10.27). In some evolutionary models that we consider q_i is independent of i, so that the Poisson process theory of Chapter 4 may be used directly for them. When q_i depends on i we may regard the process governing the number of transitions as a generalization of the Poisson process of Chapter 4.

10.7.2 Time-Dependent Solutions

By the time homogeneity property of Markov chains,

$$P_{ij}(t + h) = \sum_k P_{ik}(t)P_{kj}(h),$$

where the sum is taken over all possible states. From this, equations (10.26)–(10.28) imply that for fixed i,

$$P_{ij}(t + h) = P_{ij}(t)(1 - q_j h) + h \sum_{k \neq j} P_{ik}(t)q_{kj} + o(h). \tag{10.29}$$

From this equation and the assumptions (10.26)–(10.28) we arrive at the following system of differential equations:

$$\frac{d}{dt}P_{ij}(t) = -q_j P_{ij}(t) + \sum_{k \neq j} P_{ik}(t)q_{kj}, \quad j = 1, 2, \ldots, s. \tag{10.30}$$

These are called the *forward Kolmogorov equations* of the system. They can be solved explicitly in cases where the q_{kj} take simple forms. Some applications in the evolutionary context where these equations can be solved are given in Sections 13.3.1, 13.3.2, and 13.3.3.

10.7.3 The Stationary Distribution

A stationary distribution $\{\varphi_j\}$ has the property that if at any time t $\text{Prob}(Y(t) = j) = \varphi_j$ for all j, then for all j, $\text{Prob}(Y(u) = j) = \varphi_j$ for all $u > t$. Thus a stationary distribution, if it exists, can be found by replacing the derivatives on the left-hand sides of the system of equations (10.30) by zero and replacing $P_{ij}(t)$ and $P_{ik}(t)$ by φ_j and φ_k, respectively, to get

$$q_j\varphi_j = \sum_{k \neq j} \varphi_k q_{kj}, \quad j = 1, 2, \ldots, s. \tag{10.31}$$

This equation is used in an evolutionary context in Sections 13.3.1, 13.3.2, and 13.3.3.

10.7.4 Detailed Balance

When a stationary distribution $\{\varphi_j\}$ exists, the detailed balance conditions analogous to the discrete-time conditions (10.5) are that for all i, j, and t,

$$\varphi_i P_{ij}(t) = \varphi_j P_{ji}(t). \tag{10.32}$$

10.7.5 Exponential Holding Times

Suppose that $Y(t) = j$. Then $Y(\cdot)$ will remain at the value j for some length of time until it changes to some value other than j, and we now find the probability density function of the time until such a change occurs.

Let T be the (random) time until the value of $Y(\cdot)$ moves to some value different from j. Then

$$\text{Prob}(T \geq t + h) = \text{Prob}(T \geq t) \cdot \text{Prob}(T \geq t + h \,|\, T \geq t).$$

From the memoryless and time homogeneity properties of the process, this is

$$\text{Prob}(T \geq t + h) = \text{Prob}(T \geq t) \cdot \text{Prob}(T \geq h). \tag{10.33}$$

The probability that $T \geq h$ is the probability that the random variable has not moved from the value j before time h, and if terms of order $o(h)$ are ignored, this is the probability $1 - q_j h + o(h)$ given in equation (10.27). Thus

$$\text{Prob}(T \geq t + h) = \text{Prob}(T \geq t) \cdot (1 - q_j h) + o(h). \tag{10.34}$$

Rearrangement of terms gives

$$\frac{\text{Prob}(T \geq t + h) - \text{Prob}(T \geq t)}{h} = -q_j \, \text{Prob}(T \geq t) + \frac{o(h)}{h},$$

and the limiting operation $h \to 0$ gives

$$\frac{d}{dt} \text{Prob}(T \geq t) = -q_j \, \text{Prob}(T \geq t). \qquad (10.35)$$

Standard differential equation calculations show that

$$\text{Prob}(T \geq t) = C \cdot e^{-q_j t},$$

for some constant C. The case $t = 0$ shows that $C = 1$, so that

$$\text{Prob}(T \geq t) = e^{-q_j t}. \qquad (10.36)$$

Allowing for a change in notation, this is identical to equation (1.60), and thus T has an exponential distribution. From (1.46), the density function of T is

$$f(t) = q_j e^{-q_j t}, \quad t > 0. \qquad (10.37)$$

To summarize, if $Y(\cdot)$ has just arrived at the value j, it next moves to some other value after a random length of time having the exponential distribution given in equation (10.37). This implies that having just arrived at the value j, the mean time spent at this value before moving to some other value is q_j^{-1}.

10.7.6   The Embedded Chain

In some applications we might not be interested in the time spent by $Y(\cdot)$ at any value but only in the sequence of values that $Y(\cdot)$ assumes. In other words we are interested only in the so-called *embedded chain* of the process. This embedded chain is a discrete-time Markov chain whose transition probabilities are

$$p_{jk} = \frac{q_{jk}}{q_j}. \qquad (10.38)$$

These are conditional probabilities, derived from (1.92); given that a change occurs, the probability that it is to k is given by (10.38). Markov chain theory can be used to find properties of this embedded process. These properties can provide information about the original time-dependent process. For example, if a time-dependent process has absorbing states, the probability that the process enters a specific absorbing state is the same as the corresponding probability in the embedded chain. The latter probability might be found more easily using discrete time Markov chain theory than by the continuous time theory of this section. On the other hand it is not possible to find absorption time properties for the continuous time process from the embedded chain.

Problems

10.1. Suppose that the transition matrix P of a Markov chain is given by

$$P = \begin{bmatrix} 0.6 & 0.4 \\ 0.3 & 0.7 \end{bmatrix}. \tag{10.39}$$

Use the definitions (B.44) and (B.45) to find the (two) eigenvalues and the (two pairs of) corresponding left and right eigenvectors of P.

10.2. Use your answer to Problem 10.1 to check that equation (B.47) holds.

10.3. For the matrix P given in (10.39), find P^2 by direct matrix multiplication. Then find P^2 by using the eigenvalues and eigenvectors calculated in Problem 10.1, together with the right-hand side of equation (B.48).

10.4. Find the stationary distribution of the Markov chain whose transition matrix is given in (10.39). Find P^3 and P^4 using the spectral expansion, and thus check that P^n is approaching the matrix defined through the stationary distribution.

10.5. Use equation (10.5) to show that if the transition matrix of a finite irreducible aperiodic Markov chain is symmetric, then that Markov chain is reversible.

10.6. Suppose that a finite aperiodic irreducible Markov chain has transition probability matrix P and stationary distribution φ'. Show that if k is any constant, $0 < k < 1$, then the Markov chain with transition probability matrix $P^* = kP + (1 - k)I$ also has stationary distribution φ'. What interpretation or explanation can you give for this result?

10.7. Prove the assertion made below equation (10.8), that $p_{ij} > 0$ for all (i, j) including the case $i = j$.

11

Hidden Markov Models

We divide this brief account of hidden Markov models into three sections: (i) a description of the properties of these models, (ii) the three main algorithms of the models, (iii) applications. For a more complete account of these models, see Rabiner (1989).

11.1 What is a Hidden Markov Model?

A hidden Markov model (HMM) is similar to a Markov chain, but is more general, and hence more flexible, allowing us to model phenomena that we cannot model sufficiently well with a regular Markov chain model. An HMM is a discrete-time Markov model with some extra features. The main addition is that when a state is visited by the Markov chain, the state "emits" a letter from a fixed time-independent alphabet. Letters are emitted via a time independent, but usually state-dependent, probability distribution over the alphabet. When the HMM runs there is, first, a sequence of states visited, which we denote by $q_1, q_2, q_3, \ldots$, and second, a sequence of emitted symbols, denoted by $\mathcal{O}_1, \mathcal{O}_2, \mathcal{O}_3, \ldots$. Their generation can be visualized as a two-step process as follows:

$$\underset{q_1}{\text{initial}} \rightarrow \underset{\mathcal{O}_1}{\text{emission}} \rightarrow \underset{\text{to } q_2}{\text{transition}} \rightarrow \underset{\mathcal{O}_2}{\text{emission}} \rightarrow \underset{\text{to } q_3}{\text{transition}} \rightarrow \underset{\mathcal{O}_3}{\text{emission}} \rightarrow \cdots$$

We denote the entire sequence of q_i's by Q and the entire sequence of $\mathcal{O}_i$'s by $\mathcal{O}$, and we write "the observed sequence $\mathcal{O} = \mathcal{O}_1, \mathcal{O}_2, \ldots$" and "the state sequence $Q = q_1, q_2, \ldots$."

Often we know the sequence $\mathcal{O}$ but do not know the sequence Q. In such a case the sequence Q is called "hidden." An important feature of HMMs is that we can efficiently answer several questions about $\mathcal{O}$ and Q.

One of these questions concerns the estimation of the hidden state sequence that has the highest probability given the observed sequence. We illustrate this with a simple example. Consider the Markov chain with two states S_1 and S_2, with uniform initial distribution and transition matrix

$$\begin{bmatrix} .9 & .1 \\ .8 & .2 \end{bmatrix}.$$

Let A be an alphabet consisting only of the numbers 1 and 2. State S_1 emits a 1 or 2 with equal probability $\frac{1}{2}$, state S_2 emits a 1 with probability $\frac{1}{4}$ and a 2 with probability $\frac{3}{4}$. Suppose the observed sequence is $\mathcal{O} = 2, 2, 2$. What sequence of states $Q = q_1, q_2, q_3$ has the highest probability given $\mathcal{O}$? In other words, what is

$$\underset{Q}{\mathrm{argmax}}\, \mathrm{Prob}(Q \,|\, \mathcal{O})?$$

There are eight possibilities for Q. Each of these can be written down and its probability calculated, and from this it is found that the answer to the above question is $Q = S_2, S_1, S_1$. The sequence Q contains more S_1's, even though S_2 is more likely to produce a 2 when visited (probability $\frac{3}{4}$) than S_1 (probability $\frac{1}{2}$). The reason is because S_1 is much more likely to be visited than S_2 ($p_{11} = .9$ and $p_{21} = .8$).

We can also calculate

$$\mathrm{Prob}(\mathcal{O}) = \sum_Q \mathrm{Prob}(\mathcal{O} \,|\, Q) \cdot \mathrm{Prob}(Q). \tag{11.1}$$

This calculation is useful in distinguishing which of several models is most likely to have produced $\mathcal{O}$.

In the above example all of these calculations can be done by hand. However, models arising in practice have many states, sometimes hundreds, and an alphabet with many symbols (often 20, one for each amino acid). In these cases, calculation of the quantities above by exhaustive methods becomes impossible even for the fastest computers. Fortunately, there are dynamic programming approaches that overcome this problem, which we discuss in detail below. Before turning to the algorithms, however, it is necessary to introduce some specific notation. An HMM will consist of the following five components:

(1) A set of N states $S_1, S_2, \ldots, S_N$.

(2) An alphabet of M distinct observation symbols $A = \{a_1, a_2, \ldots, a_M\}$.

(3) The transition probability matrix $P = (p_{ij})$, where

$$p_{ij} = \mathrm{Prob}(q_{t+1} = S_j \,|\, q_t = S_i)$$

(4) The emission probabilities: For each state S_i and a in A,

$$b_i(a) = \text{Prob}(S_i \text{ emits symbol } a).$$

The probabilities $b_i(a)$ form the elements in an $N \times M$ matrix $B = (b_i(a))$.

(5) An initial distribution vector $\pi = (\pi_i)$, where $\pi_i = \text{Prob}(q_1 = S_i)$.

Components 1 and 2 describe the structure of the model, and 3–5 describe the parameters. It is convenient to let $\lambda = (P, B, \pi)$ represent the full set of parameters. We can now describe the main algorithms.

11.2   Three Algorithms

There are three calculations that are frequently required in HMM theory. Given some observed output sequence $\mathcal{O} = \mathcal{O}_1, \mathcal{O}_2, \ldots, \mathcal{O}_T$, these are:

(i) Given the parameters λ, efficiently calculate

$$\text{Prob}(\mathcal{O} \,|\, \lambda).$$

That is, efficiently calculate the probability of some given sequence of observed outputs.

(ii) Efficiently calculate the hidden sequence $Q = q_1, q_2, \ldots, q_T$ of states that is most likely to have occurred, given $\mathcal{O}$. That is, calculate

$$\underset{Q}{\text{argmax}} \ \text{Prob}(Q \,|\, \mathcal{O}).$$

(iii) Assuming a fixed topology of the model (i.e., a fixed graph structure of the underlying Markov chain, as defined in 4.8), find the parameters $\lambda = (P, B, \pi)$ that maximize $\text{Prob}(\mathcal{O} \,|\, \lambda)$.

We address these problems in turn.

11.2.1   The Forward and Backward Algorithms

We first consider problem (i). The naive way of calculating $\text{Prob}(\mathcal{O})$ is to use formula (11.1). This calculation involves the sum of N^T multiplications, each being a multiplication of $2T$ terms. The total number of operations is thus on the order of $2T \cdot N^T$.

Unless T is quite small, this calculation is computationally infeasible. For example, if $N = 4$, $T = 100$, the number of calculations is on the order of 10^{60}. It would take the life of the universe to make such a calculation.

Fortunately, there is a much more efficient and computationally feasible procedure, called the *forward algorithm*.

The forward algorithm focuses on the calculation of the quantity

$$\alpha\,(t,i) = \mathrm{Prob}(\mathcal{O}_1, \mathcal{O}_2, \mathcal{O}_3, \ldots, \mathcal{O}_t, q_t = S_i), \tag{11.2}$$

which is the joint probability that the sequence of observations seen up to and including time t is $\mathcal{O}_1, \mathcal{O}_2, \mathcal{O}_3, \ldots, \mathcal{O}_t$, and that the state of the HMM at time t is S_i. The $\alpha(t,i)$ are called the *forwards* variables.

Once we know $\alpha(T,i)$ for all i, then $\mathrm{Prob}(\mathcal{O})$ can be calculated as

$$\mathrm{Prob}(\mathcal{O}) = \sum_{i=1}^{N} \alpha(T,i). \tag{11.3}$$

We calculate the $\alpha(t,i)$'s inductively on t. The first calculation is of the *initialization* step, and uses the obvious result

$$\alpha(1,i) = \pi_i b_i(\mathcal{O}_1). \tag{11.4}$$

Next, the equation

$$\alpha(t+1,i) = \sum_{j=1}^{N} \mathrm{Prob}(\mathcal{O}_1, \mathcal{O}_2, \mathcal{O}_3, \ldots, \mathcal{O}_{t+1}, q_{t+1} = S_i \text{ and } q_t = S_j)$$

leads to the *induction* step

$$\alpha(t+1,i) = \sum_{j=1}^{N} \alpha\,(t,j) p_{ji} b_i(\mathcal{O}_{t+1}). \tag{11.5}$$

This equation gives $\alpha(t+1,i)$ in terms of the $\alpha(t,j)$, so that $\alpha(t+1,i)$ can be calculated quickly once the $\alpha(t,j)$ are known. We use (11.4) to calculate $\alpha(1,i)$ for all i; then we use (11.5) to calculate $\alpha(2,i)$ for all i and again to calculate $\alpha(3,i)$ for all i, and so on, until we have obtained the $\alpha(T,i)$ for all i, needed in (11.3).

This procedure provides an algorithm for the solution to problem (i). The algorithm requires on the order of TN^2 computations, and thus is feasible in practice, even for very large models.

Before going on to problem (ii), we consider briefly the *backward* part of the forward–backward algorithm. This provides another approach to solving problem (i), but we introduce it because we will use the "backwards" variables when we discuss problem (iii).

In the above, we calculated successively $\alpha(1,\cdot), \alpha(2,\cdot), \ldots, \alpha(T,\cdot)$, that is, we calculated forward in time. In the backward algorithm we calculate, as the name suggests, backward in time. Specifically, we calculate the probability $\beta(t,i)$, defined by

$$\beta(t,i) = \mathrm{Prob}(\mathcal{O}_{t+1}, \mathcal{O}_{t+2}, \ldots, \mathcal{O}_T \,|\, q_t = S_i), \tag{11.6}$$

starting with the value $t = T - 1$, then for the value $t = T - 2$, and so on, eventually working back to $t = 1$. The relevant equations for this procedure are

$$\beta(T - 1, i) = \sum_{j=1}^{N} p_{ij}\, b_j(\mathcal{O}_T), \tag{11.7}$$

and for $t \leq T - 1$,

$$\beta(t - 1, i) = \sum_{j=1}^{N} p_{ij}\, b_j(\mathcal{O}_T)\beta(t, j). \tag{11.8}$$

Using these two equations we can successively calculate $\beta(T - 1, i)$ for all i, $\beta(T - 2, i)$ for all $i, \ldots$, and $\beta(1, i)$ for all i. It is convenient to define $\beta(T, j)$ to be 1 for all j, so that (11.7) becomes a special case of (11.8) (with $t = T$).

11.2.2   The Viterbi Algorithm

Given some observed sequence $\mathcal{O} = \mathcal{O}_1, \mathcal{O}_2, \mathcal{O}_3, \ldots, \mathcal{O}_T$ of outputs, we want to compute efficiently a state sequence $Q = q_1, q_2, q_3, \ldots, q_T$ that has the highest conditional probability given $\mathcal{O}$. In other words, we want to find a Q that makes $\mathrm{Prob}(Q \,|\, \mathcal{O})$ maximal, that is, we want to calculate

$$\underset{Q}{\mathrm{argmax}} \;\; \mathrm{Prob}(Q \,|\, \mathcal{O}). \tag{11.9}$$

There may be many Q's that maximize $\mathrm{Prob}(Q \,|\, \mathcal{O})$. We give an algorithm that finds one of them. It can easily be generalized to find them all. However, for our applications this generalization will not be necessary.

The Viterbi algorithm carries out the efficient computation of (11.9). The algorithm is divided into two parts. It first finds $\max_Q \mathrm{Prob}(Q \,|\, \mathcal{O})$, and then "backtracks" to find a Q that realizes this maximum. This is another dynamic programming algorithm.

First define, for arbitrary t and i,

$$\delta_t(i) = \max_{q_1, q_2, \ldots, q_{t-1}} \mathrm{Prob}(q_1, q_2, \ldots, q_{t-1}, q_t = S_i \text{ and } \mathcal{O}_1, \mathcal{O}_2, \mathcal{O}_3, \ldots, \mathcal{O}_t)$$

($\delta_1(i) = \mathrm{Prob}(q_1 = S_i \text{ and } \mathcal{O}_1)$). In words, $\delta_t(i)$ is the maximum probability of all ways to end in state S_i at time t and have observed sequence $\mathcal{O}_1$, $\mathcal{O}_2, \ldots, \mathcal{O}_t$. Then

$$\max_Q \mathrm{Prob}(Q \text{ and } \mathcal{O}) = \max_i \delta_T(i).$$

The probability in this expression is the joint probability of Q and $\mathcal{O}$, not a conditional probability. Our aim is to find a sequence Q for which the maximum conditional probability (11.9) is achieved. Since

$$\max_Q \text{Prob}(Q \mid \mathcal{O}) = \max_Q \frac{\text{Prob}(Q \text{ and } \mathcal{O})}{\text{Prob}(\mathcal{O})},$$

and since the denominator on the right-hand side does not depend on Q,

$$\underset{Q}{\text{argmax}} \ \text{Prob}(Q \mid \mathcal{O}) = \underset{Q}{\text{argmax}} \ \frac{\text{Prob}(Q \text{ and } \mathcal{O})}{\text{Prob}(\mathcal{O})} = \underset{Q}{\text{argmax}} \ \text{Prob}(Q \text{ and } \mathcal{O}).$$

The first step is to calculate the $\delta_t(i)$'s inductively. Then we will "backtrack" and recover the sequence that gives the largest $\delta_T(i)$. The *initialization* step is

$$\delta_1(i) = \pi_i b_i(\mathcal{O}_1), \quad 1 \le i \le N. \tag{11.10}$$

The induction step is

$$\delta_t(j) = \max_{1 \le i \le N} \delta_{t-1}(i) p_{ij} b_j(\mathcal{O}_t), \quad 2 \le t \le T, \ 1 \le j \le N. \tag{11.11}$$

We recover the q_i's as follows. Define

$$\psi_T = \underset{1 \le i \le N}{\text{argmax}} \ \delta_T(i),$$

and put $q_T = S_{\psi_T}$. Then q_T is the final state in the state sequence required. The remaining q_t for $t \le T - 1$ are found recursively by first defining

$$\psi_t = \underset{1 \le i \le N}{\text{argmax}} \ \delta_t(i) p_{i\psi_{t+1}},$$

and then putting $q_t = S_{\psi_t}$. If the argmax is not unique, we arbitrarily take one value of i giving the maximum.

11.2.3   The Estimation Algorithms

We now address problem (iii). Suppose we are given a set of observed data from an HMM for which the topology is known (by topology we mean the graph structure of the underlying Markov model). We wish to try to estimate the parameters in that HMM. The parameter space is usually far too large to allow exact calculation of a set of parameter estimates that maximizes the probability of the data. Instead, we employ algorithms that find "locally" best sets of parameters. This partial solution to the problem has proven to be useful in many applications.

The focus on local estimation means that the procedure is heuristic. Therefore, the efficacy of the procedure must evaluated empirically by using benchmarks and test sets for which there are known outcomes. This matter is discussed further below.

Some further comments are in order. It is not necessary to assume that the data come from an HMM. Instead, it is usually more accurate to assume that the data are generated by some random process that we try to "fit" with an HMM. Sometimes it might be possible to achieve a tight fit with an HMM and sometimes it might not.

The discussion above shows that we should use the term "estimation" of parameters cautiously in this section. Our aim is to "set" parameters at values providing a good fit to data rather than to estimate parameters in the sense of Chapter 8.

We now describe the Baum–Welch method of parameter estimation. This is a difficult algorithm, so we do not provide proofs of the claims made but instead indicate the intuition behind the method.

We assume that the alphabet A and number of states N is fixed at the outset, and that the parameters π_i, p_{jk}, and $b_i(a)$ are unknown and are to be "estimated." The data we use to estimate the parameters constitute a set of observed sequences $\{\mathcal{O}^{(d)}\}$. Each observed sequence $\mathcal{O}^{(d)} = O_1^{(d)}, O_2^{(d)}, \ldots$ has a corresponding hidden state sequence $Q^{(d)} = q_1^{(d)}, q_2^{(d)}, \ldots$.

The procedure starts by setting the parameters π_i, p_{jk}, and $b_i(a)$ at some initial values. These can be chosen from some uniform distribution or can be chosen to incorporate prior knowledge about them. We then calculate, using these initial parameter values,

$$\overline{\pi}_i = \text{the expected proportion of times in state } S_i \text{ at} \qquad (11.12)$$
$$\text{the first time point, given } \{\mathcal{O}^{(d)}\},$$

$$\overline{p}_{jk} = \frac{E(N_{jk} \mid \{\mathcal{O}^{(d)}\})}{E(N_j \mid \{\mathcal{O}^{(d)}\})}, \qquad (11.13)$$

$$\overline{b}_i(a) = \frac{E(N_i(a) \mid \{\mathcal{O}^{(d)}\})}{E(N_i \mid \{\mathcal{O}^{(d)}\})}, \qquad (11.14)$$

where N_{jk} is the (random) number of times $q_t^{(d)} = S_j$ and $q_{t+1}^{(d)} = S_k$ for some d and t; N_i is the (random) number of times $q_t^{(d)} = S_i$ for some d and t; and $N_i(a)$ equals the (random) number of times $q_t^{(d)} = S_i$ and it emits symbol a, for some d and t. The expected values in (11.13) and (11.14) are conditional expected values, as defined in (2.56).

We show how to calculate these efficiently below. These are the "reestimation" parameter values that then replace π_i, p_{jk}, and $b_i(a)$. These values follow the form of estimation used, for example, in equation (3.10). The algorithm proceeds by iterating this step.

It can be shown that if $\lambda = (\pi_i,\ p_{jk},\ b_i(a))$ is replaced by $\overline{\lambda} = (\overline{\pi}_i, \overline{p}_{jk}, \overline{b}_i(a))$, then $\text{Prob}(\{\mathcal{O}^{(d)}\} \mid \overline{\lambda}) \geq \text{Prob}(\{\mathcal{O}^{(d)}\} \mid \lambda)$, with equality holding if and only if $\overline{\lambda} = \lambda$. Thus successive iterations continually increase the probability of the data, given the model. Iterations continue until either a local maximum of the probability is reached or until the change in the probability becomes negligible.

In order to discuss the calculations needed for (11.13)–(11.12), define $\xi_t^{(d)}(i,j)$ by

$$\xi_t^{(d)}(i,j) = \text{Prob}(q_t^{(d)} = S_i, q_{t+1}^{(d)} = S_j \,|\, \mathcal{O}^{(d)}), \qquad (11.15)$$

where $i,j = 1, \ldots, N$, and $t \geq 1$. The conditional probability formula (1.92) shows that this is equal to

$$\frac{\text{Prob}(q_t^{(d)} = S_i, q_{t+1}^{(d)} = S_j, \mathcal{O}^{(d)})}{\text{Prob}(\mathcal{O}^{(d)})}.$$

The denominator is $\text{Prob}(\mathcal{O}^{(d)})$ and is thus calculated efficiently using the methods of Section 11.2.1. The numerator is calculated efficiently by writing it in terms of the forwards and backwards variables discussed in Section 11.2.1,

$$\text{Prob}(q_t^{(d)} = S_i, q_{t+1}^{(d)} = S_j, \mathcal{O}^{(d)}) = \alpha_t(i) p_{ij} b_j(\mathcal{O}_{t+1}^{(d)}) \beta_{t+1}(j). \qquad (11.16)$$

Let $I_t^{(d)}(i)$ be the indicator variables defined by

$$I_t^{(d)}(i) = \begin{cases} 1, & \text{if } q_t^{(d)} = S_i, \\ 0, & \text{otherwise.} \end{cases}$$

The number of times S_i is visited is then $\sum_d \sum_t I_t^{(d)}(i)$. The expected number of times S_i is visited, given $\{\mathcal{O}^{(d)}\}$, is then

$$\sum_d \sum_t E(I_t^{(d)}(i) \,|\, \mathcal{O}^{(d)}). \qquad (11.17)$$

Now $E(I_t^{(d)}(i) \,|\, \mathcal{O}^{(d)})$ is $\text{Prob}(q_t^{(d)} = S_i \,|\, \mathcal{O}^{(d)})$, which is

$$\sum_{j=1}^{N} \xi_t^{(d)}(i,j). \qquad (11.18)$$

Thus the expected number of times S_i is visited, given $\{\mathcal{O}^{(d)}\}$, is

$$\sum_d \sum_t \sum_{j=1}^{N} \xi_t^{(d)}(i,j).$$

Similarly, the expected number of transitions from S_i to S_j given $\{\mathcal{O}^{(d)}\}$ is

$$\sum_d \sum_t \xi_t^{(d)}(i,j).$$

These expressions give efficient formulae to calculate all the quantities in equations (11.12)–(11.14) except the numerator of (11.14). This is calculated as follows.

Define the indicator random variables $I_t^{(d)}(i, a)$ by

$$I_t^{(d)}(i, a) = \begin{cases} 1, & \text{if } q_t^{(d)} = S_i \text{ and } \mathcal{O}_t^{(d)} = a, \\ 0, & \text{otherwise.} \end{cases}$$

Then $E(I_t^{(d)}(i, a) \,|\, \mathcal{O}^{(d)})$ is the expected number of times the dth process is in state S_i at time t and emits symbol a, given $\mathcal{O}^{(d)}$. The numerator of (11.14) is equal to $\sum_d \sum_t E(I_t^{(d)}(i, a) \,|\, \mathcal{O}^{(d)})$, which is

$$\sum_d \sum_t \sum_{\mathcal{O}_t^{(d)} = a} \sum_{j=1}^{N} \xi_t^{(d)}(i, j).$$

11.3   Applications

We sketch here the applications of HMMs in several different areas of computational biology. Only a brief outline of each application is given: further details may be found in the references provided.

11.3.1   Modeling Protein Families

In this section we develop an HMM to model protein families, and we will use the model for two purposes: to construct multiple sequence alignments and to determine the family of a query sequence. These applications were first presented in Krogh et al. (1994). In order to present the main ideas we simplify many of the details.

Figure 11.1 gives an example of the basic type of HMM we will use. This example has "length" five; any length is possible. The underlying Markov model is presented in graphical form (as in Section 4.8). The states are the squares, diamonds, and circles labeled $m_0, m_1, \ldots, m_5, i_0, i_1, \ldots, i_4$, and $d_1, d_2, \ldots, d_4$, respectively. The squares are called the *match* states, the diamonds the *insert* states, and the circles the *delete* states. The edges not shown have transition probability zero. State m_0 is the *start* state, so that the process always starts in state m_0. A transition never moves to the left, so that as time progresses the current state gradually moves to the right, eventually ending in match state m_5, the *end* state. When this state is reached the process ends. A match or delete state is never visited more than once.

The alphabet A consists of the twenty amino acids together with one "dummy" symbol representing "delete" (denoted δ). Delete states output δ with probability one. Each insert and match state has its own distribution over the 20 amino acids, and cannot emit a δ. That is, only a delete state can emit a δ, and each delete state emits *only* δ.

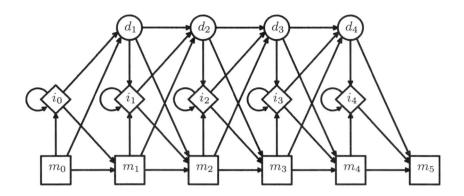

FIGURE 11.1. Hidden Markov model for a protein family.

If the emission probabilities for the match and insert states are uniform over the 20 amino acids, the model will produce random sequences that do not have much in common except possibly their lengths. At the other extreme, if each state emits one specific amino acid with probability one, and if further the transitions from m_i to m_{i+1} have probability one, then the model will always produce the same sequence. Somewhere in between these two extremes the parameters of the model can be set so that it produces sequences that are similar, thus producing what can be thought of as a "family" of sequences. Each choice of parameters produces a different family. This family can be rather "tight," meaning all sequences in it are very similar, or can be "loose," so that there is little similarity between the sequences produced. It is also possible that the similarity is high in some positions of the sequences produced and low in others. This will happen if some match states have distributions concentrated on a few amino acids while the others have distributions in which all amino acids are approximately equally likely. By contrast, the dynamic programming sequence alignment algorithms and BLAST allow one gap open penalty and use one substitution matrix uniformly across the entire length of the sequences compared. Allowing gap penalties and substitution probabilities to vary along the sequences reflects biological reality better. Alignments of related proteins generally have regions of higher conservation and regions of lower conservation. The regions of higher conservation are called functional domains, because their resistance to change indicates that they serve some critical function. Dynamic programming alignment and BLAST are essential for certain applications, such as pairwise alignments, or aligning a small number of sequences. But for modeling large families of sequences, or constructing alignments of many sequences, HMMs allow for efficiency, and at the same time exploit the larger data sets to increase flexibility.

In the HMM model of a protein family the transition (arrow) from a match state to an insert state corresponds to the gap open penalty, and the arrow from an insert state to itself corresponds to the gap extension penalty. Loosely speaking, the distribution over the amino acids for any state takes the place of a substitution matrix. The probabilities in the model can differ from position to position in the sequence, since each arrow has its own probability and each match and insert state has its own distribution. Thus the HMM model is sufficiently flexible to model the varying features of a protein along its length. While the model can be made even more flexible by adding further parameters, more data are needed to estimate these parameters effectively. The model described has proven in practice to provide a good compromise between flexibility and tractability. Such HMM models of are called *profile* HMMs.

All applications start with *training*, or estimating, the parameters of the model using a set of training sequences chosen from a protein family, such as the set of all globins in GenBank. This estimation procedure uses the Baum–Welch algorithm. The model is chosen to have length equal to the average length of a sequence in the training set, and all parameters are initialized by using uniform distributions (i.e., amino acids are given probability $\frac{1}{20}$, and transitions of the same type are given $\frac{1}{2}$ equal probabilities).[1]

11.3.2   Multiple Sequence Alignments

In this section we describe how to use the theory described above to compute multiple sequence alignments for a family of sequences. The sequences to be aligned are used as the training data, to train the parameters of the model. For each sequence the Viterbi algorithm is then used to determine a path most likely to have produced that sequence. These paths can then be used to construct an alignment. Amino acids are aligned if both are produced by the same match state in their paths. Indels are then inserted appropriately for insertions and deletions.

We illustrate this with an example. Consider the sequences CAEFDDH and CDAEFPDDH. Suppose the model has length 10 and their most likely paths through the model are

$$m_0 m_1 m_2 m_3 m_4 d_5 d_6 m_7 m_8 m_9 m_{10}$$

---

[1] What Krogh et al. (1994) do is somewhat more complicated. They allow the length of the model to change along with the parameters after each iteration of the reestimation algorithm. They also must adjust the Baum–Welch algorithm from how we have described it, in order to handle the delete states. The reader interested in implementing these applications should refer to the literature for further details.

and

$$m_0 m_1 i_1 m_2 m_3 m_4 d_5 m_6 m_7 m_8 m_9 m_{10},$$

respectively. Then the alignment induced is found by aligning positions that were generated by the same match state:

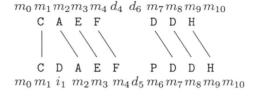

This leads to the alignment

C$-$AEF$-$DDH
CD AEF P DDH   .

More generally, suppose we have the five sequences

CAEFTPAVH,
CKETTPADH,
CAETPDDH,
CAEFDDH,
CDAEFPDDH,

and the corresponding paths returned by the Viterbi algorithm are

$$m_0 m_1 m_2 m_3 m_4 m_5 m_6 m_7 m_8 m_9 m_{10},$$
$$m_0 m_1 m_2 m_3 m_4 m_5 m_6 m_7 m_8 m_9 m_{10},$$
$$m_0 m_1 m_2 m_3 d_4 m_5 m_6 m_7 m_8 m_9 m_{10},$$
$$m_0 m_1 m_2 m_3 m_4 d_5 d_6 m_7 m_8 m_9 m_{10},$$
$$m_0 m_1 i_1 m_2 m_3 m_4 d_5 m_6 m_7 m_8 m_9 m_{10}.$$

Then the induced alignment is

C$-$AEF T P AVH
C$-$KET T P ADH
C$-$AE$-$ T P DDH   .
C$-$AEF $-$$-$DDH
CD AEF $-$ P DDH

This technique can give ambiguous results in some cases. For example, if the model has length two and the sequences ABAC and ABBAC had

$$m_0 m_1 i_1 i_1 m_2 m_3 \text{ and } m_0 m_1 i_1 i_1 i_1 m_2 m_3$$

as paths, then the leading A's and trailing C's will be aligned, but it is not clear how to align the BA from the first sequence to the BBA from the second.

In such cases Krogh et al. (1994) represent the ambiguous symbols with lowercase letters and do not attempt to give alignments of these regions.

This technique can be used to align many sequences with relatively little computing power. By contrast, dynamic programming algorithms cannot in practice align 50 or 100 long sequences. This is the value of a heuristic approach. Another advantage of the method is that it allows the sequences themselves to guide the alignment, rather than having a precomputed substitution matrix and gap penalties. Thus less bias should be introduced.

Krogh et al. (1994) tested this method on a family of 625 globin sequences. They used a published alignment of seven of these sequences constructed using knowledge of the three-dimensional structure of the sequences (Bashford et al. (1987)). (An alignment using a three-dimensional structure is considered reliable and so serves as a benchmark for testing multiple alignment algorithms.) They chose 400 of the 625 globins to train the model and then used the Viterbi algorithm to align all 625. These alignments were then compared to the induced alignment on the seven sequences from the Bashford alignment. The alignments agreed extremely well.

11.3.3   Pfam

Pfam is a web-based resource maintained by the Sanger Center (web URL (http://www.sanger.ac.uk/Pfam/). Pfam uses the basic theory described above to determine protein domains in a query sequence. A protein usually has one or more functional domains, namely portions of the protein that have essential function and thus have low tolerance for amino acid substitutions. Proteins in different families often share high homology in one or more domains. Entire protein families can be characterized by HMMs, as in the previous section, or one can characterize just functional domains. Pfam focuses on the latter.

Suppose that a new protein is obtained for which no information is available except the raw sequence. We wish to "annotate" this sequence. Annotation is the process of assigning to a sequence biologically relevant information, such as where the functional domains are, what their homology is to known domains, and what their function is. The typical starting point is a BLAST search. This will return all sequences in the chosen databases that have significant similarity to the query sequence. BLAST can return many such sequences. Though this is an important step in the annotation process, it is also desirable to have a database not of protein sequences themselves, but of protein domains. Pfam is not the first such database; however, previous domain databases do not use methods as flexible as HMMs and consequently tend not to model entire domains, but rather only the most highly conserved "motifs" that can be put in *ungapped* multiple sequence alignments. The use of HMMs allows for more effective characterization of full domains.

The domains in Pfam are determined based on expert knowledge, sequence similarity, and other protein family databases. Currently, Pfam contains 2008 protein domains. For each domain a set of examples of this domain is selected. The sequences representing each domain are put into an alignment, and the alignments themselves are used to set the parameters; that is, Baum–Welch is not used. Recall that an alignment implies for each sequence in the alignment a path through the HMM, as described in the previous section. The proportion of times these paths take a given transition is used to estimate the transition probabilities, and likewise for the emission probabilities. These alignments are called "seed alignments" and are stored in the database. Given the HMMs for all of the domains, a query sequence is then run past each one using the forward algorithm. When a portion of the query sequence has probability of having been produced by an HMM above a certain cutoff, the domain corresponding to that HMM is reported. Furthermore, the sequence can be aligned to the seed alignment using the Viterbi algorithm as described above. For more details, see the Pfam web site.

11.3.4   Gene Finding

Genomic sequences with lengths on the order of many millions of bases are now being produced, and the sequences of entire chromosomes are becoming available. Such sequences consist of a collection of genes separated from each other by long stretches of nonfunctional sequence. It is of central importance to find where the genes are in the sequence. Therefore, computational methods that quickly identify a large proportion of the genes are very useful. The problem involves bringing together a large amount of diverse information, and there have been many approaches to doing this. Currently, a popular and successful gene finder for *human* DNA sequences is GENSCAN (Burge et al. (1997)), which is based on a generalization of hidden Markov models. We sketch below an algorithm similar in spirit to that in GENSCAN in order to illustrate the basic concept of an HMM human gene finder. To increase the accuracy of the procedure it is necessary to introduce many details that we do not describe here. The interested reader is encouraged to read Burge et al. (1997) and Burge (1997).

Semihidden Markov Models

In an HMM, suppose that p is the probability of the transition from any state to itself. The probability that the process stays in this state for n steps is $p^{n-1}(1 - p)$, so that the length of time the process stays in that state follows a geometric distribution. For the gene model we construct, it is necessary to allow other distributions for this length. In a *semihidden HMM* (semiHMM) all transition probabilities from a state to itself are zero, and when the process visits a state it produces not a single symbol from the alphabet but rather an entire sequence. The length of the sequence can follow any distribution, and the model generating the sequence of that length can be any distribution. The positions in the sequences emitted from a state need not be iid.

The model is formulated more precisely as follows. Each state S has associated with it a random variable L_S (L for "length") whose range is a subset of $0, 1, 2, \ldots$, and for each observable value ℓ of L_S there is a random variable $Y_{S,\ell}$ whose range consists of all sequences of length ℓ. When state S is visited a length ℓ is determined randomly from the distribution for L_S. Then the distribution for $Y_{S,\ell}$ is used to determine a sequence of length ℓ. Then a transition is taken to a new state and the process is repeated, generating another sequence. These sequences are concatenated to create the final output sequence of the semiHMM.

The algorithms involved in this model are an order of magnitude more complex than for a regular HMM, since given an observed output sequence not only do we not know the path of states that produced it, we also do not know the division points in the sequence indicating where a transition was made to a new state. The gene-finding application requires a generalization of the Viterbi algorithm. There is a natural generalization. However, since one is generally working with very long sequences, the natural generalization does not run in reasonable time. In practice, further assumptions must be made. Burge (1997) observed that if the lengths of the long intergenic regions can be taken as having geometric distributions, and if these lengths generate sequences in a relatively iid fashion, then the algorithm can be adjusted so that practical running times can be obtained. These assumptions are not unreasonable in our case, and so they should not greatly affect the accuracy of the predictions. We will omit the technical details surrounding this issue. Our goal is to convey the main idea of how an HMM gene finder works.

A *parse* ϕ is a sequence of states $q_1, q_2, \ldots, q_r$ and a sequence of lengths $d_1, d_2, \ldots, d_r$. Given an observed sequence s from a semiHMM, the Viterbi algorithm finds an optimal parse ϕ_{opt} such that $\text{Prob}(\phi_{\text{opt}} \mid s) \geq \text{Prob}(\phi \mid s)$ for all parses ϕ. In other words, ϕ_{opt} is a parse that is most likely to have given rise to the sequence s. As we will see, the optimal parse gives the gene predictions.

Gene Structure

We now outline the basic properties of human genes that are to be captured in the model. The statistical aspects arise because (1) characteristics shared by genes have similar but not identical properties and (2) signals that genes share can also exist randomly in the nongene sequence. This issue has also been discussed in Sections 5.2 and 5.3.

A gene consists mainly of a continuous sequence of the DNA that is copied, or "transcribed," into RNA, called "premessenger" RNA or pre-mRNA. This pre-mRNA consists of an alternating sequence of exons and introns. After transcription the introns are edited out of the pre-mRNA, and the final molecule, called "messenger RNA" or mRNA, is translated into protein. There can be some other editing and processing of an mRNA before translation. However, that will not be important for our purposes.

The region of the DNA before the start of the transcribed region is called the *upstream region*. This is where the *promoter* of the gene is, the region where certain specialized proteins bind and initiate transcription. There are different definitions of what constitutes the promoter region; often it is taken to be the 500 bases before the start of transcription. Here we will be interested in only about 40 bases upstream from the start of transcription, since specific signals in the promoter region are extremely complex and are not well characterized. Our model, and the model used by Burge (1997), uses the so-called TATA box, which is a fairly common signal (approximately 70% of genes contain this signal), which is located 28–34 bases upstream from the start of transcription. We do not try to capture any other signal in the promoter region. For those genes without a TATA box, we bank on identifying the gene by the other signals in its transcribed region. The 5′ *untranslated region* (5′UTR) follows the promoter. This is a stretch of DNA that does not get translated into protein. We call the first 8 bases of this region the *cap end* of the 5′UTR. Near the other end of the 5′UTR, just before the start codon in the first exon, is a signal that indicates the start of translation, called the translation initiation signal. We will refer to the 18 bases just before the start codon as the *translation initiation end* (TIE) of the 5′UTR. This is followed either by a single exon or by a sequence of exons separated by introns. An intron may break a codon anywhere between its three nucleotides. Each intron has signals indicating its beginning and end. Modeling these signals well is crucial for correctly predicting the intron/exon structure. Following the final exon is the 3′ *untranslated region* (3′UTR), which is another stretch of sequence that is transcribed but not translated. Near the end of the 3′UTR are one or more *Poly-A* signals signaling the end of transcription. A Poly-A signal is 6 bases long with the typical sequence AATAAA.

The Training Data

Each state of the model we construct is a model in its own right. It is necessary to train each state to produce sequence that models the corresponding part of an actual gene. To do this we start with a large set of training data consisting of long stretches of DNA where the gene structures have been completely characterized. Burge et al. (1997) compiled 2.5 million bases (Mb) of human DNA with 380 genes, consisting of 142 single-exon genes and a total of 1492 exons and 1254 introns. Many of these are complete genes consisting of both the upstream and downstream regions. In addition to this they included the coding region only (no introns) of 1619 human genes.

The Model

We model a 5′ to 3′ oriented gene with a 13 state semiHMM as shown in Figure 11.2. The first row represents the intergenic region. The second row represents the promoter. The third row is the 5′UTR. The fourth row of five states represents the introns and exons. The final row is the 3′UTR and the Poly-A signal.

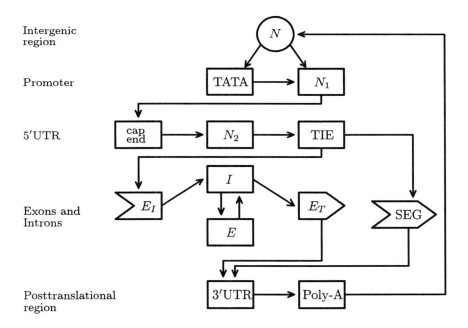

FIGURE 11.2.

We first describe how each state is trained by giving the distributions of L_S and $Y_{S,\ell}$. We then discuss how the transition probabilities are set. The intergenic region between genes is labeled N in Figure 11.2. For completely uncharacterized sequence it is reasonable to assume that genes are randomly distributed. The number of genes in any given stretch of this uncharacterized sequence is therefore modeled by a Poisson distribution, and the distance between genes is modeled by an exponential distribution.

Hence we model the length L_N with a geometric distribution (since it is the discrete analogue of the exponential) with mean equal to the size of the human genome (approximately 3 billion nucleotides) divided by the number of genes (approximately 60,000). Given an observed value ℓ of L_N, the sequence of length ℓ is generated by a fifth-order Markov model (as described in Section 10.3). The parameters of this model are set using the noncoding portions of the genes in the training set. This fifth-order model has $3 \cdot 4^5 = 3072$ parameters. Training so many parameters is possible only because there are millions of bases of DNA in the training set. There is known to be at least fifth-order Markov dependence in such noncoding DNA. A fifth-order Markov chain model is used in particular because it provides improved performance over lower-order models and because a higher-order model cannot be trained with the current training sets available. This model for producing sequence will be used for several of the states (those denoted by N_1 and N_2, as well as the entire 3'UTR), and will be referred to as the *intergenic null model*.

The TATA box is modeled with a 15-base weight matrix similar to that described in Section 5.3.2. The path can bypass the TATA box, which is necessary because only about 70% of genes have them.

The state following the TATA box produces a sequence following the null intergenic model. The length of the sequence L_{N_1} follows a uniform distribution from 28 to 34 bases.

The cap end state models the signal at the start of transcription. There has been shown to be information content in at least the first few bases of this sequence, and it is modeled with an 8-base weight matrix derived from the training data.

State N_2 is modeled with the intergenic null model with length given by a geometric distribution with mean 735 bases. This state is the main part of the 5'UTR.

The state labeled TIE contains the translation initiation signal. This state is modeled by a 18-base weight matrix derived from the training data.

The states on the next row are the exons and introns. There are two paths that can be taken through this row. The state on the right corresponds to single exon genes (SEG). It is necessary to give single exon genes their own state because the distribution of their lengths is quite different from the lengths of the exons of multiexon genes. The generation of codons is not, however, different from the multiexon genes. They both use a fifth-order Markov model, but in contrast to the intergenic region, this Markov model is nonhomogeneous. For each position in the codon there is a separate model. So in effect there are three times as many parameters as in the homogeneous intergenic model. This is partly why the training set was supplemented with the coding region of 1619 complete genes (with no introns), which gave approximately 2.3 million bases of coding sequence. The SEG generates a sequence starting with the start codon *atg* and ending with one of the three stop codons *taa, tag, tga*, chosen in accordance with their observed frequencies in the training set. The intervening codons are produced according to the fifth-order nonhomogeneous Markov model. The length of the single-exon gene is taken from an empirical distribution from the training set.

The situation with multiexon genes is more complicated. One complication arises from the fact that a single codon may be split between two exons. To handle this, the exon states in the model only produce a sequence that has length that is a multiple of three. The intron state produces one codon and then a further random number, either 0, 1, or 2, with probability $\frac{1}{3}$ of each. If this random number is 1, it places the first nucleotide at the beginning and the last two at the end; if 2, it places the first two nucleotides at the beginning and the last one at the end; and if it is 0, it does not include the codon at all.[2]

Empirical distributions from the training set are used for the lengths of the initial, internal, and terminal exons. Intron length is modeled with a geometric distribution. The codons in the exons are modeled with the same fifth-order nonhomogeneous Markov model as the single-exon genes. The intron sequence is generated by the intergenic null model, except at the ends. At the beginning of an intron is a signal called the *donor splice signal*, and at the end is the *acceptor splice signal*. Each of these tells the machinery in the cell where to splice out the intron during the editing process. It is important to model these signals well in order to predict intron/exon structure correctly. The donor signal is taken to be the first six nucleotides of the intron, and the acceptor signal is the last 20. A weight matrix or first-order Markov approach is generally insufficient to capture these signals effectively. Ideally, we would like to have a complete joint probability distribution for these sequences. However, there is insufficient data to do this. For the donor signal we instead use the maximal dependence decomposition discussed in Section 5.3.4, and for the acceptor signal we use a second-order Markov model.[3] All of this is incorporated into the single intron state I.

The 3'UTR is modeled with the intergenic null model, with geometric length of mean approximately 450. The Poly-A signal has constant length 6 bases and is modeled with a weight matrix.

---

[2]Burge handles this in a different way, by creating three internal exon and three intron states, each corresponding to a different "phase." Due to the technical nature of the algorithms, that method may be more effective than the one described here. This method does, however, require that the model be complicated substantially in order to avoid technically violating the semiHMM assumption. We made an effort here to give a model that satisfies the definition of a semiHMM, at the likely cost of some degree of accuracy.

[3]Burge models the donor splice signal also as dependent on the last three bases of the exon. Again, this violates the definition of semiHMM, so we have not included these dependencies in our model.

We now turn to the issue of assigning probabilities to the transitions. Most of the transitions have probability one. The exception is the transitions from N, from TIE, and from I. Since approximately 70% of genes have a TATA box, the transition probability from N to TATA is 0.7 and to N_1 is 0.3. The transition probability from the state labeled TIE to the state labeled SEG is taken from the proportion of single-exon genes. The transitions from the intron state I are also taken from the appropriate proportions observed in the training data.

With the model so defined, given an uncharacterized sequence of DNA, we apply the Viterbi algorithm to obtain an optimal parse. The parse gives a list of the states visited and the lengths of the sequences generated at those states. We thus get a decomposition of the original sequence into gene predictions, as well as predictions of complete gene structure for each predicted gene.

In practice, there are many more considerations in optimizing the performance. Perhaps one of the most important is that many of the probabilities that have been estimated depend on the cg content of the region of the DNA being searched. For example, regions with high cg content tend to contain a significantly higher density of genes. Burge's model takes this and several other factors into careful consideration, and the model continues to be refined as more and different types of data become available.

Problems

11.1 Define an HMM λ with the following parameters:

Three states, S_1, S_2, S_3, alphabet $A = \{1, 2, 3\}$,

$$P = \begin{bmatrix} 0 & 1/2 & 1/2 \\ 1 & 0 & 0 \\ 0 & 1 & 0 \end{bmatrix},$$

$$\pi = \begin{bmatrix} 1 \\ 0 \\ 0 \end{bmatrix},$$

$b_1(1) = \frac{1}{2}, b_1(2) = \frac{1}{2}, b_1(3) = 0,$
$b_2(1) = \frac{1}{2}, b_2(2) = 0, b_2(3) = \frac{1}{2},$
$b_3(1) = 0, b_3(2) = \frac{1}{2}, b_3(3) = \frac{1}{2}.$

What are all possible state sequences for the following observed sequences $\mathcal{O}$, and what is $p(\mathcal{O} \mid \lambda)$?

(a) $\mathcal{O} = 1, 2, 3$.
(b) $\mathcal{O} = 1, 3, 1$.

11.2 Given an HMM λ, suppose $\mathcal{O} = \mathcal{O}_1, \mathcal{O}_2, \ldots, \mathcal{O}_T$ is an observed sequence with hidden state sequence $q_1, q_2, \ldots, q_T$. For $t = 1, 2, \ldots, T$, let $\sigma_t = S_j\mathrm{m}$ where

$$j = \operatorname*{argmax}_i \ \mathrm{Prob}(\mathcal{O}_t \,|\, q_t = S_i).$$

In other words, σ_t is the state most likely to produce symbol $\mathcal{O}_t$. Construct an HMM λ with uniform initial distribution, three states, and an alphabet of size three, and give an observed sequence of length two, $\mathcal{O} = \mathcal{O}_1\mathcal{O}_2$, for which

$$\mathrm{Prob}(\mathcal{O} \,|\, \lambda) > 0, \quad \text{but} \quad \mathrm{Prob}(q_1 = \sigma_1, q_2 = \sigma_2 \,|\, \lambda) = 0.$$

11.3 Given an observed sequence $\mathcal{O}$, we have given an efficient method for calculating

$$\operatorname*{argmax}_Q \ \mathrm{Prob}(Q \,|\, \mathcal{O}).$$

One might ask why we are interested in this Q and not

$$\operatorname*{argmax}_Q \ \mathrm{Prob}(\mathcal{O} \,|\, Q) \tag{11.19}$$

instead. The latter state sequence Q, after all, is the one that if given has the highest probability of producing $\mathcal{O}$. To illustrate why this is the wrong Q to find, construct an HMM where the Q given by (11.19) is such that $\mathrm{Prob}(Q) = 0$, so could not possibly have produced $\mathcal{O}$.

11.4 Prove that the reestimation parameters (11.12)–(11.14) do indeed satisfy $\sum_i \bar{\pi}_i = 1$, $\sum_k \bar{p}_{jk} = 1$, and $\sum_a \bar{b}_i(a) = 1$.

11.5 Consider the five amino acid sequences

$$\texttt{WRCCTGC, \ WCCGGCC, \ WCGCC, \ WCCCGCC, \ WCCGC.}$$

Suppose their respective paths through a protein model HMM of length 8 are

$$m_0 m_1 i_1 m_2 m_3 m_4 m_5 d_6 m_7 m_8,$$
$$m_0 m_1 m_2 m_3 m_4 m_5 m_6 m_7 m_8,$$
$$m_0 m_1 m_2 d_3 d_4 m_5 m_6 m_7 m_8,$$
$$m_0 m_1 m_2 m_3 m_4 m_5 m_6 m_7 m_8,$$
$$m_0 m_1 m_2 m_3 d_4 m_5 d_6 m_7 m_8.$$

Using the theory of Section 11.3.2, give the alignment of the sequences that these paths determine.

12

Computationally Intensive Methods

12.1 Introduction

An important trend in statistical inference over the last twenty years has been the introduction of computationally intensive methods. These have been made possible by the availability of convenient and greatly increased computing power, and these methods are useful in bioinformatics and computational biology. Aspects of some computationally intensive methods used for both estimation and hypothesis testing are outlined in this chapter. Computationally intensive methods arise in both classical and Bayesian inference: We concentrate here on computationally intensive methods in classical inference.

It is convenient to start by defining "plug-in" probability distributions and statistics. Suppose that X is a random variable having an unknown probability distribution and that $x_1, x_2, \ldots, x_n$ are the observed values of n iid random variables $X_1, X_2, \ldots, X_n$ having the same probability distribution as X. These data encompass all the information that is available about this unknown distribution, and provide an empirical estimation of the probability distribution for X, namely

$$\mathrm{Prob}_{\mathrm{C}}(X = x) = \frac{m_x}{n}, \qquad (12.1)$$

where m_x is the number of values in the collection $x_1, x_2, \ldots, x_n$ equal to x and the suffix "C" indicates that this probability is conditional on the observed values in the data.

Suppose now that the distribution given in (12.1) is the actual probability distribution of X. Then from equation (1.16) the mean of X would be $\sum_{i=1}^{n} x_i/n = \bar{x}$, and from equation (1.20) the variance of X would be $\sum_{i=1}^{n} (x_i - \bar{x})^2/n$. The two respective estimates,

$$\bar{x} \quad \text{and} \quad \frac{\sum_{i=1}^{n}(x_i - \bar{x})^2}{n}, \tag{12.2}$$

are then called the "plug-in" estimates of the mean and variance, respectively, of the actual distribution of X. The former is an unbiased estimate of the actual mean of X, but the latter is a biased estimate of the actual variance, having expected value $\sigma^2(1-n^{-1})$ with respect to the distribution of X.

In general, a "plug-in" estimate of a parameter is calculated from the empirical distribution in the same way as the parameter itself is defined from the true, but unknown, distribution of the random variable X. Thus the observed proportion of successes in a binomial distribution is the plug-in estimate of the probability p of success, and for any probability distribution the sample median is the plug-in estimate of the median of that distribution. Plug-in estimates are clearly reasonable, but they are not necessarily unbiased, as the above example of the variance estimate, and the result of Problem 12.1, show. In both of these cases, however, the bias approaches 0 as $n \to \infty$.

12.2   Estimation

12.2.1   Classical Estimation Methods

We start with an example of a problem that arises in assessing the accuracy of a classical estimator as measured by its standard deviation. Suppose that X is some continuous random variable and the aim is to estimate the probability p_A that X is less than or equal to A, for some given number A. How can such an estimate be found? Conversely, given some probability p, how can the value A_p corresponding to this be estimated? How may the accuracy of any estimate of A_p be assessed?

The first problem is straightforward. Given the observed values of n iid random variables, each having the same distribution as X, the plug-in estimate of p_A is the proportion m/n, where m is the observed number of data values less than or equal to A. The number of observations less than A has a binomial distribution. Equation (3.12) then shows how an approximate 95% confidence interval for p_A can be found, using the estimate of the variance of m/n given by (3.11).

The converse problem is to estimate, for a continuous random variable X with unknown density function $f_X(x)$, the value A_p such that $\text{Prob}(X \leq A_p) = p$, where p is some given probability, and to assess the accuracy of this estimate. In practice, p is usually some frequently used probability such as 0.05, and it is convenient to assume that there is some integer i_0 such that $p = i_0/(n + 1)$, where n is the number of observations on which the estimate is based.

The estimation of A_p is based on the observed values $x_{(1)}, x_{(2)}, \ldots, x_{(n)}$ of the order statistics $X_{(1)}, X_{(2)}, \ldots, X_{(n)}$, where $X_1, X_2, \ldots, X_n$ are n iid random variables each having the unknown density function $f_X(x)$. If the distribution function corresponding to $f_X(x)$ is $F_X(x)$, then from (1.120), the random variables U_j, defined by $U_j = F_X(X_j)$, $j = 1, 2, \ldots, n$, have uniform distributions in $(0, 1)$. Further, since $X_1, X_2, \ldots, X_n$ are iid, so are $U_1, U_2, \ldots, U_n$. Equation (2.148), applied to $U_1, U_2, \ldots, U_n$, with $L = 1$, then shows that the random variable $U_{(i_0)} = F_X(X_{(i_0)})$ has a beta distribution with parameters i_0 and $n - i_0 + 1$, and thus from equation (1.72) has mean $i_0/(n + 1) = p$. That is to say, $E(F_X(X_{(i_0)})) = p$. Equation (B.32) then shows that to a first order of approximation, $E(X_{(i_0)}) = F_X^{-1}(p)$. Since $F_X^{-1}(p) = A_p$, this implies that $X_{(i_0)}$ is a reasonable estimator of A_p. We denote this estimator $\hat{A}_p$. However, distributional properties of $\hat{A}_p$, and thus its accuracy, depend on the unknown distribution of the X_i's. This is now illustrated by considering of the standard deviation of $\hat{A}_p$.

Since $X_{(i_0)} = F_X^{-1}(U_{(i_0)})$ and the variance of $U_{(i_0)}$ is approximately $p(1 - p)/n$, equation (B.33) shows that

$$\text{standard deviation of } \hat{A}_p \approx \sqrt{\frac{p(1 - p)}{n(f_X(A_p))^2}}. \tag{12.3}$$

In the particular case $p = \frac{1}{2}$, A_p is the median M of the probability distribution of the random variable X, so that if $\hat{M}$ is the sample median, then to a close approximation,

$$\text{standard deviation of } \hat{M} \approx \sqrt{\frac{1}{4n(f_X(M))^2}}. \tag{12.4}$$

However, neither this formula nor the more general formula (12.3) is of immediate value, since they both rely on knowledge of the unknown density function $f_X(x)$. Further, no other classical approach resolves this problem when estimating A_p, as well as other parameters, when the density function $f_X(x)$ is unknown. We now turn to an approach that attempts to overcome this problem.

12.2.2   Bootstrap Estimation and Confidence Intervals

Two frequently applied computationally intensive methods of data analysis are the jackknife and the bootstrap techniques, the latter method in particular having been developed with computational power in mind. Bootstrapping and jackknifing both operate through the empirical probability distribution (12.1). There is a voluminous literature on both procedures, particularly the bootstrap, and we consider only this procedure here. For detailed accounts of both techniques see Efron and Tibshirani (1993), and for further material see Davison and Hinkley (1997), Efron (1982), Hall (1992), Manly (1997), Sprent (1998), and Chernick (1999).

Suppose we wish to estimate the mean μ of the probability distribution of some random variable X, and we also wish to find a confidence interval for the mean, using the observed values $x_1, x_2, \ldots, x_n$ of n iid random variables $X_1, X_2, \ldots, X_n$ each having this probability distribution. The theory in Section 3.3.1 shows that $\bar{X}$ is an unbiased estimator of μ, and equation (3.8) shows how an approximate confidence interval for μ can be calculated. The form of this confidence interval is based on the assumption that $\bar{X}$ has an approximately normal distribution. These calculations are not applicable if one wishes find a confidence interval for θ using an estimator that does not have an approximately normal distribution.

Suppose that $\hat{\theta} = t(X_1, X_2, \ldots, X_n)$ is an estimator of θ, based on n iid random variables $X_1, X_2, \ldots, X_n$ having a distribution depending on θ; this might or might not be the plug-in estimator. The estimate $\hat{\theta}$ corresponding to this estimator is

$$\hat{\theta} = t(x_1, x_2, \ldots, x_n), \qquad (12.5)$$

where $x_1, x_2, \ldots, x_n$ are the observed values of $X_1, X_2, \ldots, X_n$. We abuse notation here and use the same symbol for the estimator and the estimate.

In the bootstrap procedure we regard the observations $x_1, x_2, \ldots, x_n$ as being given. These give the "plug-in" empirical distribution of X. In the case of discrete random variables, when the sample size n is large, each possible value of the random variable can be expected to occur in the sample with frequency close to its probability, while for continuous random variables the proportion of observations in any interval should be close to the probability that the random variable X lies in that interval. This observation forms the basis of the bootstrap procedure described below. When the sample size is small, greater caution is needed in applying bootstrap methods.

In the first step of the bootstrap procedure we sample from the n observations n times *with replacement*. From the comments in the previous paragraph, this may be taken, when n is large, as an approximation to iid sampling from the actual distribution of X. Some of the observations in the original data might not appear in this *bootstrap sample*, some might appear once, some twice, and so on. From this bootstrap sample we calculate an estimate $\hat{\theta}_B$ of the parameter θ of interest, replacing $x_1, x_2, \ldots, x_n$ in (12.5) by the n bootstrap sample values. Here the suffix "B" signifies "bootstrap."

This procedure is repeated a large number R of times, leading to R "bootstrap" estimates $\hat{\theta}_{B_1}, \hat{\theta}_{B_2}, \ldots, \hat{\theta}_{B_R}$. These can be regarded as describing an empirical distribution of $\hat{\theta}$. The average of the bootstrap estimates, denoted by $\hat{\theta}_{B(\cdot)}$, is the *bootstrap estimate* of θ.

Our main aim is to use bootstrap methods to find confidence intervals when we have insufficient information about the distribution of the observations to employ classical methods. In particular this overcomes the problems outlined in the example of the preceding section. There are various approaches to this problem, of which we describe two. The question of the best choice of confidence intervals is a complex one and is taken up in detail by Efron and Tibshirani (1993); see also Schenker (1985).

The first approach is based on the bootstrap estimate of the variance of the estimator $\hat{\theta}$, namely

$$\text{estimated variance of } \hat{\theta} = \frac{(\sum_{j=1}^{R} \hat{\theta}_{B_j}^2) - R\hat{\theta}_{B(\cdot)}^2}{R-1}. \tag{12.6}$$

An approximate 95% confidence interval for θ follows from this and the two standard deviation rule of Section 1.10.2, page 26, provided that one is confident that the data come from an approximately normal distribution.

The second approach to estimating a confidence interval for θ is based on quantile methods: If 2.5% of the values $\hat{\theta}_{B_i}$ lie below the value A and 2.5% lie above the value B, then (A, B) can be taken as an approximate 95% confidence interval for θ. We call this a 95% *quantile* bootstrap confidence interval. The quantile approach is used in the example below.

Example

Methods to measure gene expression intensities in a given cell type were discussed in Section 3.4.3. Golub et al. (1999) present a data set with gene intensities measured by Affymetrix microarrays. In this data set the intensities of approximately 7000 genes are measured in 72 different experiments. As an example of the data from one such gene, Figure 12.1 shows the intensity of the gene with Genbank accession number X03934, a T-cell antigen receptor gene T3-delta, in 46 acute lymphoblastic leukemia (ALL) cells. Suppose we want to estimate a 95% confidence interval for the mean of the distribution of intensities for this gene in this type of cell. By performing 5000 bootstrap estimates, we obtained a bootstrap estimate of the mean equal to 5.408, very close to the sample average of 5.407. The full range of the 5000 bootstrap estimates of the mean was $(4.30, 6.86)$, with 95% of the samples lying between $(4.77, 6.09)$. Therefore we estimate the 95% quantile confidence interval for the mean to be $(4.77, 6.09)$. For comparison, the 95% confidence interval found using (3.9) gives $(4.73, 6.09)$, which is virtually identical to the bootstrap interval. Therefore even for a highly irregular distribution that is very far from normal, such as that shown in Figure 12.1, the approximation (3.9) is apparently very accurate even with as few as 46 observations.

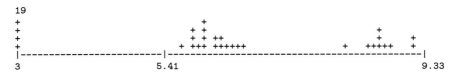

FIGURE 12.1. Intensities (in ln scale) of the gene with GenBank accession X03934 (T-cell antigen receptor gene T3-delta), measured in 46 acute lymphoblastic leukemia cells by Golub et al. (1999). There are 19 with intensity equal to 3. The sample mean is 5.41.

There is no simple analogue to (3.9) for estimating a confidence interval for the standard deviation of the expression intensity of this gene. The distribution in Figure 12.1 is apparently trimodal, and not of any standard type. Therefore it is not clear how to derive a theoretical confidence interval. We can however easily compute a bootstrap estimate. Using 5000 bootstrap replicates, we obtained a bootstrap estimate of the standard deviation equal to 2.28. The full range of the 5000 bootstrap estimates was $(1.63, 2.69)$, with a 95% quantile bootstrap confidence interval of $(1.97, 2.55)$.

More examples of applications of bootstrap methods are discussed in Chapter 14.

12.3   Hypothesis Testing: The Bootstrap Alternative to the Two-Sample t-Test

It was remarked in Section 3.5 that one disadvantage of the permutation test alternative to the two-sample t-test is that it relies on the assumption that the variances of the random variables in the two groups are equal. The bootstrap procedure overcomes this problem, and thus provides an alternative to the approximate "unequal variance" two-sample t-test discussed in Section 8.4. We describe this procedure in this section. Before doing so it is convenient to first discuss the bootstrap alternative to the "equal variance" two-sample t-test.

When the variances in the two groups are equal, the bootstrap approach is similar to the classical permutation procedure which uses R randomly chosen permutations, as described in Section 3.5. A permutation can be thought of as sampling without replacement: Once an observation has been chosen by the permutation procedure to go into one group, it cannot be chosen again. This implies that each observation is placed exactly once into one or the other group.

The bootstrap permutation method differs from the classical method in three important respects. First, in the bootstrap procedure sampling is *with* replacement, rather than without replacement, in line with the general bootstrap concepts described in Section 12.2.2. The second follows from this, namely that unlike the situation arising in the permutation procedure, there is no longer an automatic monotonic relation between the t statistic (3.19) and $\bar{x}_1 - \bar{x}_2$. The choice of test statistic then requires some discussion. Efron and Tibshirani (1993) show that use of the t statistic is preferable to use of the difference $\bar{x}_1 - \bar{x}_2$.

The third difference is perhaps the most important. The classical permutation method applies only to the case where the null hypothesis claims identical probability distributions for the observations in the two groups, since only under this assumption is the symmetry invoked by this procedure justified. This implies in particular an assumption of equal variances in the two groups. By contrast, the bootstrap method uses the plug-in principle and probabilities estimated as in equation (12.1), and can be amended to test the null hypothesis that the two group means are equal without making the assumption of equal variances. The details of this procedure are as follows.

Given the data in the two groups, one first calculates the numerical value of the "unequal variance" t' statistic defined in (8.49). The next part of the procedure involves the bootstrap operation.

The aim of the bootstrap is to estimate a null hypothesis distribution of the test statistic of interest, here t'. The null hypothesis is that the means of the two groups are equal, and to estimate the null hypothesis distribution of t', all observations are now adjusted so that the averages in the two groups are equal. This is perhaps most easily achieved by subtracting from each observation the average of the group it is in. From now on we assume that the original data have been amended in this way.

The next step is essentially conceptual, namely to think of the probability distribution of the amended data in any group as being given by the plug-in estimates found from equation (12.1) and the amended data from that group. These plug-in estimates form the basis of the bootstrap procedure.

The bootstrap procedure consists of calculating a large number R of bootstrap values of t'. In each bootstrap replication a random sample of m values is taken *with replacement* from the amended data values in the first group; these form the "data" in the first group for the replication in question. Similarly, a random sample of n values is taken *with replacement* from the amended data values in the second group; and these form the "data" in the second group for replication in question. The value of the t' statistic is then computed from each bootstrap sample. The observed value of t' is then judged to be significant with Type I error α if its value lies among the $100\alpha\%$ most extreme bootstrap t' values.

This procedure avoids the equal variance assumption by sampling within each group, using the plug-in estimates of the probability distribution for each group. It thus extends the range of possibilities for testing for equal means in two groups. On the other hand, this bootstrap procedure will involve perhaps substantial computation.

12.4   Multiple Testing Revisited

12.4.1   Introduction

The multiple testing problem was introduced in Section 3.8 in the context of expression array data analysis. We discuss it in Section 12.4.4 in the same context, focusing on an attempt to overcome the multiple testing problems raised in Section 3.8 through the use of "step-down" methods. Before doing this, we discuss general properties of step-down methods, in particular, the Westfall and Young (1993) step-down method.

12.4.2   Step-Down Methods: t-Tests

In order to indicate the essentials of the step-down procedure described below, we consider first the case of a large number g of independent two-sample t-tests, each testing a null hypothesis claiming equality of the means of two groups. We assume for purposes of illustration that the method by which each test is carried out is the equal variance normal theory t-test of Section 8.4.1.

We write the null hypothesis for test j as H_0^j. An experiment-wise Type I error of α is chosen for the procedure; this is the probability that at least one of the g null hypotheses $H_0^1, H_0^2, \ldots, H_0^g$ will be rejected if in fact all null hypotheses are true.

The test of H_0^j will lead, for each j, $j = 1, 2, \ldots, g$, to an observed value t_j of the t statistic. This in turn leads to some numerical value p_j for the P-value for the jth test. In the "*one-step*" procedure, each of these P-values is compared individually to the Šidák significance point $K(g, \alpha)$, defined from equation (2.157), with g replacing n, as

$$K(g, \alpha) = 1 - \sqrt[g]{1 - \alpha}. \qquad (12.7)$$

Those null hypotheses for cases in which the P-value is less than this significance point are rejected, while the remainder are not rejected. Although this procedure has the desired experiment-wise Type I error, each individual test is quite stringent, so that in several cases the null hypothesis might not be rejected when it is in fact false. This implies a loss of power, and it is the aim of a step-down approach to provide a more powerful testing procedure. The step-down procedure to be discussed here is that of Westfall and Young (1993).

The Westfall and Young step-down procedure is usually described in terms of so-called "amended P-values." However it is equivalent and more convenient to describe the method in terms of unamended P-values and significance points.

First, the observed unamended P-values are written in increasing order as $p_{(1)}, p_{(2)}, \ldots, p_{(g)}$, so that $p_{(j)} \leq p_{(k)}$ if $j < k$. The various null hypotheses are then correspondingly ordered as $H_0^{(1)}$, $H_0^{(2)}$, ..., $H_0^{(g)}$. These ordered P-values are now examined in turn in a sequence of steps.

The initial step in the step-down procedure is to compare the smallest observed P-value $p_{(1)}$ with the Šidák significance point $K(g, \alpha)$ defined in (12.7), where α is the chosen experiment-wise Type I error. If $p_{(1)} > K(g, \alpha)$, all null hypotheses are accepted and no further testing is done. If $p_{(1)} < K(g, \alpha)$, the null hypothesis $H_0^{(1)}$ is rejected and the sequential procedure moves to step 2.

In step 2, if reached, the null hypotheses $H_0^{(2)}$, $H_0^{(3)}$, ..., $H_0^{(g)}$ are considered further. This is done by comparing the P-value $p_{(2)}$ to the Šidák significance point $K(g - 1, \alpha)$, defined as in (12.7) but with g replaced by $g - 1$. If $p_{(2)} > K(g - 1, \alpha)$, the null hypotheses $H_0^{(2)}$, $H_0^{(3)}$, ..., $H_0^{(g)}$ are all accepted and no further testing is done. If $p_{(2)} < K(g - 1, \alpha)$, the null hypothesis $H_0^{(2)}$ is rejected and the step-down procedure moves to step 3.

The procedure in step 3 is identical to that in steps 1 and 2, with the obvious amendments. In general, if step j is reached, the null hypotheses $H_0^{(j)}$, $H_0^{(j+1)}$, ..., $H_0^{(g)}$ are considered further. If $p_{(j)} > K(g - j + 1, \alpha)$, all these hypotheses are accepted and no further testing is done. If $p_{(j)} < K(g - j + 1, \alpha)$, hypothesis $H_0^{(j)}$ is rejected and the procedure moves to step $j + 1$.

The rule at step 1 ensures that the experiment-wise Type I error is indeed α, the value chosen. This can be seen in two ways. First, the event that the smallest P-value is less than $K(g, \alpha)$ is identical to the event that at least one P-value is less than $K(g, \alpha)$. If all null hypotheses are true, the probability of the latter event, and hence of the former, is (from (12.7)) $1 - (1 - K(g, \alpha))^g = \alpha$. Since at least one null hypothesis is rejected when the smallest P-value is less than $K(g, \alpha)$, the desired Type I error α is achieved.

The second way in which this conclusion can be reached, and one more useful for examining the further steps in the procedure, is to recall from Section 3.4.2 that if all null hypotheses are true, then each P-value has a uniform distribution in $(0, 1)$. In this case the smallest P-value has the distribution of the smallest of g independent uniform $(0, 1)$ random variables as given in (2.156). The probability that such a random variable takes a value less than $K(g, \alpha)$ is found by appropriate integration of (2.156), leading to (2.157).

If step 2 of the procedure is reached, the null hypothesis $H_0^{(2)}$ is rejected if $p_{(2)} < K(g - 1, \alpha)$. The discussion in the previous paragraph indicates that this procedure is adopted because $p_{(2)}$ is regarded as the smallest of the $g - 1$ P-values remaining at this step. That is, it is the smallest of $g - 1$ independent random variables each having a uniform distribution in $(0, 1)$ when the null hypotheses $H_0^{(2)}$, $H_0^{(3)}$, ..., $H_0^{(g)}$ are true. The hypothesis

$H_0^{(1)}$, rejected in step 1, does not play a role in the procedure since it is assumed at step 2 that this hypothesis is incorrect.

The argument justifying the following steps, if reached, is parallel to that for step 2. If any step j is reached, the smallest P-value of the hypotheses considered at that step is regarded as the smallest of the $g - j + 1$ P-values considered in that step, and is thus compared to the value $K(g - j + 1, \alpha)$ for significance.

The condition (that $p_{(1)} > K(g, \alpha)$) determining whether all null hypotheses are accepted is identical in both the one-step approach and the step-down approach. However, given that $H_0^{(1)}$ is rejected, the condition $(p_{(2)} < K(g - 1, \alpha))$ that $H_0^{(2)}$ is rejected is less stringent in the step-down procedure than the corresponding condition $(p_{(2)} < K(g, \alpha))$ in the one-step procedure. A similar remark holds for all further hypotheses. This observation justifies the claim for the increased power of the step-down procedure: For the same experiment-wise Type I error, it is easier in the step-down method than in the one-step method to reject null hypotheses that are not correct. This implies that, for the same experiment-wise Type I error, the step-down procedure is more powerful than the one-step procedure.

It is possible to formulate procedures that have the same experiment-wise Type I error as both the one-step and step-down approaches, and which are more powerful than the step-down procedure. As an extreme case it might be decided to reject *all* null hypotheses if $p_{(1)} < K(g, \alpha)$. This procedure is clearly undesirable, because it does not control the false-positive rate after the first step. It is necessary to introduce some reasonable criterion for the testing procedure to satisfy. The criterion adopted in the step-down context is that of *strong control* of the experiment-wise Type I error. This requirement is that, no matter how many of the null hypotheses are true, the probability that none of the true null hypotheses are rejected is $1 - \alpha$. It can be shown that the Westfall and Young method satisfies this criterion, and clearly the extreme procedure described above does not.

This is not to say that the step-down method has uniformly better properties than the one-step method. For example, when all null hypotheses are true, the probability that a specified null hypothesis is accepted under the one-step approach is larger than the corresponding probability under the step-down approach: see Problem 12.10. This observation is in line with the fact that in statistics it is seldom the case that one procedure is better than another in all respects: Under some restrictions, or following some criteria, one procedure will be better, while under other restrictions, or following other criteria, the other will be better.

As a second example of an undesirable property of the step-down procedure, consider the case where $g = 2$, $\alpha = 0.05$, $K(2, \alpha) = .02532$, and the unamended P-values are $p_{(1)} = 0.0254$ and $p_{(2)} = 0.0255$. In this case, neither $H_0^{(1)}$ nor $H_0^{(2)}$ would be rejected, whereas if the unamended P-values were $p_{(1)} = 0.0253$, $p_{(2)} = 0.049$, both $H_0^{(1)}$ and $H_0^{(2)}$ would be rejected. It might be hard to justify the rejection of $H_0^{(2)}$ with a P-value of 0.049 and the nonrejection of $H_0^{(2)}$ with a P-value of 0.0255 because of a trivial change in the P-value of an independent test.

The above discussion assumes a continuous test statistic and thus P-values having continuous uniform distributions in $(0, 1)$ if the various null hypotheses are true. In the case of a permutation procedure the test statistic is discrete, and in the case of the bootstrap procedure an exact calculation of a P-value cannot be made. Nevertheless, to a sufficiently close approximation, the analysis described above holds in those cases also, as is demonstrated in Section 12.4.3.

12.4.3 Step-Down Methods: A Permutation Test

It was noted in Section 3.5 that the permutation procedure described in that section has properties very similar to those of the equal variance two-sample t-test. This observation leads to an examination of the permutation analogue of the step-down procedure based on two-sample t-tests discussed in Section 12.4.2. We continue to assume that the data used in any of the g tests are independent of the data used in any other test. The permutation procedure allows a relaxation of this assumption: This is discussed in Section 12.4.4.

The data used to test any of the g hypotheses fall into two groups. For reasons to be discussed in Section 12.4.4 we assume that the number of observations in the first group is the same for all hypotheses tested, and denote the common value by m. Similarly we assume that the number of observations in the second group is the same for all hypotheses tested, and denote the common value by n. The entire data set can then be arranged in a matrix with g rows and $m + n$ columns, with the entries in row j corresponding to the data $x_{j11}, x_{j12}, \ldots, x_{j1m}, x_{j21}, x_{j22}, \ldots, x_{j2n}$ used to test hypothesis H_0^j. It is convenient to think of this matrix as a collection of column vectors $x_{11}, x_{12}, \ldots, x_{1m}, x_{21}, x_{22}, \ldots, x_{2n}$. The reason for adopting this column vector notation is that in the expression array analysis discussed in Section 12.4.4, the data in any column are all taken from the same individual and give expression levels for g genes in that individual.

The permutation procedure is conducted as described in Section 3.5, with the added feature that entire columns (of length g) are permuted together. This involves no loss of generality and, for reasons to be discussed in Section 12.4.4, allows an immediate generalization to the analysis of dependent data. The various permutations are labeled as permutations 1, 2, ..., N, where $N = \binom{m+n}{m}$, with the actual data assigned to permutation 1.

For each null hypothesis H_0^j, $j = 1, 2, \ldots, g$, the N permutation values of the chosen test statistic (for example the t statistic (3.19)) are calculated. A permutation P-value associated with the observed value of this statistic is then calculated, as described in Section 3.5. When the null hypothesis is true, any such P-value is equally likely to take the values $1/N, 2/N, \ldots, 1$. This is the discrete uniform distribution analogue of the continuous uniform distribution used in Section 12.4.2. The g P-values found in this way are now ranked, and the step-down procedure carried out as in Section 12.4.2.

There are two points concerning this procedure that require further discussion. First, if g is large enough and m and n are not large, it might be impossible to achieve significance with any normally chosen Type I error. For example, if $m = n = 8$, $g = 6384$, an example discussed in Section 12.4.4, $N = \binom{8}{4} = 12{,}870$, so that the lowest achievable P-value for any given hypothesis is $1/12{,}870$. The lowest achievable experiment-wise P-value is then $1 - (1 - 1/12{,}870)^{6384} = .39$, far higher than any value accepted in any testing procedure.

The second point helps to overcome the first, and leads to a method for testing nonindependent data. Under the procedure described above, the smallest of the permutation P-values is compared to the Šidák significance point $K(g, \alpha)$, defined in (12.7), with further P-values possibly compared to further Šidák significance points. An essentially equivalent approach derives from the permutation procedure itself. For each of the N permutations of the original data there will be a minimum experiment-wise P-value, the minimum being taken over all g tests. The minimum experiment-wise P-value of the observed data (permutation 1) is then declared to be significant, under a Type I error of α, if it is among the lowest $100\alpha\%$ of the N permutation experiment-wise P-values. The similarity of this approach to that depending on Šidák significance points follows from the fact that the empirical experiment-wise P-values provided by the N permutations is very close to the probability distribution (2.156) of the minimum of g independent uniform $(0,1)$ random variables if in (2.156) n is replaced by g. If significance is achieved in this way, the second smallest observed P-value is referred to the array of N second smallest P-values, and so on.

This approach, unlike the approach using Šidák significance points, does allow for significance to be achieved in the case $g = 6384$, $n = m = 8$ discussed above.

12.4.4   Step-Down Methods Applied to Expression Arrays

In the analyses of Sections 12.4.2 and 12.4.3 it was assumed that the data used to test any hypothesis are independent of the data used to test any other hypothesis. This is not a reasonable assumption when the g hypotheses concern the expression levels of g different genes in an expression array. Genes frequently act in correlated dependent ways, and this implies potential nonindependence of data relating to different genes.

In a paper which motivates much of the discussion in 12.4.3, Dudoit et al. (2000) consider the dependent case. The data involve expression levels in 6384 genes, with 16 observations being taken for each gene, 8 in each of two groups. The method of analysis is as described in Section 12.4.3, with the proviso that an observed minimum P-value not be referred to a Šidák significance point, since such a point is calculated under the independence assumption. Instead, the minimum of the g observed P-values is declared to be significant only if it is among the lowest $100\alpha\%$ of the N permutation experiment-wise P-values. The permutation procedure automatically allows for dependence in the data.

Perhaps paradoxically, the multiple testing problem is less important in the dependent case than in the independent case. This may be seen by considering the extreme example where all g genes are perfectly correlated in their expression levels. In this case, testing all g genes is equivalent to testing just one gene, and the multiple testing problem does not arise.

We conclude with two observations. First, in the case $m = n = 8$ there are 12,870 permutations of the data into two groups. The procedure has better statistical properties the larger the number of permutations. Increasing the values of m and n to 10 would increase the number of permutations to 184,756. This implies more than an order of magnitude increase in computing time, but with sufficient computing power such an increase pays off from the statistical point of view.

Second, even with the step-down procedure described above, quite stringent individual Type I errors are needed to obtain the desired experiment-wise Type I error. It is possible to argue that these requirements are unnecessarily stringent in practice, leading to the acceptance of hypotheses that should be rejected, and that other approaches might allow for a more satisfactory testing procedure in practice. This matter was discussed further at the end of Section 3.8.

Problems

12.1 Use the result of Problem 2.22 to discuss the bias of the plug-in estimator of the median of the exponential distribution when n, the number of observations, is odd (so that $n = 2k + 1$ for some integer k).

12.2 Suppose that the observed values of n continuous iid random variables are $x_1, x_2, \ldots, x_n$, where the x_j are all distinct. A bootstrap sample (with replacement) is taken from these. Show that there are $\binom{2n-1}{n}$ different possible bootstrap samples. *Hint:* A classical result of probability theory (see Feller (1968)) is that if r indistinguishable objects are placed into n distinguishable cells, the number of different occupancy arrangements is $\binom{n+r-1}{r}$. Now put $r = n$ and make the analogy that if n_j objects are put in cell j, the value x_j arises n_j times in the bootstrap sample.

12.3 *Continuation.* Calculate the number of distinct bootstrap samples for the cases $n = 3$, $n = 4$, and $n = 5$, and display all possible samples for the case $n = 3$.

12.4 *Continuation.* Find the probability of each bootstrap sample listed in Problem 12.3 for the case $n = 3$.

12.5 *Continuation.* Rewrite the possible samples for the case $n = 3$ in Problem 12.3 using the observed values $x_{(1)}$, $x_{(2)}$, and $x_{(3)}$ of the order statistics. From this, find the sample average and the sample median of each of the possible samples. Note: If a bootstrap sample is of the form $(x_{(i)}, x_{(i)}, x_{(j)})$ for $i \neq j$, the sample median is defined to be $x_{(i)}$.

12.6 *Continuation.* Use the result of Problem 12.4 to find the conditional mean of the bootstrap sample average and of the bootstrap sample median. (The word "conditional" emphasizes that these are functions of $x_{(1)}$, $x_{(2)}$, and $x_{(3)}$.)

12.7 *Continuation.* In the case where the three random variables have the exponential distribution (1.59), find the unconditional mean of the bootstrap sample average and of the bootstrap sample median. (For the latter, use the result of Problem 2.20.) Compare these with the corresponding means of the classical plug-in estimators.

12.8 Let X_1 and X_2 be independent random variables, each having the uniform distribution on $(0, 1)$. Let $X_{(1)}$ and $X_{(2)}$ be the corresponding order statistics. Find an equation satisfied by A_1 and A_2, with $(A_1 \leq A_2)$, if we require that $\mathrm{Prob}(X_{(1)} \geq A_1, X_{(2)} \geq A_2) = 1 - \alpha$, for some given constant α, $(0 < \alpha < 1)$. Show that the choice $A_1 = A_2 = K(2, \alpha)$ (defined in (2.157)) satisfies this equation.

12.9 *Continuation.* Suppose that an experiment consists of two distinct and independent subexperiments, and that the P-values for each of these are denoted P_1 and P_2 respectively. Let $P_{(1)}$ and $P_{(2)}$ be the corresponding ordered P-values. Find an equation satisfied by two constants A_1 and A_2 $(A_1 \leq A_2)$, such that an experiment-wise Type I error of α is attained when both null hypotheses are accepted if and only if $P_{(1)} \geq A_1$ and $P_{(2)} \geq A_2$. If A_1 and A_2 are used respectively instead of $K(2, \alpha)$ and $K(1, \alpha)$ in a Westfall and Young step-down process, for what choices of A_1 and A_2 is the requirement of the strong control of the experiment-wise Type I error at α satisfied?

12.10 *Continuation.* Consider two distinct independent t-tests. Suppose the Westfall and Young step-down procedure is used to obtain an experiment-wise Type I error α. Find the probability that a specified one of the two null hypotheses is accepted, given that both null hypotheses are true. Show that this probability is less than the corresponding probability $K(2, \alpha)$ arising under the one-step procedure.

12.11 *Continuation.* Now consider g distinct independent t-tests. Suppose the Westfall and Young step-down procedure is used to obtain an experiment-wise Type I error α. Show that the probability that a specified one of the g null hypotheses is accepted, given that all null hypotheses are true, is

$$1 - \frac{\alpha}{g} - \frac{\alpha^2}{g} - \cdots - \frac{\alpha^g}{g}.$$

Compare this probability with the corresponding probability $K(g, \alpha)$ arising under the one-step procedure.

12.12 *Continuation.* It is not necessary to consider the Westfall and Young step-down procedure as being carried out in a sequential manner: which hypotheses are accepted and rejected is determined in one step by the location of the point $(p_{(1)}, p_{(2)}, \ldots, p_{(g)})$ in the $(P_{(1)}, P_{(2)}, \ldots, P_{(g)})$ space. For the case $g = 2$, and given an experiment-wise Type I error α, sketch the region R in the $(P_{(1)}, P_{(2)})$ plane having the property that both null hypotheses are rejected using this sequential procedure if $(p_{(1)}, p_{(2)})$ falls within R. Hence find the probability that both hypotheses are rejected when both null hypotheses are true. Check your answer using calculations deriving from the step-down procedure. What is the corresponding probability in the one-step procedure? Relate your answer to this last question to your answer to question 12.10.

12.13 *Continuation.* Compare the region R determined in Problem 12.12 with the region R' such that both null hypotheses are rejected using the one-step procedure if $(p_{(1)}, p_{(2)})$ falls within R'.

13
Evolutionary Models

13.1 Models of Nucleotide Substitution

The contemporary biological data from which so many inferences are made are the result of evolution, that is, of an indescribably complicated stochastic process. Very simplified models of this process are often used in the literature, in particular for the construction of phylogenetic trees, and aspects of these simplified models are discussed in this chapter. The emphasis is on introductory statistical and probabilistic aspects. A probabilistic approach has the merit of allowing the testing of various hypotheses concerning the evolutionary process. Hypothesis-testing questions in the evolutionary context are discussed in Section 14.9.

While there is clearly genetic variation from one individual to another in a population, there is comparatively little variation at the nucleotide level. For example, two randomly chosen humans will typically have different nucleotides at only one site in about 500 to 1000. To a sufficient level of approximation it is reasonable to assume, for the great majority of sites, that a single nucleotide predominates in the population. Indeed if this were not so, the concept of a paradigm "human genome" would be meaningless. Thus from now on, when we use the expression "the nucleotide at a given site in a population" we mean, more precisely, the predominant nucleotide at this site in the population.

Over a long time span the nucleotide at a given site might change. To analyze the stochastic properties of this change, we ignore the (perhaps comparatively brief) time period during which one nucleotide replaces another in the population, and imagine an effectively instantaneous change in frequency of a nucleotide from a value close to 0 to a value close to 1. We will describe this event as the *substitution* of one nucleotide by another, meaning more precisely the substitution of the predominant nucleotide by another. The time unit chosen to evaluate the properties of this substitution process is arbitrary, but is often large, perhaps on the order of tens of thousands of generations.

We focus on DNA sequences and consider the predominant nucleotide at some specified site in some population. In Section 4.4 the states of a Markov chain were labeled according to the generic convention as states $E_1, E_2, E_3, \ldots, E_s$. In analyzing substitution processes for DNA sequences it is often convenient to change notation and to identify the states with the nucleotides a, g, c, and t, taken in this order. With this notation a Markov chain would, for example, be said to be in state g if in the population considered the predominant nucleotide at the site of interest is g. If unit time in the Markov chain is taken as, for example, 200,000 generations, a change from state a to state c in one time unit means the substitution of the nucleotide a by the nucleotide c after a period of 200,000 generations.

When this notational convention is used for a discrete-time Markov chain, symbols such as p_{ag} denote the probability of a change in the predominant nucleotide from a to g after one time unit. In the continuous-time models considered in Section 13.3, again with unit time taken at some agreed value, the probability of a change in the predominant nucleotide from a to g in a time period of length t is denoted by $P_{ag}(t)$.

The comparison of genetic data from two contemporary species often relies on a model of the evolutionary processes leading to these species. Various statistical procedures used in this comparison might require tracing up the tree of evolution from one species to a common ancestor and then down the tree to the other species. If the stochastic process assumed for tracing upwards is to be the same as that for tracing downwards, the stochastic process must be reversible. This is one reason why the concept of reversibility was introduced in Sections 10.2.4 and 10.7.4. One aim of the analysis below is to check whether the various stochastic models that we introduce are indeed reversible.

13.2   Discrete-Time Models

13.2.1   The Jukes–Cantor Model

The simplest (and earliest) model of nucleotide substitution is the Jukes–Cantor model (Jukes and Cantor (1969)). The original version of this model was a continuous-time process. In this section we consider the parallel (and simpler) discrete-time version, and discuss the continuous-time process in Section 13.3.1.

The discrete-time Jukes–Cantor model is a Markov chain with four states a, g, c, and t. With the states labeled in this order, the transition matrix P for this model is given by

$$P = \begin{bmatrix} 1 - 3\alpha & \alpha & \alpha & \alpha \\ \alpha & 1 - 3\alpha & \alpha & \alpha \\ \alpha & \alpha & 1 - 3\alpha & \alpha \\ \alpha & \alpha & \alpha & 1 - 3\alpha \end{bmatrix}. \qquad (13.1)$$

Here α is a parameter depending on the time-scale chosen: If unit time were chosen as 100,000 generations, α would take a value smaller than it would be if unit time were chosen as 200,000 generations. Whatever time scale is chosen, it is clearly necessary that α be less than $\frac{1}{3}$.

The transition matrix P in (13.1) possesses several elements of symmetry, and in particular the model assumes that whatever the nucleotide in the population is at any time, the three other nucleotides are equally likely to substitute for it. Our first aim is to analyze the properties of this Markov chain by the spectral decomposition methods discussed in Section B.19.

The eigenvalue equation (B.44) shows that the matrix (13.1) has eigenvalues 1 (with multiplicity 1) and $1 - 4\alpha$ (with multiplicity 3). The left eigenvector corresponding to the eigenvalue 1 is $(.25, .25, .25, .25)$, and thus the stationary distribution is the discrete uniform distribution

$$(\varphi_a, \varphi_g, \varphi_c, \varphi_t)' = (.25, .25, .25, .25)'. \qquad (13.2)$$

This implies that after a long time has passed, the four nucleotides are essentially equally likely to predominate in the population, as might be expected from the symmetry of the model. The spectral expansion (B.48) of P^n is

$$P^n = \begin{bmatrix} .25 & .25 & .25 & .25 \\ .25 & .25 & .25 & .25 \\ .25 & .25 & .25 & .25 \\ .25 & .25 & .25 & .25 \end{bmatrix} + (1 - 4\alpha)^n \begin{bmatrix} .75 & -.25 & -.25 & -.25 \\ -.25 & .75 & -.25 & -.25 \\ -.25 & -.25 & .75 & -.25 \\ -.25 & -.25 & -.25 & .75 \end{bmatrix}. \qquad (13.3)$$

From this it follows that whatever the predominant nucleotide in the population is at time 0, the probability that this is also the predominant nucleotide at time n is

$$.25 + .75(1 - 4\alpha)^n, \qquad (13.4)$$

and the probability that some other specified nucleotide is the predominant nucleotide at time n is

$$.25 - .25(1 - 4\alpha)^n. \qquad (13.5)$$

These values confirm the stationary distribution given in equation (13.2), and also show that the rate of approach to the stationary distribution is determined by the numerical value of the eigenvalue $1 - 4\alpha$.

13.2.2   The Kimura Models

The highly symmetric assumptions implicit in the Jukes–Cantor model are not realistic. A *transition*, that is, the replacement of one purine by the other (for example of a by g) or of one pyrimidine by the other, is in practice more likely than a *transversion*, that is, the replacement of a purine by a pyrimidine or of a pyrimidine by a purine. Kimura (1980) proposed a (continuous-time) two-parameter model to allow for this. The transition matrix P for the discrete-time version of this model, with the same ordering of states as that used for the Jukes–Cantor model, is

$$\begin{bmatrix} 1-\alpha-2\beta & \alpha & \beta & \beta \\ \alpha & 1-\alpha-2\beta & \beta & \beta \\ \beta & \beta & 1-\alpha-2\beta & \alpha \\ \beta & \beta & \alpha & 1-\alpha-2\beta \end{bmatrix}. \qquad (13.6)$$

Here α is the probability of a transition in one time unit, while β is the probability that a purine is substituted by a nominated pyrimidine in one time unit and is also the probability that a pyrimidine is substituted by a nominated purine in one time unit. It is required that $\alpha + 2\beta < 1$.

The eigenvalue equation (B.44) shows that the matrix (13.6) has eigenvalues 1 (multiplicity 1), $1-4\beta$ (multiplicity 1), and $1-2(\alpha+\beta)$ (multiplicity 2). The left eigenvector corresponding to the eigenvalue 1 is $(.25, .25, .25, .25)'$, so that in this model, as in the Jukes–Cantor model, the stationary distribution is again discrete uniform. The spectral expansion (B.48) of P^n is

$$P^n = \begin{bmatrix} .25 & .25 & .25 & .25 \\ .25 & .25 & .25 & .25 \\ .25 & .25 & .25 & .25 \\ .25 & .25 & .25 & .25 \end{bmatrix} + (1-4\beta)^n \begin{bmatrix} .25 & .25 & -.25 & -.25 \\ .25 & .25 & -.25 & -.25 \\ -.25 & -.25 & .25 & .25 \\ -.25 & -.25 & .25 & .25 \end{bmatrix}$$

$$+ (1-2(\alpha+\beta))^n \begin{bmatrix} .5 & -.5 & 0 & 0 \\ -.5 & .5 & 0 & 0 \\ 0 & 0 & .5 & -.5 \\ 0 & 0 & -.5 & .5 \end{bmatrix}. \qquad (13.7)$$

This spectral expansion implies that whatever the predominant nucleotide at time 0 at any site, the probability that this is also the predominant nucleotide at time n is

$$.25 + .25(1-4\beta)^n + .5\,(1-2(\alpha+\beta))^n. \qquad (13.8)$$

If the initial nucleotide is a purine, the probability that at time n the predominant nucleotide is the other purine is

$$.25 + .25(1-4\beta)^n - .5\,(1-2(\alpha+\beta))^n. \qquad (13.9)$$

A parallel remark holds for pyrimidines. The probability that after n time units a purine has been substituted by a specific pyrimidine is

$$.25 - .25(1 - 4\beta)^n, \tag{13.10}$$

and the probability that it has been replaced by one or the other pyrimidine is

$$.5 - .5(1 - 4\beta)^n. \tag{13.11}$$

A parallel remark holds for the replacement of a pyridine by a purine. The rate of approach to the stationary distribution depends on the numerical values of α and β. If $\alpha > \beta$, as would normally be assumed, the largest nonunit eigenvalue is $1 - 4\beta$.

A generalization of the above original Kimura model is provided by the so-called Kimura 3ST model. Here the transition matrix is of the form

$$\begin{bmatrix} 1 - \alpha - \beta - \gamma & \alpha & \beta & \gamma \\ \alpha & 1 - \alpha - \beta - \gamma & \gamma & \beta \\ \beta & \gamma & 1 - \alpha - \beta - \gamma & \alpha \\ \gamma & \beta & \alpha & 1 - \alpha - \beta - \gamma \end{bmatrix}, \tag{13.12}$$

where α, β, and γ are all unknown parameters. This Markov chain also has a uniform stationary distribution. The detailed balance requirements (10.5) hold, so that the Markov chain is reversible. This also implies reversibility of the Jukes–Cantor Markov chain (13.1) and the Kimura two-parameter Markov chain (13.6), since these are special cases of this model.

13.2.3   Further Generalizations of the Kimura Models

Although the Kimura models are more realistic than the Jukes–Cantor model, they nevertheless possess symmetry assumptions that make them unrealistic. In particular, they have uniform stationary distributions, which is at variance with observation. Increasingly complex models have been proposed over the years to give a closer match between theory and observation. Each may be analyzed, in greater or lesser detail, by algebraic methods similar to those used for the Jukes–Cantor and the Kimura models. These have usually been proposed in the context of continuous-time models (discussed below), but the principles apply equally to discrete-time models, and we discuss them here in the discrete-time context.

An immediate generalization of the Kimura model (13.6) allows the purine to pyrimidine substitution probability to differ from the pyrimidine to purine substitution probability. This implies a transition matrix P of the form

$$\begin{bmatrix} 1 - \alpha - 2\gamma & \alpha & \gamma & \gamma \\ \alpha & 1 - \alpha - 2\gamma & \gamma & \gamma \\ \delta & \delta & 1 - \alpha - 2\delta & \alpha \\ \delta & \delta & \alpha & 1 - \alpha - 2\delta \end{bmatrix}. \tag{13.13}$$

The stationary distribution of the Markov chain implied by this model is

$$\left(\frac{\delta}{2(\delta+\gamma)}, \ \frac{\delta}{2(\delta+\gamma)}, \ \frac{\gamma}{2(\delta+\gamma)}, \ \frac{\gamma}{2(\delta+\gamma)} \right)'. \qquad (13.14)$$

This is not uniform if $\delta \neq \gamma$, although it does possess symmetry properties in that it allocates equal stationary probabilities to the two purines and to the two pyrimidines. The Markov chain defined by this transition matrix, like that defined by the Jukes–Cantor model and the Kimura model, is reversible.

A further generalization has been made by Blaisdell (1985), who allows different within-transition and within-transversion rates. The transition matrix P for this model is

$$\begin{bmatrix} 1-\alpha-2\gamma & \alpha & \gamma & \gamma \\ \beta & 1-\beta-2\gamma & \gamma & \gamma \\ \delta & \delta & 1-\beta-2\delta & \beta \\ \delta & \delta & \alpha & 1-\alpha-2\delta \end{bmatrix}. \qquad (13.15)$$

The stationary distribution of this Markov chain is found to be

$$\left(\frac{\delta(\beta+\gamma)}{\theta(\alpha+\beta+2\gamma)}, \ \frac{\delta(\alpha+\gamma)}{\theta(\alpha+\beta+2\gamma)}, \ \frac{\gamma(\alpha+\delta)}{\theta(\alpha+\beta+2\delta)}, \ \frac{\gamma(\beta+\delta)}{\theta(\alpha+\beta+2\delta)} \right)', \qquad (13.16)$$

where $\theta = \gamma + \delta$. This stationary distribution and the elements in the transition matrix (13.15) show that this model is not reversible. On the other hand, the elements in the stationary distribution can now all be different from each other, a property not enjoyed by any other model discussed above.

Another generalization has been made by Schadt et al. (1998), who introduce a continuous-time model whose discrete-time analogue has transition matrix P given by

$$\begin{bmatrix} 1-\alpha-\gamma-\lambda & \alpha & \gamma & \lambda \\ \epsilon & 1-\epsilon-\gamma-\lambda & \gamma & \lambda \\ \delta & \kappa & 1-\beta-\delta-\kappa & \beta \\ \delta & \kappa & \sigma & 1-\delta-\kappa-\sigma \end{bmatrix}.$$

This transition matrix is reversible if and only if

$$\beta\gamma = \lambda\sigma, \quad \alpha\delta = \epsilon\kappa. \qquad (13.17)$$

Further generalizations, steadily relaxing symmetry assumptions, have been made by Takahata and Kimura (1981) and Gojobori et al. (1982).

13.2.4 The Felsenstein Model

A different form of generalization was introduced by Felsenstein (1981), whose notation we adopt here. In the discrete-time version of this model the probability of substitution of any nucleotide by another is proportional to the stationary probability of the substituting nucleotide. This implies a transition matrix P of the form

$$\begin{bmatrix} 1-u+u\varphi_a & u\varphi_g & u\varphi_c & u\varphi_t \\ u\varphi_a & 1-u+u\varphi_g & u\varphi_c & u\varphi_t \\ u\varphi_a & u\varphi_g & 1-u+u\varphi_c & u\varphi_t \\ u\varphi_a & u\varphi_g & u\varphi_c & 1-u+u\varphi_t \end{bmatrix}, \qquad (13.18)$$

where $(\varphi_a, \varphi_g, \varphi_c, \varphi_t)$ is the stationary distribution and u is a parameter of the model. (It will be shown in Section 13.2.7, and is easily checked directly, that the stationary distribution for the model defined by (13.18) is indeed $(\varphi_a, \varphi_g, \varphi_c, \varphi_t)$.)

This model generalizes the Jukes–Cantor model, to which it reduces if $\varphi_a = \varphi_g = \varphi_c = \varphi_t = .25$. It also generalizes the Markov chain in which all rows have the form of the stationary distribution, reducing to this model when $u = 1$.

13.2.5   The HKY Model

A more complex model, introduced by Hasegawa et al. (1985), and called here the HKY model, assumes that the transition probability matrix P is of the form

$$\begin{bmatrix} 1-u\varphi_g-v\varphi_1 & u\varphi_g & v\varphi_c & v\varphi_t \\ u\varphi_a & 1-u\varphi_a-v\varphi_1 & v\varphi_c & v\varphi_t \\ v\varphi_a & v\varphi_g & 1-u\varphi_t-v\varphi_2 & u\varphi_t \\ v\varphi_a & v\varphi_g & u\varphi_c & 1-u\varphi_c-v\varphi_2 \end{bmatrix}, \qquad (13.19)$$

where $\varphi_1 = \varphi_c+\varphi_t$, $\varphi_2 = \varphi_a+\varphi_g$. This model is an amalgam of the Kimura model (13.6) and the Felsenstein model, and includes these as particular cases.

The eigenvalues and eigenvectors of these matrices follow from those given by Hasegawa et al. (1985) for the analogous continuous-time process. The eigenvalues are

$$\lambda_1 = 1, \ \lambda_2 = 1-v, \ \lambda_3 = 1-u\varphi_1-v\varphi_2, \ \lambda_4 = 1-v\varphi_1-u\varphi_2. \quad (13.20)$$

The corresponding left eigenvectors are

$$\ell_1 = (\varphi_a, \varphi_g, \varphi_c, \varphi_t)$$
$$\ell_2 = (\varphi_1\varphi_a, \varphi_1\varphi_g, \varphi_2\varphi_c, \varphi_2\varphi_t)$$
$$\ell_3 = (0,0,1,-1)$$
$$\ell_4 = (1,-1,0,0)$$

and the corresponding right eigenvectors are

$$r_1 = (1, 1, 1, 1)'$$

$$r_2 = \left(\frac{1}{\varphi_2}, \frac{1}{\varphi_2}, \frac{-1}{\varphi_1}, \frac{-1}{\varphi_1} \right)'$$

$$r_3 = \left(0, 0, \frac{\varphi_t}{\varphi_1}, \frac{-\varphi_c}{\varphi_1} \right)'$$

$$r_4 = \left(\frac{\varphi_g}{\varphi_2}, \frac{-\varphi_a}{\varphi_2}, 0, 0 \right)'.$$

These have been normalized by the requirement (B.46), and this leads immediately to the spectral expansion (B.48) for the powers of the transition matrix (13.19). The left eigenvector ℓ_1 also shows that $(\varphi_a, \varphi_g, \varphi_c, \varphi_t)$ is the stationary distribution of the Markov chain for this model, as the notation anticipated. The HKY Markov chain is reversible: See Section 13.2.7. Thus the HKY model is the most flexible of those discussed for considering evolutionary processes. This matter is discussed further in Section 14.7.

13.2.6   Other Models

All of the above models have various simplifying assumptions built into them. At the other extreme, a model having no such properties might be more appropriate in practice; the numerical model (4.25), possessing no symmetry or other simplifying features, was introduced with this possibility in mind. In contrast to the algebraically defined models described above, the rate of approach to stationarity for almost every numerical model, and the stationary distribution itself, can only be found through a numerical spectral expansion such as in Section 10.2.3.

The simpler models discussed above are unrealistic in that, for example, they imply uniform nucleotide frequency distributions. More complex models, however, might not be reversible, and thus should not be used for the reconstruction of phylogenetic trees. The choice of one model over another is thus often a difficult matter. Tests of one model against another are discussed in Chapter 14.

13.2.7   The Reversibility Criterion

The discussion of Section 10.2.4 shows that the reversibility requirement is necessary for various inferential procedures, and equation (10.5) gives the reversibility requirement for any finite Markov chain. We now discuss the reversibility criterion in the context of evolutionary models, focusing on DNA substitutions and the 4×4 transition matrices used to describe these substitutions. We assume that all transition matrices considered refer to irreducible aperiodic Markov chains, so that any model discussed has a stationary distribution, denoted as above by $(\varphi_a, \varphi_g, \varphi_c, \varphi_t)$.

The general 4×4 transition matrix has twelve free parameters, namely three free transition probabilities in each of the four rows of the transition matrix. (The fourth transition probability in each row is determined by the remaining three.) However another parameterization, using twelve different free parameters, is more useful for our present purposes. This parameterization was given by Tavaré (1986), (see Swofford et al. (1996)), and under this parameterization the transition matrix is written in the form

$$
\begin{bmatrix}
1 - uW & uA\varphi_g & uB\varphi_c & uC\varphi_t \\
uD\varphi_a & 1 - uX & uE\varphi_c & uF\varphi_t \\
uG\varphi_a & uH\varphi_g & 1 - uY & uI\varphi_t \\
uJ\varphi_a & uK\varphi_g & uL\varphi_c & 1 - uZ
\end{bmatrix} . \tag{13.21}
$$

Here A, B, ..., L are the twelve free parameters, $(\varphi_a, \varphi_g, \varphi_c, \varphi_t)$ is the stationary distribution of the Markov chain, and

$$
W = A\varphi_g + B\varphi_c + C\varphi_t, \; X = D\varphi_a + E\varphi_c + F\varphi_t,
$$

$$
Y = G\varphi_a + H\varphi_g + I\varphi_t, \; Z = J\varphi_a + K\varphi_g + L\varphi_c.
$$

The necessary and sufficient condition for the Markov chain with transition matrix (13.21) to be reversible is that the equations

$$
A = D, \quad B = G, \quad C = J, \quad E = H, \quad F = K, \quad I = L \tag{13.22}
$$

are all satisfied. When this condition holds, we call the model (13.21) the *general reversible process model*, following the terminology of Yang (1994) in the analogous continuous-time model. The general reversible model has six free parameters, which can be taken as A, B, C, E, F, and I, so one can think of paying for reversibility by losing the choice of six further parameters, or by losing six degrees of freedom.

It is easily checked that when the conditions (13.22) hold, $(\varphi_a, \varphi_g, \varphi_c, \varphi_t)$ is indeed the stationary distribution of the model (13.21): see Problem 13.4.

The Felsenstein model is the particular case of (13.21) with $A = B = \cdots = L = 1$, and the HKY model is a special case of (13.21) with $B = C = G = J = E = F = H = K$, and $A = D = I = L$. Both models are thus reversible, as claimed above. The choice of one of these models instead of the general reversible model may also be thought of as involving a loss of degrees of freedom. These degrees of freedom are relevant to the test of one these models against the general reversible model, a matter discussed in Section 14.9.5.

13.2.8   The Simple Symmetric PAM Model

The simple symmetric PAM matrix M_1 defined in in Section 6.5.4 is the "amino acid" 20×20 matrix generalization of the "nucleotide" Jukes–

Cantor matrix (13.1). Like that model it is reversible. The n-step transition probabilities $\{m_{jk}^{(n)}\}$ given in equations (6.28) and (6.29) define the spectral expansion of this matrix, which is a direct analogue spectral expansion of the Jukes–Cantor matrix (13.1).

As with the Jukes–Cantor model, this model is not realistic. We have analyzed it in some detail in Chapters 6 and 9 because it illuminates properties of more realistic PAM models.

Further transition matrices may be found as the "amino acid" analogues of the Kimura, Felsenstein and HKY models. However these have not been discussed at length in the literature and are not considered further here.

13.3   Continuous-Time Models

As justified in Section 10.7.1, it is assumed for all continuous-time evolutionary models considered that the transition probabilities are of the form described in (10.26) and (10.27).

13.3.1   The Continuous-Time Jukes–Cantor Model

The Jukes–Cantor and Kimura models were first proposed as continuous-time models, and in their continuous-time versions can be analyzed by the methods of Section 10.7. It is convenient to continue to label the states as a, g, c, and t, as in Section 13.1, and to introduce the further notation that i, j, and k are arbitrary members of the set $\{a, g, c, t\}$.

In the Jukes–Cantor model the instantaneous transition rates q_{ij} and q_i are defined by $q_{ij} = \alpha$ for all $i \neq j$, and $q_i = 3\alpha$. Replacing q_{kj} and q_j in equation (10.30) by α and 3α, respectively, we get

$$\frac{d}{dt}P_{ij}(t) = -3\alpha P_{ij}(t) + \alpha \sum_{k \neq j} P_{ik}(t). \tag{13.23}$$

Since $\sum_{k \neq j} P_{ik}(t) = 1 - P_{ij}(t)$, this equation simplifies to

$$\frac{d}{dt}P_{ij}(t) = \alpha - 4\alpha P_{ij}(t). \tag{13.24}$$

This is a linear differential equation and may be solved by standard methods. The solution involves a constant of integration, which is allocated by the boundary conditions $P_{ii}(0) = 1$, $P_{ij}(0) = 0$, $i \neq j$. Using these boundary conditions, the solution is found to be

$$P_{ii}(t) = .25 + .75e^{-4\alpha t}, \tag{13.25}$$
$$P_{ij}(t) = .25 - .25e^{-4\alpha t}, \quad j \neq i. \tag{13.26}$$

These equations show that as $t \to \infty$, both $P_{ii}(t)$ and $P_{ij}(t)$ approach 0.25. Thus in the stationary distribution all nucleotides have equal probability at any site, as is expected from the symmetry of the model. This stationary distribution can also be found by applying equations (10.31). The similarity of the expressions for $P_{ii}(t)$ and $P_{ij}(t)$ and the expressions given in equations (13.4) and (13.5) indicate why the present model is the continuous-time analogue of the discrete-time Jukes–Cantor model.

The fact that q_i is independent of i shows that if an event is defined as the substitution of one nucleotide for another, substitutions follow the laws of the homogeneous Poisson process discussed in Section 4.1, so that the probability that j substitutions occur in time t is given by the Poisson probability distribution (4.10) with $\lambda = 3\alpha$. From this it follows (see Problem 13.7) that if j transitions from one state to another have occurred in the continuous-time Markov process described by equation (13.24), the probability that the process has returned to its initial state is

$$\frac{1}{4} + \frac{3}{4}\left(-\frac{1}{3}\right)^j. \tag{13.27}$$

The probability that j transitions do occur in time t is given by (4.10) with $\lambda = 3\alpha$. Combining these results, the probability that at time t the original nucleotide is the predominant one is

$$\sum_{j=0}^{\infty} e^{-3\alpha t} \frac{(3\alpha t)^j}{j!} \left(\frac{1}{4} + \frac{3}{4}\left(-\frac{1}{3}\right)^j\right). \tag{13.28}$$

By equation (B.20), this reduces to the right-hand side in equation (13.25). Equation (13.26) is found similarly.

Suppose that two independent contemporary populations are available that descended from a common population t units of time ago, and that the evolutionary properties of these two populations since their common ancestor are described by the Jukes–Cantor model. We would like to use data from these two populations to estimate the evolutionary parameter λ in this model, describing in a sense the rate at which one nucleotide replaces another in these populations.

It is convenient to begin the analysis by considering the probability p that at a given site, the predominant nucleotides in the two populations differ. Given two DNA sequences each containing N nucleotides, and with the assumption that the replacement processes at all sites are independent and have identical stochastic properties, the theory of Section 3.3.1 shows that $\hat{p}$, the observed proportion of sites at which the predominant nucleotides differ between the two populations considered, is an unbiased estimator of p. It is thus natural to estimate p by $\hat{p}$. Further, equation (2.74) shows that the variance of this estimate is

$$\text{variance of } \hat{p} = \frac{p(1-p)}{N}. \tag{13.29}$$

The simplicity of this calculation follows largely from the iid assumption made. This assumption is unlikely to hold in practice, and we later examine some of the consequences of its not holding.

We now use this estimate, and its variance, to discuss properties of the estimates of the parameters in the Jukes–Cantor model. Poisson process theory shows that in this model, the mean number of substitutions, in t units of time, at any nucleotide site down the two lines of descent leading from the common founder to the two populations is $2\lambda t$. The mean number ν of substitutions down the two lines of descent at all N sites together is then $\nu = 2N\lambda t$. Since $\lambda = 3\alpha$, both of these are functions of the parameter αt. The way in which the parameter αt is related to p is found as follows.

The symmetry inherent in the Jukes–Cantor model implies that whatever the predominant nucleotide at time 0, the probability $I(t)$ that at time t the two descendent populations have the same predominant nucleotide is

$$I(t) = \left(P_{ii}(t)\right)^2 + \sum_{j \neq i} \left(P_{ij}(t)\right)^2 ,$$

where $P_{ii}(t)$ is given by (13.25) and $P_{ij}(t)$ is given by (13.26). These equations show that

$$I(t) = .25 + .75e^{-8\alpha t}. \tag{13.30}$$

The probability p that the nucleotides are different in the two populations is thus

$$p = 1 - I(t) = .75(1 - e^{-8\alpha t}). \tag{13.31}$$

This equation relating p to αt can be found in a second, more efficient, way. The reversibility of the Jukes–Cantor model implies that the properties of the stochastic process describing any line of descent is the same as that describing the process in reverse, that is, by considering the corresponding line of ascent. The elapsed time up the line of ascent from one of the two contemporary populations up to the founder population, and then down the line of descent from the founder population to the other contemporary population, is $2t$. Therefore, the probability that, at any nucleotide site, the same nucleotide occurs in both populations is given by equation (13.25) with t replaced by $2t$. This leads directly to the expression (13.30) and thus to (13.31).

The evolutionary parameters introduced above are all functions of the composite parameter αt. Standard practice in estimating αt is to invert the relation between p and αt in (13.31), yielding

$$\alpha t = -\frac{1}{8} \log \left(1 - \frac{4}{3}p\right),$$

and thus to estimate αt by

$$\widehat{\alpha t} = -\frac{1}{8} \log \left(1 - \frac{4}{3}\hat{p}\right). \tag{13.32}$$

This right-hand side is defined only if $\hat{p} < \frac{3}{4}$. When $\hat{p} \geq \frac{3}{4}$ the two sequences are more dissimilar than two random sequences will tend to be, and it is natural to define $t = +\infty$. Even when $\hat{p} < \frac{3}{4}$ the estimator (13.32) is a biased estimator of αt, and in fact, there is no unbiased estimator of αt. In practice, this observation is usually ignored.

Equation (13.32) and the identity $\lambda = 3\alpha$ show that unless t can be estimated extrinsically, the aim of estimating the parameter λ cannot be achieved, and all that is possible is estimation of λt. Equation (13.32) shows that

$$\widehat{\lambda t} = -\frac{3}{8} \log\left(1 - \frac{4}{3}\hat{p}\right). \tag{13.33}$$

Similarly, the estimate $\hat{\nu}$ of $\nu = 2N\lambda t$ is

$$\hat{\nu} = -\frac{3N}{4} \log\left(1 - \frac{4}{3}\hat{p}\right). \tag{13.34}$$

The issue of bias discussed above arises also for these estimators, but again we ignore them.

The estimation formula (13.34) has several interesting consequences. Perhaps the most interesting is that if $\hat{p}$ is small, the Taylor series approximation (B.25) shows that

$$\hat{\nu} \approx N\left(\hat{p} + \frac{2}{3}\hat{p}^2\right).$$

Thus $\hat{\nu}$ is marginally larger than $N\hat{p}$, the observed number of sites at which the predominant nucleotides in the two populations differ from each other. For example, if in 300 of $N = 3{,}000$ sites the predominant nucleotide differs between the two populations, then $\hat{p} = 0.1$ and $\hat{\nu} \approx 320$. Thus it is estimated that about 20 substitutions occurred but are not observed by counting the differences between the extant sequences. This is because more than one substitution can occur in the same position. The quantity ν accounts for *all* substitutions.

The statistical differentials variance formula (B.33) can be used to approximate the variances of the estimators (13.32), (13.33), and (13.34). We illustrate this by considering the variance of $\hat{\nu}$. Equation (B.33) shows that this variance is approximately

$$\left(\frac{d\hat{\nu}}{d\hat{p}}\bigg|_{\hat{p}=p}\right)^2 (\text{variance of } \hat{p}),$$

which from (13.29) is

$$\left(\frac{d\hat{\nu}}{d\hat{p}}\bigg|_{\hat{p}=p}\right)^2 \frac{p(1-p)}{N}.$$

By equation (13.34), this equals

$$Np(1-p)e^{8\nu/3N}. \tag{13.35}$$

Since maximum likelihood estimators have optimality properties as discussed in detail in Chapter 8, it is appropriate to find the maximum likelihood estimate of the parameter ν. Suppose that of the N sites sampled there are $j = N\hat{p}$ sites for which different nucleotides appear in the two populations. Under the iid assumption the likelihood of the data in terms of ν is, from equations (13.30) and (13.31), proportional to

$$\left(\frac{3}{4}\left(1 - e^{-4\nu/3N}\right)\right)^j \left(\frac{1}{4} + \frac{3}{4}e^{-4\nu/3N}\right)^{N-j}, \qquad (13.36)$$

where $8\alpha t$ has been replaced by $4\nu/3N$. The derivative of the logarithm of this expression with respect to ν is

$$\frac{1}{N}e^{-4\nu/3N}\left(\frac{j}{p} - \frac{N-j}{1-p}\right), \qquad (13.37)$$

where

$$p = \frac{3}{4}\left(1 - e^{-4\nu/3N}\right). \qquad (13.38)$$

It follows from this that the expression (13.34) for $\hat{\nu}$ in is the maximum likelihood estimate of ν if $\hat{p}$ is given by j/N.

This result also follows directly from the invariance property of maximum likelihood estimators, discussed in Section 8.2.2. The maximum likelihood estimator of p is $\hat{p} = j/N$, and the relation (13.38) between p and ν implies that the maximum likelihood estimate $\hat{\nu}$ is given by (13.34). In the following section we use this more direct approach when analyzing the continuous-time Kimura model.

A further differentiation of the log likelihood and the asymptotic variance formula (8.17) shows that the asymptotic variance of $\hat{\nu}$ is identical to the expression in (13.35).

13.3.2   The Continuous-Time Kimura Model

The analysis of the continuous-time Kimura model (13.6) is more complex than that for the Jukes–Cantor model, and the notation must be handled more carefully. The instantaneous parameters q_{ij} and q_i, introduced in equations (10.26) and (10.27), take the values

$$q_{ij} = \alpha \text{ for the } (i,j) \text{ pairs } (a,g), (g,a), (c,t), \text{ and } (t,c),$$
$$q_{ij} = \beta \text{ for the } (i,j) \text{ pairs } (a,c), (a,t), (g,c), (g,t),$$
$$(c,a), (c,g), (t,a), \text{ and } (t,g).$$

In all cases,

$$q_i = \alpha + 2\beta. \qquad (13.39)$$

If these choices are made, the system of equations (10.30) yields a set of four simultaneous differential equations typified by

$$\frac{d}{dt}P_{aa}(t) = -(\alpha + 2\beta)P_{aa}(t) + \alpha P_{ag}(t) + \beta\left(P_{ac}(t) + P_{at}(t)\right), \quad (13.40)$$

where the notation now indicates specific nucleotide types. The boundary conditions for these equations are $P_{ii}(0) = 1$ for all i, $P_{ij}(0) = 0$ for all $j \neq i$. This system of four differential equations can be solved, and for any i the value of $P_{ii}(t)$ is found to be

$$P_{ii}(t) = .25 + .25e^{-4\beta t} + .5e^{-2(\alpha+\beta)t}. \quad (13.41)$$

The expression for $P_{ij}(t)$ depends on the choice of i and j. If the initial predominant nucleotide is a specified purine (respectively pyrimidine), then the probability that at time t the predominant nucleotide is the other purine (respectively pyrimidine) is

$$.25 + .25e^{-4\beta t} - .5e^{-2(\alpha+\beta)t}. \quad (13.42)$$

If the initial predominant nucleotide is a specified purine (respectively pyrimidine), then the probability that at time t the predominant nucleotide is a pyrimidine (respectively purine) is

$$.5 - .5e^{-4\beta t}. \quad (13.43)$$

These expressions reduce to the corresponding formulae for the Jukes–Cantor model when $\alpha = \beta$. They are also the continuous time analogues of the discrete-time equations (13.8), (13.9), and (13.11). A further implication of these equations is that, as expected, in the stationary distribution for this model all nucleotides have equal probability.

The detailed balance requirements (10.32) hold for the Kimura model, so that the process is reversible. This implies that the Jukes–Cantor process is also reversible.

As in the Jukes–Cantor model, Poisson process theory is relevant to the evolutionary properties of the Kimura model. For the Kimura model q_i, defined in (13.39), is independent of i. Thus if an event is defined as a substitution from one nucleotide to another at any nucleotide site in the line of descent of any population, the probability that j events occur in time t is given by equation (4.10), with $\lambda = \alpha + 2\beta$.

The observed number of transition substitutions and the observed number of transversion substitutions can be used to estimate the parameters of the Kimura model using the reversibility property of the model and the "line of ascent and descent" argument described in Section 13.3.1. Here t is replaced by $2t$ in the right-hand sides of equations (13.41), (13.42), and (13.43). The resulting expressions then show that if

$$\gamma = 4(\alpha + \beta)t, \quad \phi = 8\beta t,$$

then the probability p_1 that at any site one purine (pyrimidine) occurs in one sequence and the other purine (pyrimidine) in the other sequence is

$$p_1 = .25 + .25e^{-\phi} - .5e^{-\gamma}.$$

The probability p_2 that at any site a purine (pyrimidine) occurs in one sequence and a pyrimidine (purine) in the other sequence is

$$p_2 = .5 - .5e^{-\phi}.$$

Suppose that in a DNA sequence of length N there are n_0 sites where the same nucleotide occurs in both populations, n_1 sites where one purine (pyrimidine) occurs in the sequence of one population and the other purine (pyrimidine) in the sequence of the other population, and n_2 sites where a purine occurs in one sequence and a pyrimidine in the other. The observed proportions $\hat{p}_1 = n_1/N$ and $\hat{p}_2 = n_2/N$ are the maximum likelihood estimators of p_1 and p_2, respectively (see Problem 8.2). The invariance property of maximum likelihood estimators discussed in Section 8.2.2 implies that the maximum likelihood estimates $\hat{\gamma}$ and $\hat{\phi}$ satisfy the equations

$$\hat{p}_1 = .25 + .25e^{-\hat{\phi}} - .5e^{-\hat{\gamma}},$$
$$\hat{p}_2 = .5 - .5e^{-\hat{\phi}}.$$

These equations lead to the maximum likelihood estimates

$$\hat{\gamma} = -\log(1 - 2\hat{p}_1 - \hat{p}_2), \tag{13.44}$$
$$\hat{\phi} = -\log(1 - 2\hat{p}_2). \tag{13.45}$$

In the length of DNA considered, the mean number ν of nucleotide substitutions through the course of evolution since the original founding population is

$$\nu = 2N\lambda t = N(.5\gamma + .25\phi).$$

From equations (13.44) and (13.45), the maximum likelihood estimator $\hat{\nu}$ of ν is

$$\hat{\nu} = -N\left(.5\log(1 - 2\hat{p}_1 - \hat{p}_2) + .25\log(1 - 2\hat{p}_2)\right). \tag{13.46}$$

This estimator is subject to the same qualifications as those made for the Jukes–Cantor estimator.

It is interesting to compare calculations derived from (13.46) with those from the parallel Jukes–Cantor model. If $N = 3,000$ and the data yield 210 transitional and 90 transversional changes, and thus 300 changes in total, then $\hat{p}_1 = 0.07$ and $\hat{p}_2 = 0.03$. The total number of changes from one predominant nucleotide to another at some time during the evolution of the two populations since their common founder population is then estimated from equation (13.46) to be about 326. This is somewhat larger than the corresponding estimate in the Jukes–Cantor model.

This implies that if a satisfactory estimate of λ is available extrinsically, the estimate of t differs in the Kimura and the Jukes–Cantor models given the same data in each. Thus if the true model is the Kimura model and the Jukes–Cantor equation (13.33) is used to estimate t, then a biased estimation procedure has been used. Since the estimate of t is often taken as proportional to an estimate of distance between two species, the estimate of this distance will also be biased in this case. Even larger biases will arise if the true model is more complex than the Kimura model.

All the above estimators can also be found by writing down the likelihood and maximizing its logarithm with respect to γ and ϕ. In this case the likelihood is a multinomial, being proportional to

$$\left(.25 + .25e^{-\phi} + .5e^{-\gamma}\right)^{n_0} \left(.25 + .25e^{-\phi} - .5e^{-\gamma}\right)^{n_1} \left(.5 - .5e^{-\phi}\right)^{n_2}.$$
$$(13.47)$$

Maximization of this function with respect to ϕ and γ leads to estimators identical to those in (13.44) and (13.45). If the values of α and β are known, a similar procedure may be used to find the maximum likelihood estimate of t. A procedure very similar to this is discussed below in Section 13.3.5.

While the procedure of maximizing (13.47) is unnecessarily long so far as finding estimates is concerned, there is one advantage to it. Variance approximations for $\hat{\gamma}$ and $\hat{\phi}$ can be found from a second differentiation of the logarithm of the likelihood, using the information matrix (8.30), together with an approximation for the covariance of $\hat{\gamma}$ and $\hat{\phi}$ parallel to those for the variances of $\hat{\gamma}$ and $\hat{\phi}$. The parameter ν is a linear combination of γ and ϕ, and the variance of $\hat{\nu}$ can be found by applying equation (2.62).

13.3.3 The Continuous-Time Felsenstein Model

The discrete-time Felsenstein model was discussed in Section 13.2.4. That model is the discrete-time version of a continuous-time model originally proposed to address the problem of phylogenetic tree reconstruction, and so we now describe this continuous-time model.

The essential assumption of the discrete-time Felsenstein model (13.18) is that the probability of the substitution of one nucleotide by another is proportional to the stationary probability of the substituting nucleotide. This assumption is maintained in the continuous-time version of the model. If, for example, the predominant nucleotide at any given time t is a, the model assumes that for small h, the probability that the predominant nucleotide at time $t + h$ is X is $u\varphi_X h$, where X stands for any of the nucleotides g, c, and t, while the probability that the predominant nucleotide continues to be a at time $t + h$ is $1 - u(\varphi_g + \varphi_c + \varphi_t)h$, where in all the above expressions terms of order $o(h)$ are ignored. Parallel assumptions are made if the predominant nucleotide at time t is g, c, or t.

These assumptions define the instantaneous transition rates q_{jk} and q_j in equations (10.26) and (10.27), and this leads to specific forms for four simultaneous differential equations of the form (10.30). The solution of these equations is

$$P_{ii}(t) = e^{-ut} + (1 - e^{-ut})\varphi_i, \qquad\qquad (13.48)$$

$$P_{ij}(t) = (1 - e^{-ut})\varphi_j, \quad j \neq i. \qquad\qquad (13.49)$$

This solution implies that the continuous-time Felsenstein model satisfies the detailed balance equations (10.32), and is thus reversible.

13.3.4   The Continuous-Time HKY Model

The continuous-time analogue of the HKY model (13.19) is discussed in detail by Hasegawa et al. (1985). Its time-dependent solution is given by the spectral expansion corresponding to the discrete-time model (13.19), with λ_j^n replaced by $e^{(\lambda_j - 1)t}$. It shares all the desirable properites of its discrete-time analogue, and is used in phylogenetic studies, as discussed in Chapter 14.

13.3.5   The Continuous-Time PAM Model

The simple symmetric discrete-time PAM model of Section 13.2.8 has a direct continuous-time analogue. This is found from the Kolmogorov equations (10.30) by putting $q_{jk} = \alpha$. Here α is chosen so that the PAM requirement that the probability that the initially predominant amino acid after one time unit is still predominant is 0.99. With these choices the solutions of equations (10.30) are

$$P_{ii}(t) = .05 + .95e^{-20\alpha t}, \qquad\qquad (13.50)$$

$$P_{ij}(t) = .05 - .05e^{-20\alpha t}, \quad j \neq i. \qquad\qquad (13.51)$$

These equations are similar in form to the Jukes–Cantor equations (13.25) and (13.26). The requirement $P_{ii}(1) = 0.99$ yields $\alpha = 0.00053$.

These calculations allow maximum likelihood estimation of t, the time (in PAM units) back to an assumed common ancestor of two contemporary amino acid sequences, provided that the simple symmetric PAM model can be assumed. Suppose, for example, that two contemporary sequences of length 10,000 are compared and that in this comparison there are 1,238 matches and 8,762 mismatches. Since the simple symmetric PAM model is reversible and the time connecting the two sequences is $2t$, the invariance property of maximum likelihood estimators and equation (13.50) show that the maximum likelihood estimate $\hat{t}$ of t satisfies the equation

$$.1238 = .05 + .95e^{-40\alpha\hat{t}}. \tag{13.52}$$

The solution of this equation (with $\alpha = 0.00053$) is $\hat{t} = 120.5$.

Further properties of PAM-type matrices where reversibility is assumed are developed by Müller and Vingron (2000).

Problems

13.1 Prove equation (13.3) by induction on n. (For a discussion of proofs by induction, see Section B.18.)

13.2 Check that the stationary distribution of the Markov chain (13.15) is the vector given in (13.16).

13.3 Check that the Markov chains (13.12) and (13.13) are reversible by showing that they satisfy the detailed balance conditions (10.3), but that the chain defined by (13.15) does not satisfy these conditions and is thus not reversible.

13.4 Show that when the conditions (13.22) hold, the stationary distribution of the model (13.21) is $(\varphi_a, \varphi_g, \varphi_c, \varphi_t)$.

13.5 The Jukes–Cantor model (13.1) is a special case of (that is, is *nested within*) the Kimura model (13.6). It is also special case of (that is, is *nested within*) the Felsenstein model (13.18). is either one of the Kimura model and the Felsenstein model nested within the other?

13.6 Show that the conditions (13.22) are necessary and sufficient for the transition matrix (13.21) to be reversible.

13.7 Derive the expression (13.27).

13.8 Check that the continuous-time Kimura model satisfies the detailed balance requirements (10.32).

13.9 Show that the sum in the expression (13.28) does reduce, as claimed, to the expression on the right-hand side of (13.25).

13.10 The aim of this question is to compare the numerical values of the two estimators $\hat{\nu}$ given respectively in equations (13.34) for the Jukes–Cantor model and (13.46) for the Kimura model. In this comparison we identify the values of $\hat{p}$ in the Jukes–Cantor model with $\hat{p}_1 + \hat{p}_2$ in the Kimura model.

(i) Use the logarithmic approximation (B.24) to show that when $\hat{p}_1$ and $\hat{p}_2$ are both small, the two estimators are close, and differ by only a term of order $\hat{p}^2$.

(ii) Use the more accurate approximation (B.25) to compare the values of the two estimators when (a) $\hat{p}_1 = \hat{p}_2$, (b) $\hat{p}_1 = 2\hat{p}_2$. Thus show that even when $\hat{p}_1 = \hat{p}_2$ the two estimates of ν differ.

13.11 Estimate the standard deviation of the estimate $\hat{t} = 120.5$ found by solving equation (13.52).

14
Phylogenetic Tree Estimation

14.1 Introduction

The construction of phylogenetic trees is of interest in its own right in evolutionary studies. It is also useful in many other ways, for example in the prediction of gene function; aspects of this area of the emerging field of *phylogenomics* are discussed by Eisen (1998).

A phylogenetic tree is *binary* if each node except the *root* connects with either one or three branches, while the root, of which there can be at most one, connects with two branches. A tree without a root is called an *unrooted* tree.

The evolutionary relationships between a set of species is represented by a binary tree, and in this book we consider only binary trees. "Species" may refer either to organisms or to sequences such as protein or DNA. It is required that the set of "species" have a common ancestor, so that while one would not construct a tree relating a hemoglobin protein to a ribosomal protein, it would be natural to construct a tree relating the various hemoglobins.

The following are examples of rooted phylogenetic trees. The first relates five different species and the second relates five different hemoglobins. The unlabeled nodes represent common ancestors of the children nodes below them.

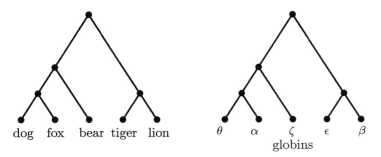

dog   fox   bear tiger lion θ α ζ ϵ β
                                                 globins

The lengths of the edges represent evolutionary time, so the longer an edge is, the longer is the evolutionary time separating the nodes at the ends of the edge. The node at the top, the root, represents a common ancestor to all species in the tree. In these examples the leaves represent extant species, so they are all at the same vertical height. This may not be so in a tree with extinct species, where the leaves are at possibly different levels, as in the following example:

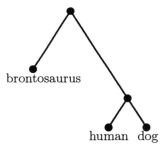

brontosaurus

human dog

An unrooted tree shows the evolutionary relationships between the species at the leaves but does not indicate the direction in which evolutionary time flowed between the internal nodes. Each node connects either one or three edges in an unrooted tree. An example of an unrooted tree (with five leaves) is:

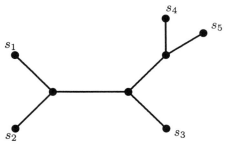

s_4

s_5

s_1

s_2

s_3

Given a set of species, there is some (unknown) phylogenetic tree connecting the members and giving their true evolutionary relationships. Our goal is to infer the tree as accurately as possible from information about the species. Before the advent of genetic information this was normally done using morphological information, such as the shape of the teeth and bones. The advent of genetic information in the form of DNA or protein sequences has fostered a more mathematical/algorithmic approach to tree reconstruction.

We discuss in turn three main types of tree-reconstruction methods, namely *distance* methods, *parsimony* methods, and *statistical* methods.

14.2   Distances

Several of the algorithmic procedures used in tree reconstruction are based on the concept of a *distance* between species. We thus introduce the concept of distance in this section and describe some properties of the distances that arise in phylogenetic reconstruction.

The most natural way to define the distance between species is in terms of years since their most recent common ancestor. If this distance were known, no other distance measure would be needed. In practice, this distance is seldom if ever known, and surrogate distances are used instead. The use of surrogates will affect the accuracy of tree reconstruction, as discussed in Sections 14.5 and 14.8. For the moment we assume that distances in years *are* known, and we call these *exact* distances.

Let S be a set of points. The standard requirements for a distance measure on S are that for all x and y in S, (i) $d(x,y) \geq 0$, (ii) $d(x,y) = 0$ if and only if $x = y$, (iii) $d(x,y) = d(y,x)$, and (iv) for all x, y, and z in S $d(x,y) \leq d(x,z) + d(z,y)$.

A distance measure $d(\cdot, \cdot)$ on a set of species is said to be *tree-derived* if there is a tree with these species at the leaves such that the distance $d(x,y)$ between any pair of leaves x and y is the sum of the lengths of the edges joining them. This distance satisfies the three requirements above. However, not all distance measures on a set S are tree derived, since a tree-derived measure must satisfy certain further requirements, as demonstrated by the unrooted tree shown:

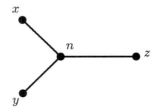

The distances in this tree must satisfy the *strict* inequality $d(x, z) < d(x, y) + d(y, z)$. For a tree-derived distance measure, this strict inequality must hold for all species x, y, and z in the tree.

For rooted trees joining extant sequences there are further requirements. In the rooted tree shown,

$$(14.1)$$

the distances between x, y and z must satisfy the equation $d(x, y) = d(x, z)$ and the further requirements $d(y, z) < d(x, y)$, $d(y, z) < d(x, z)$. This requirement motivates a further abstract distance measure concept. Any distance measure on a set S that satisfies the requirement that of the three distances between any three members of the set, two are equal and are greater than the third, is said to be *ultrametric*.[1]

Thus a tree-derived distance measure on a (rooted) tree of extant species (with exact edge lengths) is ultrametric. The converse of this statement, that any ultrametric distance is tree derived, will be proved in Section 14.3. We will show further how such a tree can be found, and that up to trivial changes this tree is unique. This allows us to go from a set of extant species to the unique *rooted* tree that relates them, as long as we know the exact distances between all pairs of species.

Not all tree-derived distance measures are ultrametric, as is the case when the tree represents only extant species and exact distances. However, if a distance measure is tree-derived it is still possible to recover a unique *unrooted* tree giving all distances between species. This is demonstrated in Section 14.4. In practice, distances used in tree reconstruction are not necessarily tree-derived, being surrogate distances that only estimate true distances. Properties of surrogate distances are discussed in Section 14.5.

14.3   Tree Reconstruction: The Ultrametric Case

In this section we prove that if an ultrametric distance measure between species is given, then there is a unique (up to trivial changes) rooted tree joining these species that gives these distances. (What "up to trivial changes" means will be made clear as we proceed.) We find this tree using a constructive method that differs from the well-known UPGMA (un-

---

[1]Ultrametric distances arise in many contexts, not only biology, and the definition varies depending on the context. For phylogenetics, this is the correct definition.

weighted pair group method using arithmetic averages) algorithm discussed below.

We assume that a set of species $s_1, s_2, \ldots, s_n$ is given with an ultrametric distance measure $d(x, y)$ between all pairs of species x and y. The proof of the above claim is by induction. We first construct the correct tree relating two species and show it is unique. We then show that if we can reconstruct the tree correctly and uniquely for m species $s_1, s_2, \ldots, s_m$, for some $m \geq 2$, we can do this also for $m + 1$ species $s_1, s_2, \ldots, s_{m+1}$.

The first step is the construction of a tree relating s_1 and s_2. Both s_1 and s_2 must be equidistant from the root, so there is clearly a unique solution:

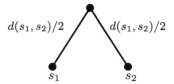

$d(s_1, s_2)/2$ $d(s_1, s_2)/2$

s_1 s_2

(In this and similar steps we ignore trivial changes such as interchanging s_1 and s_2.) Now suppose there are $m + 1$ species, $m \geq 2$. By the induction hypothesis, we can assume the tree can be uniquely reconstructed for $s_1, s_2, \ldots, s_m$. Consider the root of the tree and the two edges emanating from it. The species $s_1, s_2, \ldots, s_m$ divide into two sets, the set S_L consisting of the species down the left edge from the root and the set S_R consisting of the species down the right edge from the root. Let x be any element of S_L and y be any element of S_R.

r

$x \in S_L$ $y \in S_R$ (14.2)

Now consider species s_{m+1} and the three distances between x, y, and s_{m+1}. The ultrametric property implies that there are three possibilities, depending on which pair of these three is equidistant from the third. These possibilities are (1) $d(s_{m+1}, x) = d(s_{m+1}, y)$, (2) $d(s_{m+1}, y) = d(x, y)$, (3) $d(s_{m+1}, x) = d(x, y)$.

If (1) holds, $d(x, y)$ is strictly less than $d(s_{m+1}, x) = d(s_{m+1}, y)$. In this case, we place s_{m+1} follows:

new root

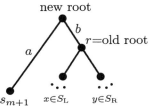

b

r=old root

a

s_{m+1} $x \in S_L$ $y \in S_R$

where $a = d(s_{m+1}, x)/2$ and $b = d(s_{m+1}, x) - d(x, y)/2$. Since $d(x, y) < d(s_{m+1}, x)$, b is positive, as required.

This gives a tree that gives the correct distances between s_{m+1}, x, and y. We must check that this placement gives the correct distance from s_{m+1} to any other species. For any species z in S_L or S_R, $d(z, \text{old root}) = d(x, y)/2$. Thus the distance from s_{m+1} to z in the new tree is $a + b + d(x, y)/2$, and it is required to show that this is equal to $d(s_{m+1}, z)$. Now, in the new tree

$$d(s_{m+1}, x) = a + b + \frac{d(x, y)}{2}, \tag{14.3}$$

so it is sufficient to show that $d(s_{m+1}, x) = d(s_{m+1}, z)$ for all z in S_L and S_R. Suppose $z \in S_L$. By the induction hypothesis, the tree does give the correct distance for x, y, and z, so $d(x, z) < d(x, y)$. Therefore, $d(x, z) < d(x, y) < d(s_{m+1}, x)$, the second inequality following from (14.3). Therefore, by the ultrametric property, $d(s_{m+1}, x) = d(s_{m+1}, z)$ as desired. The corresponding argument when z is in S_R is symmetric to this case.

In case (2), $d(s_{m+1}, y) = d(x, y)$. The ultrametric property then shows that $d(s_{m+1}, x) < d(x, y)$. Since x is in S_L and y is in S_R,

$$d(x, y) = d(x, z) \text{ for any } z \text{ in } S_R. \tag{14.4}$$

Therefore, $d(s_{m+1}, x) < d(x, z)$ for all z in S_R, and

$$d(s_{m+1}, z) = d(x, z) \text{ for all } z \text{ in } S_R. \tag{14.5}$$

Thus s_{m+1} is equidistant from every member of S_R. There are two possibilities: Either s_{m+1} is also equidistant from every member of S_L or it is not. If s_{m+1} is equidistant from every member of S_L, we add s_{m+1} to the tree, as shown:

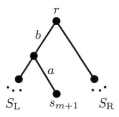

Here $a = d(s_{m+1}, x)/2$ and $b = d(s_{m+1}, y)/2 - a$. Since $d(s_{m+1}, x) < d(s_{m+1}, y)$ it follows that $b > 0$, as it must be.

We next check that this allocation gives the correct distances. For z in S_L we must show that $d(s_{m+1}, z) = 2a$, that is, that $d(s_{m+1}, z) = d(s_{m+1}, x)$. This requirement follows from the assumption that s_{m+1} is equidistant from every member of S_L. Now suppose that z is in S_R. Then the distance between s_{m+1} and z in the tree above is $a + b + \frac{1}{2}d(x, y)$. This simplifies to $\frac{1}{2}(d(s_{m+1}, y) + d(x, y))$. By (14.4) and (14.5), this is equal to $d(s_{m+1}, z)$, as required.

This leaves the case where s_{m+1} is equidistant from every member of S_R but *is not* equidistant from every member of S_L. If we remove the root and its two edges from the tree (14.2), we are left with two trees, one relating the species in S_L and one relating the species in S_R. We call these T_L and T_R, respectively. The tree T_L contains $m-1$ or fewer leaves, so we know by induction that s_{m+1} can be uniquely added to this tree and all the distances between s_{m+1} and the rest of the leaves in T_L will be correct. Call this new tree T'. We know further that since s_{m+1} is not equidistant to every leaf in T_L, T' has the same root as T_L. Therefore, we can put back together the whole tree relating all of $s_1, s_2, \ldots, s_{m+1}$ as

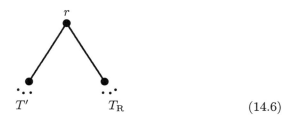

$$(14.6)$$

We now show that s_{m+1} has the correct distances to the rest of the species $s_1, \ldots, s_m$. Since we know that T' has been correctly reconstructed, this is true for all species in S_L. Suppose z is in S_R. Now x is in S_L, so the tree constructs the distance from s_{m+1} to z to be $d(x, z)$. We must therefore show that $d(x, z) = d(s_{m+1}, z)$. We know that $d(x, s_{m+1}) < d(x, y)$ (this follows from assumption 2), and we know that $d(x, y) = d(x, z)$. Therefore, $d(x, s_{m+1}) < d(x, z)$. The ultrametric property therefore gives us $d(x, z) = d(s_{m+1}, z)$, as desired.

The proof for case (3) is identical in form to this proof.

$\square$

This proof is constructive in that it provides an algorithm for reconstructing the tree. One adds the species to the tree one by one in no particular order, with the inductive step showing explicitly where each next species should be added.

We now describe the UPGMA algorithm of Sokal and Michener (1958), which is used frequently for tree reconstruction. In the case of ultrametric distances, this will reconstruct the same tree as that derived by the methods described above. This algorithm can be used for any set of distances, ultrametric or not; however, if the distance is not ultrametric, it cannot not reconstruct a tree that gives back the distance. It is often used anyway, as an approximation.

Two species are *neighbors* if the path between them contains only one node. Thus in the tree below, x and y are neighbors while x and z are not.

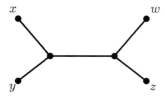

For a tree whose derived distance satisfies the ultrametric property, pairs of species that are closest together with respect to the distance metric are neighbors.

The UPGMA algorithm proceeds as follows. Let $d(i,j)$ be the distance between species i and j. This distance is assumed to be derived from data by using some method, perhaps a surrogate distance method discussed later in Section 14.5. Suppose that the two species with minimal distance between them are species x and y. These are now joined with a two-leaf rooted tree, with root "species" r_1, as shown in the diagram.

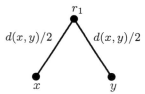

The distance from r_1 to any remaining species z is defined by

$$d(r_1, z) = \left(d(x, z) + d(y, z) - d(x, y)\right)/2. \qquad (14.7)$$

Species x and y are now replaced by the single species r_1. The next step is to find a minimal-distance pair of species in the new set of species consisting of all original species other than species x and y, together with a new species r_1. If neither of the species in this minimal pair is r_1, a new two-leaf rooted tree is constructed joining this minimal pair in the same manner as that joining species x and y, and consequently two more species will be replaced by one new species r_2. If the species r_1 is one of the minimal distance pair, a new root r_2 to the first tree is constructed, and joined to the new species r_1, as in the diagram below.

This process is repeated, building the tree in a bottom-up way until it is entirely reconstructed. If the distance is ultrametric, this procedure will build the (unique) correct tree. However, the proof of this claim is more straightforward when the first method described above is used rather than the UPGMA method.

If the distance is not ultrametric, an incorrect tree can be drawn by this process, as shown in the diagram below. In this tree, x and y are the two closest leaves in the tree, but they are not neighbors.

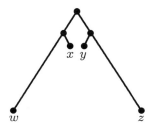

This problem leads to the tree construction method described in the following section.

14.4   Tree Reconstruction: Neighbor-Joining

In this section we show that given any tree-derived distance measure on a set of species, a tree can be constructed giving this distance measure. This method was introduced by Saitou and Nei (1987).

The proof of this claim uses the quantity $\delta(x, y)$, defined by

$$\delta(x, y) = (N - 4)d(x, y) - \sum_{z \neq x, y} (d(x, z) + d(y, z)), \qquad (14.8)$$

where N is the total number of species. The function $\delta(x, y)$ is not a distance measure, because it may take negative values. However, the following is true.

Theorem. Suppose S is a set of species and d is a tree-derived distance on S obtained from an unrooted tree. If x and y are such that $\delta(x, y)$ is minimum, then x and y are neighbors.

Proof. We follow the proof of Studier and Keppler (1988). Consider four species as in the following tree:

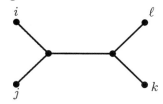

We have
$$\delta(i,j) = -d(i,k) - d(i,\ell) - d(j,k) - d(j,\ell)$$
and
$$\delta(i,k) = -d(i,j) - d(i,\ell) - d(k,j) - d(k,\ell),$$
so that
$$\delta(i,k) - \delta(i,j) = -d(i,j) - d(k,\ell) + d(i,k) + d(j,\ell).$$

The above diagram shows that the positive terms on the right-hand side in this equation include the length of the internal edge twice, whereas the negative terms do not include it. Therefore, $\delta(i,k) - \delta(i,j) > 0$, that is, $\delta(i,k) > \delta(i,j)$. Every leaf in an (unrooted) four-leaf tree has a neighbor. Therefore, any value of δ arising for a pair of nonneighbors x and y is greater than that for neighbors x and z.

Suppose now that $N > 4$. In this case not all species have neighbors in the sense defined above. Suppose i and j are such that $\delta(i,j)$ is minimum, but i and j are not neighbors. If i has a neighbor k, then

$$\delta(i,k) - \delta(i,j)$$
$$= (N-4)d(i,k) - (N-4)d(i,j)$$
$$\quad - \sum_{z \neq i,k} d(i,z) - \sum_{z \neq i,k} d(k,z) + \sum_{z \neq i,j} d(i,z) + \sum_{z \neq i,j} d(j,z)$$
$$= (N-4)d(i,k) - (N-4)d(i,j)$$
$$\quad - \left(d(i,j) + \sum_{z \neq i,j,k} d(i,z) \right) - \left(d(k,j) + \sum_{z \neq i,j,k} d(k,z) \right)$$
$$\quad + \left(d(i,k) + \sum_{z \neq i,j,k} d(i,z) \right) + \left(d(j,k) + \sum_{z \neq i,j,k} d(j,z) \right)$$
$$= (N-3)d(i,k) - (N-3)d(i,j) - \sum_{z \neq i,j,k} d(k,z) + \sum_{z \neq i,j,k} d(j,z)$$
$$= \sum_{z \neq i,j,k} \left(d(i,k) + d(j,z) - d(i,j) - d(k,z) \right). \tag{14.9}$$

The following tree shows that each quantity in the summand is negative.

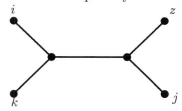

Therefore, $\delta(i,k) < \delta(i,j)$, contradicting the minimality assumption on $\delta(i,j)$. It follows that neither i nor j can have a neighbor if $\delta(i,j)$ is minimal, and the tree must be as shown:

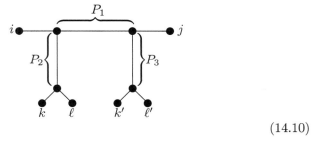

$$(14.10)$$

In this tree P_1, P_2, and P_3 are paths, not edges, but all other line segments represent edges. The path to any leaf other than those shown joins this diagram along one of the paths P_1, P_2, or P_3. The key observation is that we can assume the existence of both k and ℓ down path P_2, because if there were only one leaf down path P_2, it would have to be a neighbor of i. A similar observation holds for k' and ℓ', because neither is a neighbor of j.

We focus for the moment on $i, j, k,$ and ℓ.

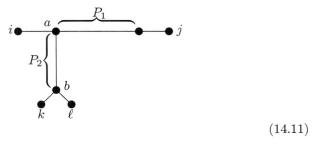

$$(14.11)$$

A calculation similar to the one that led to (14.9) gives

$$\delta(k,\ell) - \delta(i,j)$$
$$= \sum_{z \neq i,j,k,\ell} \Big([d(i,z) + d(j,z) - d(i,j)] - [d(k,z) + d(\ell,z) - d(k,\ell)] \Big).$$

$$(14.12)$$

The proof of this claim is left as an exercise (Problem 14.3).

By our assumption that $\delta(i,j)$ is minimal, the left-hand side of (14.12) is nonnegative, so that the sum on the right-hand side is nonnegative. Any leaf z different from i, j, k, and ℓ joins the diagram (14.11) at path P_1 or at path P_2. Let $p(z)$ be the point where the path from z meets this diagram above. This leads to the two possibilities shown below.

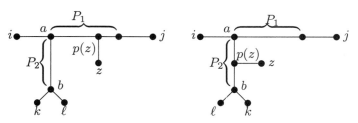

We calculate the summand in (14.12) in both cases. In the left-hand case this summand is equal to $-2d(a,b) - 2d(a,p(z))$, and in the right-hand case it is equal to $2d(a,p(z)) - 2d(b,p(z))$. Given one leaf of each type, say leaf z_1 of the type in the left-hand diagram and leaf z_2 of the type in the right-hand diagram, addition of the two summands gives

$$-2d(a,b) - 2d(a,p(z_1)) + 2d(a,p(z_2)) - 2d(b,p(z_2)),$$

and since $-d(a,b) + d(a,p(z_2)) < 0$, this is negative. It follows that there must be more leaves of type 2 than there are of type 1. Tree (14.10) then shows that there are strictly more leaves joining P_2 than P_3.

However, exactly the same argument can be applied to show that more leaves must join P_3 than P_2. This contradiction shows that the assumption that i and j are not neighbors is impossible, so that the theorem is proved.□

The above proof shows how a tree can be reconstructed when the distances $d(i,j)$ are tree-derived. Let x and y be such that $\delta(x,y)$ is the minimum of all $\delta(i,j)$ values, where $\delta(i,j)$ is defined in (14.8). We know from the above that x and y are neighbors. Let r_1 be the node on the path that joins them. Since $d(i,j)$ is assumed to be a tree-derived metric, the distance $d(r_1,z)$ for any leaf $z \neq x, y$ is calculated as defined in (14.7). And using any such leaf, $d(x,r_1)$ and $d(j,a)$ are determined by

$$d(x,r_1) = \frac{d(x,z) - d(y,z) + d(x,y)}{2},$$

and

$$d(y,r_1) = \frac{d(y,z) - d(x,z) + d(x,y)}{2}.$$

We next replace x and y in our original set of species with r_1, and repeat the process. Eventually the entire tree is reconstructed.

The "neighbor-joining" algorithm of tree reconstruction for distances that are possibly not tree-derived is based on the above procedure. For each distance $d(i,j)$, we calculate the quantity $\delta(i,j)$, and the species x, y for which this quantity is minimized are first paired. Distances from the node r_1 joining species x and y are then calculated using (14.7). Species x and y are then replaced by a new species r_1 and the procedure repeated starting from a new set of values $\delta(i,j)$ including this node. Eventually, a tree is reconstructed. If the distances between species are not tree-derived, and are given for example as surrogate distances as discussed in Section 14.5, there is no guarantee that the tree developed is the correct tree.

14.5 Surrogate Distances

Unfortunately, exact distances between species are seldom if ever known, and surrogate distances must be used instead. We now discuss some properties of these distances.

If $d(x,y)$ is the exact distance between extant species x and y and $d'(x,y)$ is a surrogate distance between x and y, and if there is a constant C such that $d(x,y) = Cd'(x,y)$ for all x and y, then since d is tree-derived, d' will be also. The relationship between trees and tree-derived distance measures discussed in Section 14.4 shows that a tree reconstructed from d' has the same topology, that is, the same branching structure and same labeling of the leaves, as that found using d.

We now discuss surrogate distances used in practice using DNA information derived from each species considered. The molecular clock need not run at the same speed along different lineages. This can happen for several reasons, including differing generation lengths in different species, different degrees of importance of the biological sequence to the particular species, and so on. Thus a distance measure reflecting the molecular clock might lead to the following tree joining mouse, elephant, and elephant shrew:

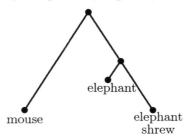

In this example the molecular clock ran faster in the mouse and shrew lineages than in the elephant lineage, so that proportions were not kept with the actual exact distances. If a surrogate distance measure perfectly captures this difference in rates, it will be tree-like, and the neighbor-joining algorithm can be used to reconstruct the tree that gave these surrogate distances. This tree will have the same branching structure as the true tree, often its most interesting feature.

In practice, such reconstruction is seldom possible. A surrogate distance between two extant species is sometimes found by using aligned DNA sequences taken from each, and using a surrogate distance measure proportional to

$$-\log\left(1 - \frac{4}{3}p\right), \tag{14.13}$$

where p is the proportion of nucleotides where the two sequences differ. This estimate derives from equations such as (13.32), and thus implicitly assumes a Jukes–Cantor model of evolution. It is thus subject to the comments made below equation (13.46), so that if some more complicated model is appropriate, these estimators might have undesirable properties.

A more complicated surrogate distance may be derived from the Kimura two-parameter model of Section 13.3.2, using the analogue of (14.13) for that model. This is done in the example of Section 14.8. If, as is almost certain to be the case in practice, a model more complex than the Kimura two-parameter model is appropriate, then surrogate distances using this model might also have undesirable properties.

There are further problems concerning surrogate distance estimates. First, proteins and protein-coding DNA sequences have subregions that are highly conserved and other regions that change more readily. After a long time the latter regions will diverge significantly between sequences, while the conserved regions remain more or less the same. Second, the rates at which these sequences evolve also depend on the generation times of the species involved. Finally, measures such as (14.13) are only as good as the alignment on which they are based. "Nonfunctional" sequences, when distantly related, are difficult to align correctly.

As a result of these problems, surrogate distances calculated from genetic data of extant species will seldom if ever give a distance proportional to the true distance, and need not even satisfy the three fundamental properties of distance measures given in Section 14.2.

Despite these problems it is still possible to apply the UPGMA and neighbor-joining algorithms. Because surrogate numbers do not necessarily satisfy the three basic properties of distances, negative distances can arise in the inferred trees. Further, the UPGMA and neighbor-joining trees need not agree, as is shown in the example of Section 14.8.

14.6   Tree Reconstruction: Parsimony

Under the parsimony approach a total "cost" is assigned to each tree, and the optimal tree is defined as that with the smallest total cost. The focus

is on finding an optimal *topology*, or shape, and not on edge lengths. The
parsimony approach to tree construction is based on the assumption that
the tree (or trees) with least cost should be used to estimate the phylogeny
of the species involved.

We describe the cost calculation for the case of DNA sequences when
unit cost is made for each nucleotide substitution. A central step in the
procedure is to allocate sequences to the internal nodes in the tree. For any
such set of sequence allocations the total cost of the tree is the sum of the
costs of the various edges, where the cost of an edge joining two nodes, or
a node and a leaf, is the number of substitutions needed to move from the
sequence at one to the sequence at the other.

In principle the optimal tree is found in two steps. First, all possible
tree topologies must be listed, including the allocation of the species from
which the data are found to the leaves. Second, for any such choice the
labeling of the internal nodes of the tree that minimizes the cost of the
tree must be found. This second step is accomplished by *Fitch's algorithm*
(Fitch (1971)). This algorithm provides the optimal set of internal nodes
for a fixed topology. Once these are found the minimum "cost" of a tree
with a given fixed topology can be found. The final tree is that which, over
all choices of topology, has overall minimum cost.

In practice, the most difficult problem associated with this procedure
concerns the listing of all possible topologies of the tree. For trees joining
a small number of species the optimal tree can be found by complete enumeration of all possible tree topologies. For three species there is only one
form for the topology, namely that shown in (14.1), with three choices for
the labeling of the species x, y and z. Figure (6.3) on page 207 shows the
five optimal allocations of internal nodes for the three aligned sequences
AA, AB, and BB, found by complete enumeration. With four species there
are two different forms of topology for rooted trees, as shown. There are
three essentially different labelings for the left-hand tree (three choices for
the species chosen as the neighbor of the left-most species) and twelve for
the right-hand tree (four choices for the rightmost species, three remaining
choices for the next to rightmost species), giving fifteen essentially different
labeled trees in all. Once again complete enumeration is straightforward.

Although complete enumeration methods are possible when the number of species is small, the number of possible trees increases extremely
rapidly as the number of species increases, and complete enumeration is
not possible with more than a small number of species. When there are s

species at the leaves of a rooted tree there are $1\times 3\times 5\times \cdots \times(2s-3)$ $= (2s-3)!/2^{s-2}(s-2)!$ essentially different tree topologies. For 20 species this is about 8×10^{21} topologies. Despite this, some researchers use parsimony methods to reconstruct trees with as many as 500 or more species, and this requires the use of heuristic methods (see Durbin et al. (1998)).

The parsimony method does not construct arm lengths. Nor does it use any explicit evolutionary model. Proponents of the method see the latter as an advantage: to the extent that any evolutionary model used in tree reconstruction might be misleading, a method not dependent on any model has some merit. Others see the lack of an evolutionary model, and the lack of constructed arm lengths, as disadvantages to the method.

14.7   Tree Reconstruction: Maximum Likelihood

In this section we discuss the maximum likelihood estimation of the phylogenetic tree leading to a number of contemporary species from some common ancestor species, given genetic data from the contemporary species and some continuous-time evolutionary model such as one of those discussed in Section 13.3. Maximum likelihood phylogenetic tree estimation was introduced by Felsenstein (1981).

Suppose first that the topology of the tree is given and that the aim is to find the maximum likelihood estimate of the various arm lengths in the tree, given some continuous-time evolutionary model. This is done by writing down the likelihood of the data in terms of these lengths as parameters, and then maximizing this likelihood with respect to these lengths. This maximization process is, in practice, usually extremely difficult, even when many simplifying assumptions are made. One such simplifying assumption often made is the iid assumption, that the substitution processes at different sites within any species are described by the same stochastic model and that these processes are independent from one site to another. The assumption of independent evolution at different sites allows an analysis of the evolution at the various sites separately, with an overall likelihood obtained by multiplying individual site likelihoods, or in practice an overall log likelihood obtained by summing log likelihoods. It is also frequently assumed that time homogeneity of the common stochastic process applies, and that different species evolve independently. These assumptions can hardly be expected to approximate reality closely, and much recent research attempts to remove them or at least to assess the biases involved when they are incorrectly made. We refer to some of these analyses in Section 14.9.5. On the other hand, in view of the comments about modeling in Section 4.9, in particular the fact that one can expect a simple initial model to be refined by further work, these assumptions form a reasonable starting point for the stochastic approach to the estimation of phylogenetic trees.

We now discuss an analysis in which the iid assumption is made. Initially, no specific assumption is made below about the evolutionary model chosen. As a specific example, suppose that we have data from five species $s_1, \ldots, s_5$

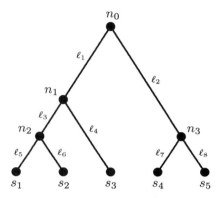

FIGURE 14.1.

and that the topology of the phylogenetic tree connecting them is as given in Figure 14.1. Then at any particular site the five respective nucleotides in these five species are known, and these comprise all the data available for this site. Denote these five nucleotides by A, B, C, D, E. We do not know the nucleotides at the internal nodes n_0, n_1, n_2, and n_3, nor do we know the lengths $\ell_1, \ldots, \ell_8$ of the various arms in the tree. The aim of the process is to estimate $\ell_1, \ldots, \ell_8$, which are lengths of time.

Suppose that the nucleotide at the root n_0 is W and is given, and that the nucleotides at the nodes n_1, n_2, and n_3 are X, Y, and Z, respectively. Then there is a certain likelihood for the arm lengths $\ell_1, \ldots, \ell_8$, deriving from whichever nucleotide evolutionary model is chosen. The stationary probability of the nucleotide W is denoted by φ_W and the probability, under the model assumed, of a substitution of nucleotide A by nucleotide B after time ℓ is denoted by $P_{AB}(\ell)$. Then the joint probability that W is indeed the nucleotide at the root of the tree, that the nucleotides X, Y and Z occur at the nodes as indicated, and that the arm lengths take the values $\ell_1, \ldots, \ell_8$ is

$$\varphi_W P_{WX}(\ell_1) P_{WZ}(\ell_2) P_{XY}(\ell_3) P_{XC}(\ell_4) P_{YA}(\ell_5) P_{YB}(\ell_6) P_{ZD}(\ell_7) P_{ZE}(\ell_8). \tag{14.14}$$

Expressions of this form are now computed for all 64 possible combinations of nucleotides at the internal nodes. The sums of the resulting expressions give the likelihood of $\ell_1, \ldots, \ell_8$ conditional on the assumption that the nucleotide at the root of the tree is W. This calculation is then made for all four possible nucleotides at the root of the tree, and the sum of these four expressions is, for the site in question, the likelihood of $\ell_1, \ldots, \ell_8$.

This procedure is now repeated over all the nucleotide sites in the data, and the overall likelihood is computed from the product of the various site likelihoods. This overall likelihood can now, at least in principle, be maximized.

This procedure can now be done, again in principle, for all possible topologies, and then the maximum likelihood tree is that which, taken over all possible topologies, has the maximum likelihood.

These comments gloss over difficult computational problems. There are, however, some computational simplifications possible. In the first part of the process, the various summations implied by the above operations can be reorganized in a way that saves considerable computational effort, as pointed out by Felsenstein (1981). A second computational savings was also noted by Felsenstein. If the stochastic process of nucleotide substitution assumed is reversible, it is impossible to estimate the location of the root of the tree. This "pulley principle" implies that in the reversible case the estimation procedure relates to unrooted trees. Since with s observed species there are $(2s-3)!/2^{s-2}(s-2)!$ rooted trees, as noted in Section 14.6, and the smaller number $(2s-5)!/2^{s-3}(s-3)!$ of unrooted trees (Edwards and Cavalli-Sforza (1964), Felsenstein (1978a)), one might think that the use of unrooted trees would imply some useful economy in computation. However, even for comparatively small s the number of unrooted trees is still prohibitively large for a complete search, and thus heuristic methods are necessary for the second part of the process.

Felsenstein (1981) suggests as one heuristic method the strategy of building up an unrooted tree starting with two species and successively adding species. If at any stage there are $k - 1$ species in the tree, there are $2k - 5$ segments to which species k can be attached. Each of these is tried in turn, the maximum of the likelihood for each possible attachment is calculated, and the topology with the highest likelihood accepted. Before the next species is tried, local rearrangements of the current tree can be carried out to see whether any rearrangements increase the likelihood. If so, the rearrangement maximizing the likelihood is accepted.

This process does not necessarily lead to the tree with the maximum likelihood, and the tree constructed by this process will depend on the initial two species chosen and the order in which species are added. Nevertheless, with reasonably self-consistent data the same tree will tend to arise whatever order of species is chosen (Felsenstein (1981)). Different orderings can be tried, and if more than one tree is obtained from the different orderings, that with the highest likelihood can be chosen. Heuristic methods such as this overcome some of the computationally difficult problems of tree estimation.

In some cases the maximum-likelihood tree can be shown to be identical to the tree found by parsimony methods. This is interesting because the first attempt at phylogenetic tree reconstruction by maximum-likelihood methods (Edwards and Cavalli-Sforza (1964)) proved to be beyond the computing power of the time, and Edwards and Cavalli-Sforza resorted to an algorithmic parsimony approach to the problem. The links between maximum likelihood estimates and parsimony estimates are examined in detail by Tuffley and Steel (1997). Identity of the two will occur only under the simplest, and probably unrealistic, evolutionary models. Further aspects of the relation between the maximum likelihood and parsimony tree reconstruction methods are given by Lewis (1998).

When the evolutionary process possesses simplifying properties, for example iid substitution processes at the various sites, time homogeneous substitution processes, and identical processes along the various arms of the phylogenetic tree, one would expect that all methods discussed above would lead to similar trees, all of which should be close to the true tree. This would be especially so if the stochastic substitution process assumed in the data analysis were the same as that generating the actual substitutions. Kuhner and Felsenstein (1994) generate data at the tips of various trees under all the simplifying assumptions listed above. The substitution process at any site used in deriving the simulated data was chosen as the Kimura two-parameter model discussed in Section 13.3.2, and this was in effect the model assumed in the data analysis. Further, the transition/transversion ratio assumed in the analysis was set equal to that in the simulation. A further feature of the analysis was that the mean number of substitutions at any one site in the entire phylogenetic tree was about 0.02 in the "low" substitution and about 0.2 in the "high" substitution model. Thus even in the high substitution model, the probability of two or more substitutions at a given site in the entire tree is, from the Poisson approximation (4.11), about $1 - e^{-.2}(1 + .2) = .0175$. This implies that the tree reconstruction process is simpler than those arising when many substitutions can be expected to occur, for which there will be an increased level of scrambling of data at the tips of the trees.

The simulation results of Kuhner and Felsenstein (1994) confirm that under these many simplifying assumptions, all standard tree estimation methods lead to similar results and to trees either identical to, or very close to, the true tree, with accurate estimates of arm lengths and other features. Kuhner and Felsenstein also carry out simulations when substitution rates are allowed to vary between branches of the tree and when substitution rates are allowed to vary between different sites. In this case inaccuracies and biases arise in the various estimation procedures considered. This indicates the danger of tree reconstruction methods using real data when the simplifying iid and constancy of rates assumptions are made, since in practice these assumptions are unlikely to hold. This is illustrated by the example of Section 14.8.

For real data sets a different picture can emerge. Yang (1994), (1996a), (1996b), (1997a), (1997b) analyzed a variety of data sets from various primate species. Although it is not easy to summarize the results of these analyses briefly, some broad conclusions do emerge. First, simple models such as the Jukes–Cantor, Kimura and Felsenstein models can result in severe estimation errors of branch lengths. The HKY model must perform better than these models since they are all special cases of this model, and the added complexity of the HKY model does appear to add significant flexibility to the modeling process. However the HKY model itself sometimes performs significantly worse than the general reversible process model (see Section 13.2.7), of which it is itself a particular case. Thus increasing the complexity of the model seems often to be necessary and to be worth the loss of degrees of freedom involved. On the other hand the completely general evolutionary model (13.21) does not seem usually to perform significantly better than the general reversible process model. Thus given the desirability of the reversibility criterion, the latter model appears often to provide a reasonable compromise between simplicity and analyzability on the one hand and complexity and reality on the other.

Further aspects of Yang's analysis refer to the question or substitution rate heterogeneity, discussed in Sections 14.9.4 and 14.9.5.

14.8 Example

The various tree construction methods outlined above use different approaches to construction and different optimality criteria. Those methods based on distances suffer from the fact that the surrogate distances normally used need not be proportional to the true distances, and in fact may not even be tree-like. It is therefore not surprising that the four methods ultrametric reconstruction, neighbor joining, parsimony, and maximum likelihood will often produce different trees even though using the same data. This is illustrated in the example in this section, which investigates the evolutionary relationships between 14 species of mammals. We will see that even though all four methods disagree to some extent, there are divisions that they all do agree on. This leads to the concept of a *consensus tree*, which tries to capture this information. We will not give the details of consensus trees, but we will illustrate the basic idea in this example. The DNA used in the tree construction is from the interphotoreceptor retinoid binding protein (Stanhope et al. (1996)). Sequences for the 14 species were taken from Genbank, aligned using CLUSTAL W (Thompson et al. (1994)), and a 532 nucleotide ungapped subalignment extracted by eye and used as input to the various tree algorithms. The species are

Marsupial Mole	Whale	Insectivore
Wombat	Dolphin	Human
Rodent	Pig	Sea Cow
Elephant Shrew	Horse	Hyrax
Elephant	Bat	

and the alignment is

```
gctccagcaaatgatcaagtaccaggtattggagggcaatgtgggttacctaagagtggactacatccctggccag
gctccagcaaatgatcaagtaccaggtactggagggtaatgtgggttacctgagagtggactacatccctggccag
gctacagaggaatattcaccatgaggttctggagggcaacttgggttacctatgggtggacgatctcttgggccag
gctggagagaagcatgagctacaggattctggatggtaatgtgggctacttgcagatagacaacatcccaggccag
gctgcagacaagcatgagctacaaggttctggagggcaacgtgggctacctgcgggtagacaacatcccaggccag
gctgcagaacggcctccgccatgaggttctggaaggcaatgtgggctacctgcgggtggacgacatcccaggccag
gctgcagaacggcttccgccatgaggttctggaaggcaatgtgggctacctgcgggtggacgacatcccaggccag
gctgcacaatagtctccgccatgaggttctggaaggcaatgtgggctacctgcgggtggacgacatcccaggccag
gctgcaggagggcatccgctatgacattctggagggcgacgtgggctacttgcgagtggacaacatcccgggccag
gctgcaaaaaggccatccactacaatgttctggagggcaacgtgggctactttcgggtggacgacatcccgagccag
gctgcagagggccatccgctaccaggttctggcggccaatgtgggctacctggggagggataacctccccggtcag
gctgcaaagggggcctccgccatgaggttctggagggtaatgtgggctacctgcgggtggacagcgtcccgggccag
gctgcagaccagcatgagctacaaggttctggatggcaatgtgggctacctgcgggtagacaacatccctggccag
actgcagacaagcatgagctacaaggttctggagggcaacgtgggttacctgcgggtagacaacattccgggtcaa
```

```
gaggtagtagaaaaagtcggggagttcctggtgaatgacatctggaagaagctcatggggacatcctctctagtgc
gaggtggtagagaaaagtcggggagttcctggtgaatgatgtctggaagaagctcatggggacctcttctctggtgt
gaggtactgagtaagctcgggggattcctggtggcccacatgtggggggcagctcatgaatacctctggcttggtgc
gaggtactgagccgactaggggccttcctggtggcccatgtctggagacagctcatgggcacctctgctttggtgt
gaggtgctgaaccagctgggggccttcctggtgactcacgtctggaagcagcttatgggctcctctgccttagtgc
gaggtgatgagcaagctgaggagcttcctggtggccaacgtctggaggaagctcatgggcacctctgccttggtgc
gaggtgatgagcaagctgaggagcttcctggcggccaacgtctggaggaagctcatgggcacctctgccttggtgc
gaggtgatgaacaagctggggagcttcctggtagtcaacgtctgggaaaagctaatgggcacctctgccttggtgc
gaggtggtgagcaagctggggggcttcctggtggacaatgtctggaggaagctcatgggcacctctgccttggtgc
gaggtggtgagcaatcttgggggcttcctcgtggacaatttctggaggaagctcctgggcacctctgccttggtgc
gaggtggtgaccatactgggggctctcctggtggccaatgtctggggggaagctcatagccacctctcccttggtgc
gaggtgctgagcatgatgggggagttcctggtggcccacgtgtggggggaatctcatgggcacctccgccttagtgc
gaggtgctgagccgtctgggggggcttcctggtgactcacatctggaagcagctcatgggctcctctgccttagtcc
gatgtgctgaaccagctgggggggcttcctggtgactcatgtgtggaagcagctcatgggctcctctgccttagtgc
```

```
tagatctccagcacagcacaggggggtgaagtttcgggaatcccctttgtcatttcctatctacatcaggggggatat
tggatctccagcacagcacgggaggcgaagtttcaggaatcccgtttgtcatttcctacctacaccaggggggataa
tagatctccggcactgtactggggggggcatgtttctggtattccctatgtcatctcctacttgcaccccgggaacac
tggacctgcgcgcagtgcacaggaggccatgtttccagcatcccttaccttatttcctacctgcacccagcgggcac
tggacctgcgcgacactgcacaggggggccatgtctccagcatcccttacctcatttcctacctgcacccgggcggcac
tggacctccgccattgcactgggggccacatttctggcatcccctatgtcatctcctacctgcacccggggaacac
tggacctccgccactgcactggcggccacatttccggcatcccctatgtcatctcctacctgcacccaggggaacac
tagacctccggcactgcaccaggggccacgtttctggcatcccctatgtcatctcctacctgcacccaggggaacac
tggacctccggcactgcactggggggccacgtttccggcatcccctatatcatctcctacctgcacccaggaaacac
tagacctcccacactgcactggggggcacgtttctgggatctcctatgtcatctcctacttgcaccgagggaacac
tggacctccgacactgcactggggggccatgtctctgggatccctacgtcatctcctacctgtacccaggaaacac
tggatctccggcactgcacaggaggccaggtctctggcattccctacatcatctcctacctgcacccagggaacac
tggacctgcggcactgtatgggtggccatgtctccagcatcccttacatcatctcctacctacaccccggaggagc
```

tggacctaaggcactgcacgggggccatgtctccagtatcccttacctcatctcctacctgcatccagggagcac

cctgctccatgtagacacagtttatgaccggccatcaaacactaccacagagatctggacccagcctcaggtgctg
tctgctgcatgtagacacagtttatgaccggccatcaaacaccaccacagagatctggaccctgccccaggtgttg
aatcatgcatgtgaacaccatctatgatcggccctctaataccaccacagagatctggaccttggccaaggtcctg
ggtcctgcacgttgacaccatttacaaccgtccctctaacacaaccactgagctctggactttgcctcaggtgctt
cgtgctgcacgtggacaccatttacaaccgcccctccaatacgactacggagctctggaccttgccccaggtgctg
agtcctgcacgtggataccatctatgatcgcccctctaatacgaccactgagatctggacctgcccgaagtcta
agtcctgcatgtggataccatctacgatcgcccctctaatacgaccactgagatctggaccctccccgaagtccta
ggtcctgcacgtggacaccatctatgaccgtccctccaatacgaccactgagatctggaccctgcccgaagtcctg
ggtcctgcacgtggacaccatctacgaccgcccctccaatacgaccactgagatctggaccctgcccgaggtcctg
cgtcctgaatgtggacccactctatgaccccccctccaacacgaccacagagatctggaccctgccccaggtcctg
ggtcctgcatatggacaccatctatgaccgcccctccaatatcaccactgagctctggaccctgccccagctccag
catcctgcacgtggacactatctacaaccgcccctccaacaccaccacggagatctggaccttgccccaggtcctg
agtgctgcatgtggacaccatttacaaccgcccctccaatacgactactggggtctggaccttgccccaagtgctg
tgtgctgcacgtggacaccatttacaaccgcccctccaatacaactactgagctctggaccttgccccaggtgctg

ggtgagaggtatggagggagaaggacatggtggttctcaccagccatcatactgtaggggtagctgaggatatcg
ggtgagaggtacggtggggagaaggacgtggtggtcctcaccagccatcacacggtcggggtagcagaggatattg
ggggagaggtacagtgctgacaaggatgtggtggtcctcaccagtggccacactggaggagtgggtgaggacattg
ggggagagatacagtgctgagaaggatgtggtggtcctcaccagtggtcaaacccggggtgtggctgaggacattg
ggggagaggtatagcgccgacaaggatgtggtggtcctcaccagtggccacaccaggggcgtggccgaggacatcg
ggagagaactacggtgccgataaggatgtggtggtcctcaccagtggtcgcaccgggggtgtggctgaggacatcg
ggagacaactacggtgccgataaggatgtggtggtcctcaccagtggtcgcacggggggtgtggctgaggacatct
ggagacaggtacagtgcggataaggacgtggtggtcctcaccagcagccacacaggggcgtggctgaggacatcg
ggagagaggtacagtgccgacagggatgtggtggtcctcaccagtggccacaccggggcgtggccgaggacattg
ggagagaggtacagtgctgacaaggatgttgtggtcctcaccagtggccacactggaggagtggctgaggacattg
ggagagcggtacggtgcagacaaggatgtggtggtcctcatcagcgaccacactgggggtgtggctgaggacatta
ggagaaaggtacggtgccgacaaggatgtggtggtcctcaccagcagccagaccaggggcgtggccgaggacatcg
ggagaaaggtacagtcccaacaaggatgtggtggtcctcaccagtggccacaccaggggcgtggccgaagacatcg
ggggagagatacagtgctgacaaggatgtggtggtcctcaccatgggccacaccaggggtgtggccgaggacatcg

cctatattctcaagaagatgcgccgggccattgtggtggggagagcagactctgggaggggccctagatctccggaa
cctacatcctcaagaagatgcgccgggccattgtggtggggagagcagactctgggaggggccctagatctccggaa
cctatatcctcaaacagatgcgcagggccatcatggtgggtgagcagactgaaggtggtgccctggacctccagaa
tctacatcctcaagcagatgggcagggccatagtggtgggtgaacgtactggggggggtctccctggacctccagaa
tctacatcctcaagcagatgggcagggccatcgtggtgggcgagcggactgagggtggtgccctggacctccagaa
cttatatcctcaaacaaatgcgcagggccattgtggtgggcgagcggactgtggggggggggccttggacctccagaa
cttatatcctcaaacagatggacagggccatcgtggtggacgaacggactgtggggggggggccttggacctccagaa
cctacatcctcaaacagatgcgcagggccattgtggtcggcgagcgaactgtgggggggtgccctggacctccagaa
cttacatcctcaaacagatgcgcaggaccatcgtggtgggtgagcggaccgtgggaggtgccctggacctccagaa
cttacatcctcaaacagatgcgcagggccattgtggtgggtgagcagactgtgggggggtgccctggacctccagaa
cttacatcctcaaacagatgcgccgggctattgtggtgggcgagcagactgtgggggctgctctggacctccagaa
cgcacatccttaagcagatgcgcagggccatcgtggtgggcgagcggactggggagaggggccctggacctccggaa
ttcacatccttaagcagatgggcagggccatagtggtgggcgagaagacggaggcaggtgccctgcacctccagaa
tctacatcctcaagcagatgggcagggccattgtggtaggcgagcggaccgagggtggtgccctggacctccagaa

gctgcgcatcggtcagtcagacttttttcatcactgtgcccgtgtcacgctccctgagcccccttggtggggggagt
gcttcgtattggtcagtcagacttttttcatcactgtgcccgtgtcccgttctctgagcccccctcagtgggggggagc
actgaggataggccagtccaacttcttcctcacagtgcctctggcgatgtctctggggccgatgggtggaggtggc
gctaaggatagccaactctgacttcttcctcactctacctgtgtccaggtccttggggcctctgggtggaggcacc
gataggccactctgacttcttcctcactctgcctgtgtctaggtccttaggcccccctgggcgggggaagccagaca
gataggccagtctgacttcttttctcaccgtgcccgtgtccaggtccctggggcccctgggcaagggcagtcagact

gataggccagtctgagttctttctcacagtgcccgtgtccaggtccctggggcccctgggcaagggcagccagact
gataggccagtccgacttctttctcaccgtgcctgtgtccaggtccctggggcccctgggtgagggcagccagaca
gataggccagtccgacttcttcctcaccgtgcccgtgtccaggtccctgggtctgcgcgaggtcctcatgcataac
gataggccagtctgacttcttcctcactgtgcctgtgtctaggtccctgggggctctgggtgggggcaggcagaca
gataggccagtctgacttcttcatcactctgcctgtctccaggtctctggggactctgggcggggggcagccagaca
gataggcgagtctgacttcttcttcacggtgcccgtgtccaggtccctggggcccccttggtggaggcagccagacg
gataggtcactctgatttctttctcactctgcctgtgtccaggtccttggggcctttgggcaggggaagccagaca
aataggtcactcagacttcttttttcactctgcctgtgtccaggtcactgggcccccttaggcaggggaagccagaca

Tree reconstruction using the parsimony, maximum likelihood, UPGMA, and neighbor-joining methods was carried out using PHYLIP (Phylogeny Inference Package, Felsenstein (1980–2000)). We give below the output as it is given from the package. The distances used in the UPGMA and neighbor-joining methods are those given by the distance matrix in PHYLIP, and are based on the maximum likelihood estimates of the divergence times between any two species under the Kimura two-parameter model of Section 13.3.2. These distances are:

```
mars.mole   .00 .11 .41 .44 .39 .37 .40 .37 .41 .36 .40 .37 .42 .39
wombat      .11 .00 .39 .40 .36 .33 .35 .33 .35 .32 .33 .33 .38 .34
rodent      .41 .39 .00 .33 .30 .24 .25 .22 .28 .23 .25 .23 .32 .31
elph.shrew  .44 .40 .33 .00 .20 .26 .26 .25 .28 .28 .28 .26 .20 .21
elephant    .39 .36 .30 .20 .00 .22 .23 .22 .25 .25 .24 .21 .11 .12
whale       .37 .33 24  .26 .22 .00 .03 .10 .16 .17 .18 .17 .22 .24
dolphin     .40 .35 .25 .26 .23 .03 .00 .11 .16 .19 .18 .17 .22 .25
pig         .37 .33 .22 .25 .22 .10 .11 .00 .17 .18 .19 .17 .24 .24
horse       .41 .35 .28 .28 .25 .16 .16 .17 .00 .17 .21 .20 .25 .26
bat         .36 .30 .23 .28 .22 .17 .19 .18 .18 .00 .15 .20 .27 .27
insectivor  .40 .33 .25 .28 .26 .18 .18 .19 21  .15 .00 .19 .26 .26
human       .37 .33 .28 .26 .21 .17 .17 .17 .23 .20 .19 .00 .22 .23
sea cow     .42 38  .32 .20 .11 .22 .22 .24 .25 .27 .26 .22 .00 .14
hyrax       .39 .34 .31 .21 .12 .24 .25 .24 .26 .27 .26 .24 .14 .00
```

Because these distances are not tree-derived, the trees found from the UPGMA and neighbor-joining methods need not be identical (and, as shown in Figure 14.2, are not in this case), and will only approximate the real tree. These two trees are shown in Figure 14.2, and the trees derived from the parsimony and maximum likelihood trees are shown in Figure 14.3.

It is interesting to note that some relationships are preserved in all four trees, while others vary from tree to tree. The human species has the most ambiguous placement. The UPGMA tree groups human with horse, pig, dolphin, and whale; the neighbor-joining tree places human further away from those species; the parsimony tree places humans with the group containing elephant, sea cow, hyrax, and elephant shrew; and the maximum likelihood tree places humans with rodents. These wide differences highlight the effects of the different modeling assumptions, the different optimality criteria, and the different broad approaches to tree construction. The problems associated with modeling and hypothesis testing in tree construction are thus significant, and are discussed further in Section 14.9.

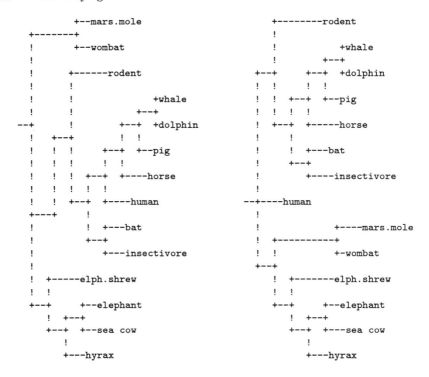

FIGURE 14.2. UPGMA and neighbor-joining trees

14.9   Modeling, Estimation, and Hypothesis Testing

14.9.1   Estimation and Hypothesis Testing

Suppose that a phylogenetic tree has been estimated by one of the methods discussed in Section 14.8 or by some other reasonable method. How much confidence can we place in the various features that this tree possesses? We illustrate approaches to this question by discussing the concept of monophyletic groups.

Given any internal node in a phylogenetic tree, there is a set of leaves, that is, contemporary species, that descend from that node. We call any such a set of leaves a *monophyletic group*. A tree reconstructed from a set of species can be thought of as a set of predictions of monophyletic groups. Suppose that a given tree estimation process leads, from certain data, to some monophyletic group that is of interest to the investigator. How much confidence can be placed in the claim that this represents a true monophyletic group?

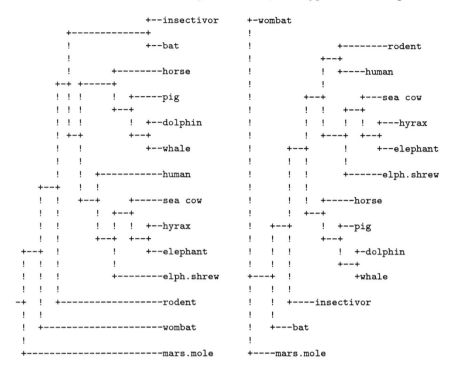

FIGURE 14.3. Parsimony and Maximum Likelihood trees

First, any tree estimation method implicitly or explicitly assumes certain
properties of the evolutionary process leading to the data at the leaves
of the tree from which the evolutionary reconstruction is made. If these
assumptions are incorrect, a misleading assessment of the tree topology can
arise. In particular, an estimation procedure that makes incorrect implicit
assumptions about the evolutionary process may well lead to estimation
of incorrect monophyletic groups. If several different estimation procedures
all rely on similar but nevertheless incorrect implicit assumptions about
the evolutionary process, consistency of a monophyletic group from one
estimation procedure to another does not add confidence to the claim that
this represents a true monophyletic group. Thus the fact that {Elephant,
Elephant Shrew, Hyrax, Sea Cow} forms a monophyletic group in all four
of the tree estimation examples in Section 14.8 does not mean that we can
assign any particular measure of confidence to the statement that Elephant,
Elephant Shrew, Hyrax, and Sea Cow really do form a monophyletic group.

Even if an estimation procedure correctly assumes the properties of the
evolutionary process leading to the data at the leaves of the tree, it is not
certain that the correct tree will be estimated from these data. This follows

from the stochastic nature of the evolutionary process and the consequent possibility that the data from which the tree estimation is made lead to an incorrect tree estimate. This is another example of the properties of estimation as a statistical process.

Given data from a set of contemporary species and a tree estimated by a given method from these data, how may the accuracy of this tree be assessed? How much confidence may we place in some monophyletic group defined by the estimated tree? To address this question, Hillis and Bull (1993) define the concepts of the *precision*, *repeatability*, and *accuracy* of estimating phylogenetic trees.

The investigator is most interested in accuracy, defined (in the case of monophyletic groups) as the probability that an estimated monophyletic group is a real monophyletic group. Accuracy cannot be estimated from data directly. Repeatability is concerned with the extent to which independent samples of sites of the same size, and from the same evolutionary process, will all estimate the monophyletic group of interest.

The bootstrapping procedure applied in the phylogenetic context is an attempt to address these problems. The concept of precision relates to the bootstrapping procedure, and relates to properties of R bootstrap samples of a given data set. We now discuss this procedure in more detail.

14.9.2 Bootstrapping

There is some confusion in the literature about what a bootstrap procedure can do and what the properties of the procedure are. For example, bootstrap methods are often said to lead to confidence intervals, or at least bootstrap confidence intervals, for some monophyletic group. However this use of terminology is misleading and is avoided here. Some investigators misinterpret bootstrapping as referring to accuracy or repeatability. It is therefore necessary to consider the nature and the properties of the bootstrap procedure in the phylogenetic context.

We now describe how the bootstrapping procedure attempts to gauge repeatability. Suppose that a phylogenetic tree has been estimated by some given method, using as data the nucleotides at n aligned nucleotide sites in each of the m species at the leaves of the tree. We can think of the m-tuple of m nucleotides at site i in these species as "data" x_i, and the entire collection of the data at n aligned sites as a set of data values $(x_1, x_2, \ldots, x_n)$. In the bootstrap procedure of Section 12.2.2, n "observations" are drawn from $(x_1, x_2, \ldots, x_n)$ without replacement. Similarly, here n sets of m-tuples are drawn with replacement from the data. Some sites might not be represented at all in this sample, some might be represented once, some twice, and so on. Given this bootstrap sample, a phylogenetic tree can be estimated by the method used for the estimation using the original data. This bootstrap procedure is now repeated R times, for some large number R. This results in R estimated trees, possibly not all equal.

Recall from Section 12.2.2 that the bootstrap procedure assumes that the data $(x_1, x_2, \ldots, x_n)$ are the observed values of iid random variables. Thus in applying the bootstrap procedure as above one assumes implicitly that the evolutionary processes at the various sites are independent and have identical stochastic properties.

Those monophyletic groups that do not appear in most of the R trees cannot be considered reliable. What can be said about those monophyletic groups that do occur in all, or almost all, of the R trees? In this case we have gained confidence only in the *precision* of the bootstrap procedure, that is that the inherent variability in the reconstruction method being employed is minimal. What is the relation between this precision and the concepts of accuracy and repeatability?

Hillis and Bull (1993) show by simulation of a process with independent evolution at different sites, and the same stochastic properties at all sites, that the precision properties of bootstrap samples are different from those of the repeatability properties of independent samples. In particular, while the means of bootstrap and independent estimators are similar, as expected, the variance of bootstrap estimates is much larger than that of estimates found from independent samples, and indeed so large that little reliance can be placed on any single bootstrap estimate when used to estimate repeatability. They also show that bootstrap estimates when taken as estimates of accuracy are biased and often quite conservative. This conclusion was also reached in laboratory experiments with bacteriophage T7. Thus the bootstrap procedure may not be used as an assessment of accuracy. In particular, even if a certain group is monophyletic in (say) 95% of the R bootstrap samples, the value 95% should not be used in assessing accuracy. This problem is made worse if the evolutionary processes at the various sites are dependent. Felsenstein and Kishino (1993) reinforce this point.

14.9.3 An Example

In some simple cases, the question of the relation between accuracy and precision can be addressed theoretically. In this section we outline an example of Zharkikh and Li (1992a) in which this is done.

Zharkikh and Li consider a simple example of a Kimura two-parameter evolutionary process that is identical at all times and in all branches of the tree. The processes at different sites are also assumed to be independent and all have the same stochastic properties. Their example considers a tree with four leaves, or contemporary species, with one species (called here species 4) known to be an out-group. There are thus four possible phylogenetic trees, as shown in Figure 14.4, and the correct tree is tree A.

With the evolutionary model assumed it is possible to calculate various probabilistic properties of the data, which consist of the nucleotides at n aligned sites in each of the four species. These properties depend on the

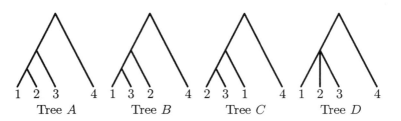

FIGURE 14.4.

model chosen and the time that the evolutionary process is allowed to run, and their details are not important for the analysis below.

Consider now the estimation of the phylogenetic tree, given these data, using the parsimony approach. Under this approach, the data at any given nucleotide site are said to be *informative* for choosing between trees A, B, and C if exactly two nucleotides types are observed at that site, one type in two species and the other type in the remaining two. This definition of "informative" depends on the fact that parsimony is the method used for tree estimation. Thus the data at a given site are informative and support tree A if the nucleotide type at that site is the same in species 1 and 2, and is also the same in species 3 and 4, but different from that in species 1 and 2. Similar conventions apply for informativeness and support of trees B and C.

We define p_X as the probability that the nucleotides at a given site support tree X, $X = A, B, C$. Under the evolutionary model assumed, $p_B = p_C$, and this is the only property of the evolutionary model that is necessary for the analysis. Our focus is on p_A, the probability that the data at a given site support the correct tree A, that is, correctly suggest that species 1 and 2 form a monophyletic group. Note that $p_A + p_B + p_C < 1$, since it is quite possible that the data at a given site do not support either tree A, tree B, or tree C.

Define $Y_X/n = \hat{p}_X$ as the proportion of sites (out of n) that support tree X. Then under the parsimony approach, the data in the sample of n sites leads to the choice of tree A if

$$\hat{p}_A > \max(\hat{p}_B, \hat{p}_C), \tag{14.15}$$

or equivalently if

$$Y_A > \max(Y_B, Y_C). \tag{14.16}$$

Write $Y_0 = n - Y_A - Y_B - Y_C$, the number of noninformative sites. Then under the simplifying assumptions made, the vector (Y_A, Y_B, Y_C, Y_0) has the multinomial distribution (2.6), with index n and parameters $(p_A, p_B, p_C, 1 - p_A - p_B - p_C)$. For small n and given values of p_A, p_B, and p_C, the probability of the event defined in (14.16) can be found, after some computation, from multinomial distribution probabilities. For large n it is convenient to approximate the multinomial distribution by the multivariate normal distribution (2.61), with a choice of parameters determined by the (binomial) means and variances of the Y_X and the covariance formula (2.60). The multivariate normal distribution is assumed as a sufficiently close approximation in the following analysis.

Our first aim is to find an expression for the probability P of the event defined by (14.16), that is, of the probability that the correct tree is inferred from the data by the parsimony approach. The event (14.16) can be written as the event $Y^* > 0$, where

$$Y^* = Y_A - \left(\frac{Y_B + Y_C}{2} + \frac{|Y_B - Y_C|}{2} \right). \qquad (14.17)$$

From (2.66), the mean of Y^* is the sum of the means of the three terms on the right-hand side of (14.17). Since $p_B = p_C$, the mean of $(Y_B + Y_C)/2$ is np_B. Furthermore, $Y_B - Y_C$ has mean 0 and, from (2.60) and (2.62), has variance $2np_B(1 - p_B) + 2np_B^2$, the second term in this expression arising from the covariance between Y_B and Y_C. This covariance is small when p_B is small and is ignored. With this approximation, the mean of $|Y_B - Y_C|/2$ is, from (1.58), $\sqrt{np_B(1 - p_B)/\pi}$. Thus the mean of Y^* is, approximately,

$$np_A - np_B - \sqrt{np_B(1 - p_B)/\pi}. \qquad (14.18)$$

By again ignoring small covariance terms, the variance of Y^* is found to be, approximately,

$$np_A(1 - p_A) + np_B(1 - p_B) \left(1 - \frac{1}{\pi} \right). \qquad (14.19)$$

Thus the probability P of the event defined by (14.16) is approximated by the probability that a normal random variable with mean (14.18) and variance (14.19) exceeds 0.

This probability depends only on n, p_A, p_B, and p_C. It can thus be estimated by using the sample frequencies $\hat{p}_A$, $\hat{p}_B$, and $\hat{p}_C$ in the relevant calculations in place of p_A, p_B, and p_C. The resulting estimate, denoted by $\hat{P}$, is a random variable whose probability distribution can be found in principle by transformation methods from the approximately multivariate normal distribution of Y_A, Y_B, and Y_C. This enables an assessment of the properties of $\hat{P}$, in particular its variance and the probability that $\hat{P}$ exceeds certain interesting values. For example,

$$\text{Prob}(\hat{P} > P) \approx \frac{1}{2}, \quad \text{Prob}\left(\hat{P} > \frac{1}{2} \right) \approx P. \qquad (14.20)$$

Calculations similar to those above are not possible when more than four species occur at the leaves of the tree, and in such cases estimation of P

is sometimes carried out using a bootstrap procedure. The main purpose of the above calculations is to provide the background for comparison with the parallel calculations of the bootstrap procedure. This will allow us to assess the validity of the bootstrap estimation process.

In any bootstrap resample there will be some number Y_{BX} of sites that support tree X, $X = A, B, C$, the additional suffix "B" indicating that this number is found from a bootstrap sample. It is natural to mimic the procedures above leading to (14.18), (14.19), and (14.20) with the data from any bootstrap resampling, replacing Y_X by Y_{BX} and $\hat{p}_X$ by $\hat{p}_{BX}$, where $\hat{p}_{BX} = Y_{BX}/n$. The event that the bootstrap sample supports tree A is, from (14.16), the event that

$$Y_{BA} > \max(Y_{BB}, Y_{BC}). \tag{14.21}$$

Following (14.17), this can be written as the event that $Y_B^* > 0$, where

$$Y_B^* = Y_{BA} - \left(\frac{Y_{BB} + Y_{BC}}{2} + \frac{|Y_{BB} - Y_{BC}|}{2} \right). \tag{14.22}$$

Write $Y_{B0} = n - Y_{BA} - Y_{BB} - Y_{BC}$. Then the vector $(Y_{BA}, Y_{BB}, Y_{BC}, Y_{B0})$ has the multinomial distribution (2.6), with index n and parameters $(\hat{p}_A, \hat{p}_B, \hat{p}_C, 1 - \hat{p}_A - \hat{p}_B - \hat{p}_C)$, and one is thus tempted to claim that to a close approximation, the mean and variance of Y_B^* are given by (14.18) and (14.19), respectively, with p_X replaced by $\hat{p}_X$. However, this claim is not correct. Whereas $p_B = p_C$, so that the mean of $Y_B - Y_C$ is 0, it is not necessarily true that $\hat{p}_B = \hat{p}_C$, so that the mean of $Y_{BB} - Y_{BC}$ is not necessarily 0. Thus the calculation leading to (14.18) does not apply to the bootstrap sample, since this relies on the assumption that the mean of $Y_B - Y_C$ is 0. Further, the variance of Y_B^* is not given by (14.19) with p_X replaced by $\hat{p}_X$. The reason for this is that Y_B^* has a compound distribution: The variance of Y_B^* has two components, one of which is the variance of Y_B and the other coming from the bootstrap procedure. This problem persists if Y_B^* is replaced by an average value over the R bootstrap samples, and disappears only in the limit as $R \to \infty$.

It follows that the bootstrap resamples do not have the same probability distribution as the original sample. Thus a procedure made with bootstrap data leading to the estimation of P parallel to that above can lead to biased estimates of P, that is, to biased estimates of the probability that species 1 and 2 form a monophyletic group. Zharkikh and Li show further that the bootstrap estimate of P tends to be too large for small values of n and too small for large values of n. If a tree is accepted as being the true tree only if it arises from at least 95% of the bootstrap replicates, there can be a high probability, sometimes exceeding 50%, of failing to accept the true tree, even when P is as high as 0.95. That is, the bootstrap process can be quite conservative in estimating the reliability of an inferred tree. This agrees with the conclusion of Hillis and Bull (1993).

These problems arise despite the very simple evolutionary model assumed. When the methods of analysis do not correspond as well to the actual evolutionary process, even further deviations of bootstrap inferences from the true phylogenetic tree should be expected.

14.9.4 Assumptions and Problems

The fact that in the example of Section 14.8 the four tree construction methods lead to four quite different trees should be a cause for concern. The differences arise because of the different approaches that the four methods adopt. Further, the methods based on distance measures assumed the Kimura two-parameter stochastic evolutionary model. This is an oversimplified model, which cannot be expected usually to describe reality well. As discussed in Chapter 13, distances estimated under one model when another more complex model is appropriate are biased. Of course, it is difficult in practice to know what a reasonable model might be, so that it is important to assess the problems that arise if a tree is estimated under one model when some other model is appropriate. In this section we discuss aspects of modeling problems in the phylogenetic context and in the following section discuss hypothesis testing involving models.

The concept of a consistent estimator of a parameter was introduced in Section 8.2.1. Steel (1994) demonstrated that if the frequencies of the various nucleotides (for DNA sequences) or amino acids (for protein sequences) can be taken as known, then under an iid assumption of the substitutions at different sites, a consistent estimator of a phylogenetic tree can be found. However, Steel et al. (1994) show that if the frequencies of the various nucleotides (for DNA sequences) or amino acids (for protein sequences) are unknown, then with site-to-site variation in substitution rates it can be impossible in principle to estimate a phylogenetic tree consistently, even with the most extensive data.

Similarly, Chang (1996b) demonstrated that when the correct evolutionary model is used, and under further mild restrictions, a consistent estimate of an evolutionary tree topology, together with the various branch lengths, can be achieved. These models assume identical substitution processes at all sites. However, when different substitution processes apply at different sites, for example with site-to-site variation in substitution rates, inconsistent tree estimation can occur (Chang (1996a)).

Yang (1993) approached the problem of site-to-site variation in substitution rates in the Felsenstein model by allowing the rate parameter u to be a site-to-site random variable following a gamma distribution with mean 1 and variance $1/\beta$, for which, from Section 1.10.5,

$$f_U(u) = \frac{\beta^\beta u^{\beta-1} e^{-\beta u}}{\Gamma(\beta)}, \quad u \geq 0. \tag{14.23}$$

Yang's analysis assumes the Felsenstein model of substitution, and a similar analysis can be made for other models. Yang tests whether the site-to-site variation model provides a significantly better fit than the "fixed u" model by using a $-2 \log \lambda$ testing procedure. This procedure is discussed further in Section 14.9.5.

Lake (1987) also addressed the problem of site-to-site variation in substitution rates and developed a model allowing this variation for which consistent estimation is possible. However, this model is specialized and may not apply widely. In general, the problem of site-to-site variation in substitution rates appears to be difficult to overcome. Questions concerning the reconstruction of phylogenetic trees by algorithmic methods in the presence of site-to-site variation in substitution rates are discussed by Steel et al. (1994).

The analysis described in Section 14.7, and further analyses of phylogenetic tree estimation in the literature, assume that even if rates differ from site to site, at least the evolutionary processes at the various sites are independent. However, population genetics theory shows that very close sites evolve in a dependent manner. This matter has been addressed by various authors, in the context of both maximum likelihood estimation and algorithmic estimation. Dixon and Hillis (1993) discuss the use of weights to accommodate nonindependence in a parsimony analysis. Others have examined substitution processes in regions rather than at nucleotide sites when various correlated and other independent substitution processes structures are assumed (Muse (1995), Schöniger and von Haeseler (1994), (1995)). Since correlated substitutions are known to be the case, more research on this matter is necessary.

A further interesting fact is that there can be more than one maximum likelihood tree, and that the topologies of the various trees having maximum likelihood can be different. This was demonstrated by Steel (1994), and disproved a claim (Fukami and Tateno (1989)) that a unique maximum likelihood tree exists. It can be argued that while Steel's result is true in principle, it is very unlikely in practice that multiple maximum likelihood trees will arise, especially with extensive data sets. The simulations of Rogers and Swofford (1999) suggest that the multiple maximum phenomenon will rarely occur in practice. Nevertheless, the very possibility of multiple maxima should cause concern, especially when the various optimality properties of maximum likelihood estimators are invoked in phylogenetic tree estimation.

14.9.5 Phylogenetic Models and Hypothesis Testing

In this section we discuss aspects of hypothesis testing in the context of phylogenetic tree construction. The derivation of tests of hypotheses in this area, and the analysis of properties of these tests, is an active area of current research and we refer to aspects of this later. Here we consider comparatively simple aspects of the hypothesis-testing question.

The testing vehicle that we discuss is the $-2\log\lambda$ test statistic discussed in Section 8.4.2. It is tempting to use this statistic in cases where various parameters of the phylogenetic tree of interest have been estimated by maximum likelihood, since this test statistic is a function of the ratio of two maximum likelihoods. However, inappropriate use of this statistic can cause problems, as we discuss below.

Suppose that a set of aligned DNA sequences from a collection of species is given, and we wish to test the (alternative) hypothesis that these data are significantly better explained by the Kimura model of Section 13.3.2 than by the Jukes–Cantor model of Section 13.3.1. The null hypothesis is that the Kimura model does not explain the data better than the Jukes–Cantor model. In statistical terms, this is a test of the null hypothesis that in the Kimura model, $\alpha = \beta$.

It is important to note two aspects of this test. First, the two hypotheses are nested, since the Jukes–Cantor model is a special case of the Kimura model. Thus the "nesting" requirement of the $-2\log\lambda$ testing method is met. Second, if the actual topology of the phylogenetic tree, which can loosely be thought of as an unknown parameter, is also estimated as part of the maximizing procedure, then the requirement that the parameters of interest be real numbers is not met. Because of this, the asymptotic null hypothesis distribution of $-2\log\lambda$ might not be chi-square. Thus for the moment we assume that the tree topology is given, and that the estimation procedure relates only to the various arm lengths.

Even this restriction does not guarantee that the asymptotic null hypothesis chi-square distribution of $-2\log\lambda$ can be assumed. Part of the estimation procedure relates to the unknown DNA sequences at the internal nodes of the tree, and if we take these as unknown parameters, the question of estimating discrete parameters arises. It is therefore important to assess what the asymptotic null hypothesis distribution of $-2\log\lambda$ is. It appears that this is very difficult to do theoretically, and that it is necessary to resort to computer simulation. Here we present the results of Whelan and Goldman (1999), who follow the simulation approach.

Whelan and Goldman take the topology of the phylogenetic tree as given, and carry out a random procedure of DNA substitution at the various sites in a DNA sequence, assuming that the substitution process is described by the Jukes–Cantor model. They then find the maximum of the likelihood of the derived data in the species at the leaves of the tree under both the Jukes–Cantor model and the Kimura model, and from this they compute $-2\log\lambda$. This procedure is then replicated a large number of times, and from these replicates an accurate assessment of the null hypothesis distribution of $-2\log\lambda$ can be found. It is found that this distribution is very close to that of chi-square with one degree of freedom.

It is therefore reasonable to assume in this case, where the topology of the tree is given, that the $-2\log\lambda$ statistic has an approximate chi-square distribution with one degree of freedom under the null hypothesis. Suppose that for a given set of data we do not reject the null hypothesis when carrying out this test. It is important to be aware of what this implies. It does not necessarily imply that the Jukes–Cantor model provides a satisfactory explanation of the data. What it does imply is that the Kimura model, with one further free parameter, does not give a significantly better fit to the data than the Jukes–Cantor model. Thus the test is a comparative one rather than an absolute one, and we have not shown in any absolute sense that the Jukes–Cantor model provides an adequate fit to the data. It is important to make this point, since claims of absolute fit rather than relative fit are sometimes made by carrying out the $-2\log\lambda$ procedure in similar cases.

Whelan and Goldman (1999) carry out a variety of similar assessments, and we outline here three of their results. First, if the stationary probabilities φ_a, φ_g, φ_c, and φ_t in the Felsenstein model of Section 13.3.3 are all equal to .25, then that model becomes the Jukes–Cantor model, so that the Jukes–Cantor model is nested within the Felsenstein model. Since $\varphi_a + \varphi_g + \varphi_c + \varphi_t = 1$, there are three more free parameters under the Felsenstein model than under the Jukes–Cantor model. It then becomes reasonable to hope that, if the null hypothesis is the Jukes–Cantor model and the alternative hypothesis is the Felsenstein model (with arbitrary stationary probability values), and where the topology of the tree is given, the null hypothesis distribution of $-2\log\lambda$ is approximately chi-square with three degrees of freedom. The simulations of Whelan and Goldman (1999) support this conclusion.

Second, suppose that the null hypothesis is the Jukes–Cantor model and the alternative hypothesis is the Felsenstein model with stationary probability values equal to the observed values in the data. Then neither model is nested within the other and we have no theoretical support for the claim that the null hypothesis distribution of $-2\log\lambda$ is chi-square. Whelan and Goldman (1999) show that the null hypothesis distribution of $-2\log\lambda$ is not close to a chi-square with three degrees of freedom, and that in fact negative values of $-2\log\lambda$ can arise, an impossibility for a random variable truly having a chi-square distribution. This matter is referred to again in Problem 14.6.

Third, in Section 14.9.4 we discussed models that allow unequal substitution rates at various sites. Since the assumption of equal rates leads to simpler models, it is important to find a test for it. Such tests are important because, for example, an unjustified assumption of equal substitution rates can lead to biases in substitution matrix estimation (Kelly and Churchill (1996)).

Whelan and Goldman (1999) and Goldman and Whelan (2000) discuss this matter. They consider the test of a null hypothesis Jukes–Cantor model, with substitution rate 1 at all sites, against an alternative hypothesis Jukes–Cantor model for which the substitution rate at any site is a random variable having the gamma distribution (14.23) with mean 1 and variance $1/\beta$. They also consider the parallel test for the Felsenstein model.

Although in both models the relevant null and alternative hypotheses differ by one free parameter, and the null hypothesis is nested within the alternative, the simulations show that the null hypothesis distribution of $-2\log\lambda$ is not chi-square with one degree of freedom. This occurs because the null hypothesis value of β is $+\infty$, corresponding to a variance of zero. The value $+\infty$ is not covered by the asymptotic $-2\log\lambda$ theory, since that theory requires β to be a real number. The $-2\log\lambda$ testing procedure has been misused in the literature to assess whether significant site-to-site variation in substitution rates occurs within the Felsenstein model.

Self and Liang (1987) suggest that in this case the null hypothesis distribution of $-2\log\lambda$ is approximately that of a random variable taking the value 0 with probability $\frac{1}{2}$ and having a chi-square distribution with one degree of freedom also with probability $\frac{1}{2}$, and their simulations support this suggestion. On an associated point, Ota et al. (1999) show that boundary maximum likelihood estimates such as those arising in Example 3 of Section 8.4.2 of parameters cannot have asymptotic normal distributions, and the theory of Chapter 8 then implies that the asymptotic $-2\log\lambda$ testing theory cannot hold. A simple example of such a boundary estimate, and its clearly nonnormal asymptotic distribution, is provided by Example 3 of Section 8.4.2.

The null hypothesis chi-square mixture of Self and Liang (1987) is not always applicable. Ota et al. (2000) show that when there are two or more boundary parameter estimates, the null hypothesis distribution of $-2\log\lambda$ cannot be expressed as a linear combination of any number of chi-square distributions. This remark is relevant because models with two or more boundary maximum likelihood estimates are beginning to appear (Huelsenbeck and Nielsen (1999)).

These observations indicate the dangers in unthinking application of tests using $-2\log\lambda$. A further and more serious problem arises, as discussed above, when the maximum likelihood procedure involves estimating the topology of the phylogenetic tree as well as its branch lengths, a case not covered by the $-2\log\lambda$ theory. Here the null hypothesis distribution of $-2\log\lambda$ is far from chi-square, essentially because the tree topology is not a real number able to take values in some interval. This matter is discussed further by Goldman (1993), who notes that several authors have made an inappropriate chi-square assumption in the literature when the tree topology is estimated as part of the testing procedure.

The bootstrap testing procedure in Section 12.3 is often thought of as one that overcomes problems of an unknown null hypothesis distribution. However Andrews (2000) has shown that when a maximum likelihood estimate is a boundary value, problems can arise even with the bootstrap operation.

Much of the discussion above concerning the problems of hypothesis testing in the phylogenetic tree context has used as an example the test of the (null hypothesis) Jukes-Cantor model against the (alternative hypothesis) Kimura model, within which it is nested. These problems also arise in other tests. Yang (1994) discusses tests of the Felsenstein and the HKY models against the general reversible model, within which they are nested. He also considers the test of the general reversible process model itself against a completely arbitrary model. The number of degrees of fredom used in these tests are derived from the number of degrees of freedom by which the respective models differ, as discussed in Section 13.2.7. The extent to which a more restrictive model can be accepted depends on the data analyzed, so that general conclusions are not easily obtained. Yang does however claim that, broadly speaking, the general reversible process model can be recommended and that the completely arbitrary model is not recommended.

It is clear that the hypothesis tests considered above barely scratch the surface of what is possible and necessary. A list of possible tests, and the outcomes of these tests, is given by Huelsenbeck and Rannala (1997). Among the many results that they present, Huelsenbeck and Rannala confirm that the distribution of $-2 \log \lambda$ is far from chi-square when the tree topology is estimated as part of the testing procedure. They also give the results of several tests where $-2 \log \lambda$ is correctly applied to real data sets, to assess what modeling assumptions may reasonably be made in practice. As one important example, they show that substitution rates vary significantly from one lineage to another, thus rejecting the "molecular clock hypothesis" favored by many population geneticists that the substitution rate per generation is constant in all branches of the phylogenetic tree. This hypothesis is further discussed by Tourasse and Li (1999). They show that the Kimura model fits observed data significantly better than the Jukes–Cantor model, so that transition substitution rates differ significantly from transversion substitution rates. Models allowing different substitution rates between sites fit observed data significantly better than models that assume the same rate at all sites. Models that allow different substitution rates in different regions of the genome fit observed data significantly better than models that assume the same rate in all regions, a result confirmed by the recent chromosome 22 data (Dunham et al. (1999)). Liò and Goldman (1998) discuss further aspects of model testing, and the use of hidden Markov models, in analyzing models with different substitution rates at different sites.

Testing for monophyly provides an important case of hypothesis testing involving tree structure. Here the null hypothesis to be tested is that some group of species is monophyletic. The maximum likelihood tree can be constructed under the (null) hypothesis of monophyly of a certain set of species and can also be constructed without this restriction, and the two likelihoods compared. Huelsenbeck et al. (1996) and Huelsenbeck and Crandall (1997) point out that the null hypothesis distribution of $-2\log\lambda$ is far from chi-square when the null hypothesis claim of monophyly is correct. This again arises because tree topology is not a real number taking values in some interval. Indeed, it is not even clear how many degrees of freedom $-2\log\lambda$ would have if it did have a null hypothesis chi-square distribution. If $-2\log\lambda$ is to be used in a test of monophyly, its null hypothesis distribution must be found empirically, using simulation methods.

An excellent description of these problems is given by Huelsenbeck and Crandall (1997), who also provide a summary of maximum likelihood estimation procedures used in phylogenetic analysis, and their properties. A further excellent description of statistical problems in phylogenetic analysis is provided by Holmes (1999), who gives a table listing "translations" of expressions frequently used in biological articles into the corresponding standard statistical terminology.

Problems

14.1 Prove that a tree-derived distance satisfies the four properties of a distance given on page 387.

14.2 Prove that for a tree whose derived distance satisfies the ultrametric property, pairs of species with the smallest distance between them are neighbors.

14.3 Prove equation (14.12).

14.4 For five species a, b, c, d, and e, with distance given by

	a	b	c	d	e
a	0	2	8	8	8
b		0	8	8	8
c			0	4	4
d				0	2
e					0

reconstruct the tree using both algorithms in Section 14.3. For the first algorithm, use the order a, b, c, d, e. Show the partial tree for each step, together with edge lengths.

14.5 For five species a, b, c, d, and e, with distance given by

	a	b	c	d	e
a	0	9	8	7	8
b		0	3	6	7
c			0	5	6
d				0	3
e					0

reconstruct the tree using the neighbor joining algorithm of Section 14.4. Show the partial tree in each case, together with edge lengths. Now reconstruct the tree, together with edge lengths, using the UPGMA algorithm of section 14.3. Compare your answers for the two reconstructions.

14.6 In what circumstances will negative values of $-2 \log \lambda$ arise in testing the Jukes–Cantor model against the Felsenstein model in which the stationary probabilities are set to the observed values?

Appendix A
Basic Notions in Biology

We outline here the basic notions from biology that are needed in the book. Deoxyribonucleic acid (DNA) is the basic information macromolecule of life. It consists of a polymer of nucleotides, in which each nucleotide is composed of a standard deoxyribose sugar and phosphate group unit, connected to a nitrogenous base of one of four types: adenine, guanine, cytosine, or thymine (abbreviated here a, g, c, and t, respectively). Because of similarities in the chemical structure of their nitrogenous bases, adenine and guanine are classified as purines, while cytosine and thymine are classified as pyrimidines. Adjacent nucleotides in a single strand of DNA are connected by a chemical bond between the sugar of one and the phosphate group of the next. The classic double-helix structure of DNA is formed when two strands of DNA form hydrogen bonds between their nitrogenous bases, resulting in the familiar "ladder" structure. Under normal conditions, these hydrogen bonds form only between particular pairs of nucleotides (referred to as base pairs): Adenine pairs only with thymine, and guanine pairs only with cytosine. Two strands of DNA are complementary if the sequence of bases on each is such that they pair properly along the entire length of both strands (see Figure A.1). The sequence in which the different bases occur in a particular strand of DNA represents the genetic information encoded on that strand. By virtue of the specificity of nucleotide pairing, each of the two strands of any DNA molecule contains all of the information present in the other. There is also a chemical polarity to polynucleotide chains such that the information contained in the a, g, c and t bases is synthesized and decoded in only one direction.

```
...   ------------------------------------------   ...
      a c c g t a t a a c g a t c c t c t g a
      : : : : : : : : : : : : : : : : : : : :
      t g g c a t a t t g c t a g g a g a c t
...   ------------------------------------------   ...
```

FIGURE A.1. Example of a portion of DNA sequence. The dashed line represent the sugar-phosphate backbone, and the letters represent the nitrogenous bases. Dotted lines connecting base pairs denote hydrogen bonds. Note the specificity of base-pairing (a to t, c to g).

In the cell, DNA is organized into chromosomes, each of which is a continuous length of double stranded DNA that can be hundreds of millions base pairs long. Most human cells contain 23 pairs of chromosomes, one member of each pair paternally inherited and the other maternally inherited. The two chromosomes in a pair are virtually identical, with the exception of the sex chromosome, for which there are two types, X and Y. Nearly every cell in the body contains identical copies of the full set of 23 pairs of chromosomes. An organism's total set of DNA is referred to as its genome; the human genome contains more than three billion base pairs.

A human chromosome consists mostly of so-called "junk DNA," whose function, if any, is not well understood. Interspersed in this junk DNA are genes, the classic unit of genetic information. Genes themselves are often organized into exons, which are the sequences that will eventually be used by the cell, alternating with introns, which will be excised and discarded. The human genome is currently thought to contain approximately 30,000–40,000 genes (International Human Genome Sequencing Consortium (2001)). The information in these genes will go on to be encoded in RNA (ribonucleic acid), and in many cases ultimately in proteins.

The first step in this process is transcription, the creation of an RNA molecule using the DNA sequence of a gene as a template. Transcription is initiated at non-coding sequences called promoters, located immediately preceding the gene. Like DNA, RNA is made up of a series of nucleotides, but with several important differences: RNA is single-stranded, contains the sugar ribose, and substitutes the nitrogenous base uracil for thymine. After post-transcriptional modification, which includes the removal of introns, the RNA will go on to various fates within the cell. Of particular interest is mRNA (messenger RNA), which will be translated into protein.

A protein is comprised of a sequence of amino acids. There are twenty amino acids which commonly appear in proteins. Each of these amino acids is represented by one or more sequences of three RNA nucleotides known as a codon; for example, the RNA sequence aag encodes the amino acid lysine. The combination of four possible nucleotides in groups of three results in 4^3 or 64 codons, meaning that most amino acids are coded for by more than one codon. An organelle known as the ribosome performs the translation of mRNA into protein. The ribosome pairs each codon in the RNA sequence with the appropriate amino acid, and then adds the amino acid onto the growing protein. The process of translation is mediated by two special types of codon: start codons signal the location on the RNA molecular where translation should begin, while stop codons signal the location where translation should terminate. Once the sequence of amino acids that make up a particular protein is assembled, the protein dissociates from the ribosome and folds into a specific three-dimensional form. The function of a protein ultimately depends on both its three-dimensional structure and its amino acid sequence. Proteins go on to perform a variety of functions in the cell, covering all aspects of cellular functions from metabolism to growth to division.

Currently, functions have been assigned to only a small proportion of the genes in even the best understood of model organisms. In order to assign function to the remaining genes, it is helpful to examine the expression patterns of these genes in various tissues. Microarray technology developed over the past several years now allows the measurement of mRNA levels for tens of thousands of genes simultaneously. This provides an efficient and convenient way to determine the expression patterns of genes in many different types of tissues, but at the same time provides new challenges in information cataloging and statistical analysis.

Appendix B
Mathematical Formulae and Results

B.1   Numbers and Intervals

It is a nontrivial matter to define the real numbers carefully, and for our purposes this is not necessary. We will instead take the real numbers as a starting point. Intuitively, they represent all the numbers that correspond to lengths, together with their negatives and zero. The real numbers can be put in one-to-one correspondence with the points on a line.

The *integers* are all the numbers $\dots, -3, -2, -1, 0, 1, 2, 3, \dots$. The *positive* integers, sometimes called the *natural numbers*, are the numbers 1, 2, 3, $\dots$, whereas the *nonnegative* integers are the numbers $0, 1, 2, 3, \dots$. The *rational* numbers are those that can be written as ratios of integers, i.e., in the form $\frac{a}{b}$, where a and b are integers. The *irrational* numbers are the real numbers that cannot be so written. It can be shown, for example, that $\sqrt{2}$, π, and e are all irrational.

We use four kinds of intervals in this book. The first are called *open* intervals and are denoted by round brackets:

$$(a, b) = \text{ set of all real numbers } x \text{ such that } a < x < b.$$

When we write such an interval, we will always assume $a < b$; otherwise, the interval would be empty. We also allow a to be $-\infty$ and/or b to be ∞. The second kind of intervals are the *closed* intervals, denoted by square brackets:

$$[a, b] = \text{ set of all real numbers } x \text{ such that } a \le x \le b.$$

In this case we do allow $a = b$, but we do not allow a or b to be infinite. The final kind of intervals are the *half-open* intervals

$$(a, b] = \text{ set of all real numbers } x \text{ such that } a < x \le b$$

and

$$[a, b) = \text{ set of all real numbers } x \text{ such that } a \le x < b.$$

For the first kind a can be $-\infty$ and in the second b can be ∞. We require $a < b$ in all such intervals.

B.2   Sets and Set Notation

The sets we need to consider in this book are subsets of n-dimensional space, for $n = 1, 2, \ldots$. By n-dimensional space we mean the set of n-tuples $(x_1, x_2, \ldots, x_n)$ where $x_1, x_2, \ldots, x_n$ are real numbers. Other sets we consider are subsets of these sets. For example, S might be the set of all n-tuples $(x_1, x_2, \ldots, x_n)$ where x_n is positive. If $n = 2$, then this is the half-plane above the horizontal axis. We refer to the tuples in S as *members* or *elements* of S. If s is a member of S, we write $s \in S$, and if s is not a member of S, we write $s \notin S$. If S_1 and S_2 are two sets in the same space, their *union* is the set of elements that are in S_1 or S_2 (or both), written $S_1 \cup S_2$. Their *intersection* is the set of elements that are in both S_1 and S_2, written $S_1 \cap S_2$.

Suppose S_1 is a subset of m-dimensional space and S_2 is a subset of n-dimensional space. Then the Cartesian product of S_1 and S_2 is the subset of $(m + n)$-dimensional space consisting of $(m + n)$-tuples whose first m components form an element of S_1 and whose last n components form an element of S_2. The Cartesian product of S_1 and S_2 is written $S_1 \times S_2$. If $S_1 = [a, b]$ and $S_2 = [c, d]$, then $S_1 \times S_2$ is a rectangle.

B.3   Factorials

Let n be a positive integer. Then we define $n!$ (pronounced "*n factorial*") to be

$$n! \stackrel{\text{def}}{=} n(n-1)(n-2) \cdots 3 \cdot 2 \cdot 1.$$

It will be apparent later that it is convenient to define $0! = 1$. Note that $n! = n(n-1)!$, and thus $\frac{n!}{(n-1)!} = n$.

B.4   Binomial Coefficients

Let r be *any real number* and let k be a positive integer. We define the *binomial coefficient* $\binom{r}{k}$ by

$$\binom{r}{k} \stackrel{\text{def}}{=} \frac{r(r-1)(r-2)(r-k+1)}{k!},$$

and for $k = 0$, put $\binom{r}{k} = 1$. The reason why it is called a "coefficient" will be explained in B.5.

When n is a positive integer and $k \leq n$, $\binom{n}{k} = \frac{n!}{(n-k)!k!}$ takes on a combinatorial meaning, which we will discuss shortly.

B.5   The Binomial Theorem

In this section we describe the basic version of the binomial theorem, and in a later section we will revisit it in more generality. The basic version says that for any real numbers a and b, and any positive integer n,

$$(a+b)^n = \sum_{k=0}^{n} \binom{n}{k} a^k b^{n-k}. \tag{B.1}$$

In particular,

$$(x+1)^n = \sum_{k=0}^{n} \binom{n}{k} x^k. \tag{B.2}$$

We provide a proof of (B.1) later, but first we need to discuss some *combinatorial* issues.

B.6   Permutations and Combinations

Consider n distinguishable balls in an urn, numbered 1 to n. Let $k \leq n$ be a positive integer. We are interested in the following two "combinatorial" quantities:

(1) In how many orders can we choose k of the balls *without replacement*? Equivalently, how many k-tuples of distinct integers $(n_1, n_2, \ldots, n_k)$ are there, where $1 \leq n_i \leq n$?

(2) In how many ways can we choose k of the balls from the urn, without replacement?

The key distinction between these two counting problems is that the first considers the order in which the objects were chosen, whereas the second does not. The first quantity is called the number of *permutations* of n objects taken k at a time, and we denote it by $_nP_k$. The second quantity is called the number of *combinations* of n objects taken k at a time and is often referred to as "n choose k." We will not introduce a notation for "n choose k," since we will soon show that it is equal to $\binom{n}{k}$.

We can easily list all the possible permutations of 3 things taken 2 at a time as

$$(1,2), \ (1,3), \ (2,3), \ (2,1), \ (3,1), \ (3,2),$$

so that $_3P_2 = 6$. On the other hand, there are 3 combinations of 2 things,

$$\{1,2\}, \quad \{1,3\}, \quad \{2,3\}.$$

In general, there can be n choices for the first element in the permutation, and for each of those there are $n-1$ choices for the second element, etc., down to $n-k+1$ choices for the kth element in the permutation. Therefore,

$$_nP_k = n(n-1)(n-2)\cdots(n-k+1) = \frac{n!}{(n-k)!}.$$

It should be clear that for each permutation of n objects taken k at a time, there will be $k!$ that involve the same k objects, each arranged in a different order. Therefore, it follows that the number of combinations of n things taken k at a time (i.e., the answer to question 2 above) is equal to

$$\frac{_nP_k}{k!} = \binom{n}{k}.$$

Thus, it is natural that $\binom{n}{1} = n$ and $\binom{n}{n} = 1$. The motivation behind the definition of $\binom{n}{0} = 1$ should now be clear.

We can now give a combinatorial proof of the binomial theorem. Consider the product

$$(a+b)^n = \overbrace{(a+b)(a+b)\cdots(a+b)}^{n \text{ times}}.$$

The right-hand side can be expanded by repeated application of the distributive property. We end up with one term of the form $a^k b^{n-k}$ for every way we can choose a's from k of the terms in the product. Therefore, the coefficient of $a^k b^{n-k}$ is $\binom{n}{k}$. This explains why $\binom{n}{k}$ is called a binomial *coefficient*.

B.7   Limits

We cannot give a rigorous introduction to limits, since the subject has many subtleties and complexities that would take us unreasonably far afield. There are two kinds of limits we consider in this book: *discrete* and *continuous* limits. Continuous limits arise for functions defined on the real numbers. Discrete limits are for functions defined on the natural numbers. The latter kind of functions are usually called *sequences* and are usually denoted using subscripts, typically with a notation of the form $a_1, a_2, a_3, \ldots, a_n, \ldots$, instead of the usual functional notation $f(1), f(2), f(3), \ldots, f(n), \ldots$. When we want to address the entire function (sequence) at once, we write $\{a_n\}_{n=m}^{\infty}$, where m is the first index in the sequence, or as $\{a_n\}$ when the domain of n is clear. In the discrete case we are mainly interested in the limit as n goes to infinity:

$$\lim_{n \to \infty} a_n.$$

An important discrete limit, proven in most basic calculus books by an application of L'Hospital's rule, is that for any fixed t,

$$\lim_{n \to \infty} \left(1 + \frac{t}{n}\right)^n = e^t. \tag{B.3}$$

Certain uniform approximations can be derived from this. For example, for all $|t| \leq 1$, the approximation

$$\left(1 + \frac{t}{n}\right)^n \approx e^t$$

is accurate to within 1% for all $n \geq 50$, to within .1% for all $n \geq 500$, and to within .01% for all $n \geq 5,000$.

For functions of a real variable $f(x)$, we consider limits as x approaches any real value (in the domain of the function), or $\pm\infty$. A limit of importance for us is the limit of the function $x \log x$ as $x \to 0$ from the right, usually written

$$\lim_{x \to 0^+} x \log x.$$

It can be shown that this limit is equal to 0.

B.8   Asymptotics

Computational issues often require an analysis of the asymptotic behavior of functions. To be more precise, let $f(t)$ and $g(t)$ be functions that are defined on the nonnegative real numbers and takes values in the positive real numbers. We are interested in methods for describing how $f(t)$ and $g(t)$ behave with respect to each other as t approaches some limit (possibly infinity). To this end, four basic relations between such functions are defined.

(1a) $f = O(g)$ at ∞, if there are constants $C, K \geq 0$ such that $\frac{f(t)}{g(t)} \leq C$ for all $t > K$.

(1b) $f = O(g)$ at a, $a < \infty$ if there are constants $C, h \geq 0$ such that $\frac{f(t)}{g(t)} \leq C$ for all t within h of a.

(2) $f = o(g)$ at a, if $\lim_{t \to a} \frac{f(t)}{g(t)} = 0$.

(3) $f \sim g$ at a, if $\lim_{t \to a} \frac{f(t)}{g(t)} = 1$.

(4) $f \, \Omega \, g$ at a, if $f = O(g)$ and $g = O(f)$ at a.

Conditions 1a and 1b are read "f is big 'oh' of g as t approaches a," condition 2 is read "f is little 'oh' of g as t approaches a," condition 3 is read 'f is asymptotic to g as t approaches a," and condition 4 is read "f is 'omega' of g as t approaches a."

The intuition behind 1a is that if it holds, then for sufficiently large values of t, $f(t)$ is no bigger than some fixed constant times $g(t)$. The intuition behind 1b is similar, replacing "sufficiently large" with "sufficiently close."

Even though the function $f(t) = 2t$ becomes arbitrarily larger than $g(t) = t$, as t gets larger, $f = O(g)$ at infinity. When $a = \infty$, and the functions involved are polynomials, the relevant terms are the largest powers of t that occur in each polynomial. If the largest power in g is at least as large as the largest power in f, then $f = O(g)$ at infinity, and if the largest power in f is at least as large as the largest power in g, then $f = O(g)$ at zero. On the other hand, if f is a polynomial, and g is an exponential function with base greater than 1, such as e^t, then it is never the case that $g = O(f)$ at infinity. These statements are usually verified by repeated applications of L'Hospital's rule.

If $f = o(g)$ at a, then f is in some sense "eventually" (i.e., for t sufficiently close to a) smaller than g. For example, $t^2 + t = o(t^3)$ at infinity. However, even though $t^2 - t$ is smaller than t^2 by more than a constant factor, $t^2 - t$ is *not* $o(t^2)$ at infinity (in fact $t^2 - t \sim t^2$ at infinity).

If $f \sim g$, then f and g grow at roughly the same rate. This does not, however, mean that the values $f(t)$ and $g(t)$ get closer to each other as $t \to \infty$. For example if $f(t) = t^2$ and $g(t) = t^2 + t$, then $f \sim g$; however, $g(t) - f(t) = t \to \infty$. One must therefore be very careful to keep the exact definitions close at hand when using these asymptotic concepts.

For the purposes of bioinformatics, the most important asymptotic classes of functions are the ones that are asymptotic to either polynomials (or more generally, to any power of t), logarithms, or exponentials. We say that a function asymptotic to some polynomial as $t \to \infty$ has *polynomial growth*, one asymptotic to a logarithm as $t \to \infty$ has *logarithmic growth*, and one asymptotic to an exponential as $t \to \infty$ has *exponential growth*. Obviously, other types of growth exist, such as "doubly exponential," "factorial," etc. One should note that a function with polynomial growth is not necessarily a polynomial, for example $f(t) = t + \sqrt{t}$ is not a polynomial, but $f(t) \sim t$ at infinity, and $g(t) = t$ is a polynomial.

An order of magnitude relation used often in this book is of the form $f(h) = \lambda h + o(h)$ for h small. This means that if we write $f(h) = \lambda h + g(h)$, then $g(h)/h \to 0$ as $h \to 0^+$. Terms of order $o(h)$ are always written as $+o(h)$ in such cases, even though they may well be negative.

B.9 Stirling's Approximation

The concepts of the previous section were phrased in terms of continuous limits, but they can all be translated into the discrete case by substituting sequences for the functions of real variables. Our next two results are discrete asymptotic equations for $n!$ and $\binom{n}{k}$. The first is Stirling's approximation:

$$n! \sim \sqrt{2\pi}\, n^{n+1/2} e^{-n}. \tag{B.4}$$

From this it follows that

$$\binom{n}{k} \sim \frac{1}{\sqrt{2\pi}} \frac{\sqrt{n}}{\sqrt{k(n-k)}} \left(\frac{k}{n}\right)^{-k} \left(1 - \frac{k}{n}\right)^{-(n-k)} \tag{B.5}$$

B.10 Entropy as Information

Suppose that some number between 1 and 64 (inclusive) is chosen at random by person A, and person B is required to find this number. By asking questions using a strategy in which the number of possibilities is halved with the answer to each question (the first being, perhaps, "is the number 32 or less?"), it is clear that six questions are sufficient to determine unambiguously which number was chosen. The precise form of the question is not fixed: For example, the first question might be, "is the number odd or even?" but the form of the questions, in which the number of possible cases is halved after each question, is clear. We can say that six bits of information are sufficient to determine the number chosen if this form of question is used. Further, no other system of questions can consistently outperform this one, in the sense that the mean number of questions asked by any other system must be at least 6.

It is clear that when a person must determine one of 2^k numbers, k questions, or k bits of information, are needed. A more complicated calculation arises in other cases. Suppose, for example, that one of the 20 amino acids is chosen at random. If the amino acids are numbered $1, 2, \ldots, 20$ in some agreed order, the first two questions, and their answers, might be, "is the number of the amino acid 10 or less?" (yes), "is the number of the amino acid 5 or less?" (no). At this point the next question might be "is the number 6, 7, or 8?" If the answer is "no," only one further question is needed to determine whether the number is 9 or 10. If the answer is "yes," the next question might be, "is it either 6 or 7?". If the answer is "no," no further questions are needed to determine the number. If the answer is "yes," one further question is needed. Thus often four but in some cases five questions are needed. (It is an interesting exercise to prove that the mean number of questions that need to be asked is 4.4.) In general, if a number is chosen from a large number N of numbers, the mean number of questions needed to determine which one was chosen is approximately $\log_2 N$. This is confirmed by the fact that $\log_2 20 = 4.32$. The number of questions is exactly $\log_2 N$ when N is of the form 2^k, for some positive integer k.

If the random variable Y discussed in the entropy formula (1.107) takes one of N possible values, each with equal probability, and 2 is used as the base of the logarithms in the formula, then the entropy defined is $\log_2 N$. Thus in this case the entropy measures the number of bits of information needed to find any given value of this random variable. The more general formula (1.107) allows for an extension of this argument to the case where Y does not have a uniform distribution.

B.11   Infinite Series

We present the definition of what a series is and give a survey of the few most relevant results for us.

Let $\{a_k\}_{k=1}^{\infty}$ be a sequence of real numbers. Using the a_k's we define a new sequence

$$s_n = a_1 + a_2 + \cdots + a_n,$$

and s_n is called the nth *partial sum* of the sequence $\{a_k\}$. A shorthand notation is $s_n = \sum_{k=1}^{n} a_k$. We define the *infinite series* $\sum_{k=1}^{\infty} a_k$ by

$$\sum_{k=1}^{\infty} a_k \overset{\text{def}}{=} \lim_{n \to \infty} s_n,$$

as long as the limit on the right-hand side exists. Thus the symbol $\sum_{k=1}^{\infty} a_k$ is the limit of the partial sums s_n as $n \to \infty$. If the limit does not exist, we say that the series diverges. There are many ways in which a series can diverge, but if in particular the limit on the right-hand side tends to infinity, we say that the series diverges to infinity. As an example of a series that diverges, but not to infinity, consider $\sum_{k=1}^{\infty}(-1)^k$; for this series $s_n = 0$ if n is even and $s_n = -1$ if n is odd.

A basic fact about infinite series is that

$$\sum_{k=m}^{\infty} a_k \text{ converges} \Rightarrow \lim_{k \to \infty} a_k = 0.$$

The essence of the proof is that $a_k = s_k - s_{k-1}$, but $\lim s_k = \lim s_{k-1}$, so the limit of a_k as $k \to \infty$ is zero.

The converse is not, however, true. The classic example is the harmonic series

$$\sum_{k=1}^{\infty} \frac{1}{k}.$$

This series diverges, albeit very slowly. If s_n is the sum of the first n terms in this series, then it can be shown that $s_n \sim \log n$, so s_n "diverges like" $\log n$. In fact, more can be shown. The limit

$$\lim_{n \to \infty} \left(\sum_{k=1}^{n} \frac{1}{k} - \log n \right) \tag{B.6}$$

exists, and is approximately equal to .5772156649.... This is the famous *Euler's constant*, denoted by γ, that has appeared in Chapter 2 as the mean of the extreme value distribution (2.123). Because of the limiting relationship (B.6), it is common practice, for large n, to use the approximation

$$\sum_{k=1}^{n} \frac{1}{k} \approx \log n + \gamma, \tag{B.7}$$

since the right-hand side is easier to calculate than the left-hand side. More important, use of this approximation often simplifies and clarifies the implication of some mathematical formulae involving $\sum_{k=1}^{n} 1/k$ when n is large (see, for example, Problem 2.21).

While the harmonic series diverges, the series

$$\sum_{k=1}^{\infty} \frac{1}{k^p} \tag{B.8}$$

for $p > 1$ converges. For some values of p, the sum of the series is known. For example, when $p = 2$

$$\sum_{k=1}^{\infty} \frac{1}{k^2} = \frac{\pi^2}{6}. \tag{B.9}$$

This surprising fact is used in Chapter 2.

Another important kind of series used often in this book is the *geometric series*. These are any series of the form

$$\sum_{k=m}^{\infty} r^k, \tag{B.10}$$

where $-1 < r < 1$. In contrast to the series in (B.8), there is a simple formula for the sum of a geometric series, namely

$$\sum_{k=m}^{\infty} r^k = \frac{r^m}{1 - r}. \tag{B.11}$$

In particular,

$$\sum_{k=0}^{\infty} r^k = \frac{1}{1-r}. \tag{B.12}$$

The most basic example is

$$1 + \frac{1}{2} + \frac{1}{4} + \frac{1}{8} + \frac{1}{16} + \cdots = \sum_{k=0}^{\infty} \frac{1}{2^k} = \frac{1}{1-\frac{1}{2}} = 2.$$

B.12   Taylor Series

Until now we have been discussing series of fixed terms. It is also necessary to consider series of variable terms. For example, we can consider the left side of equation (B.12) as a function of r, where r ranges over the (open) interval from -1 to 1, and thus consider equation (B.12) as an equation of functions. To emphasize this point we rewrite (B.12), replacing r with x, as

$$\sum_{k=0}^{\infty} x^k = \frac{1}{1-x}, \quad \text{for } -1 < x < 1. \tag{B.13}$$

This is an example of a *Taylor* series, and in this section we state without proof various relevant results from the theory of Taylor series. There is a class of functions, called *analytic functions*, that have various important properties. In particular, they are infinitely differentiable. That is, the first, second, third, etc. derivatives exist at every point in their domain. We denote the kth derivative of $f(x)$ at the point a by $f^{(k)}(a)$.

Definition. A function $f(x)$ defined on an open interval containing zero is analytic at zero if $f^{(k)}(0)$ exists for all k, and for all x in some open interval containing zero (this interval is possibly smaller than the domain of $f(x)$),

$$f(x) = \sum_{k=0}^{\infty} \frac{f^{(k)}(0)}{k!} x^k. \tag{B.14}$$

In the expression the zeroth derivative of the function $f(x)$ is taken as the function itself. Equation (B.14) is called the *Taylor series expansion of $f(x)$ at zero*.

The motivation for the expression (B.14) is as follows. The linear polynomial $p_1(x) = f(0) + f'(0)x$ is the tangent line to the graph of $f(x)$ at the point $(0, f(0))$. This linear function has the same value and same derivative at zero as $f(x)$. This is the best linear approximation to $f(x)$ at zero (see figure B.1). The polynomial $p_2(x) = f(0) + f'(0)t + \frac{f'(0)}{2}x^2$ has the same value, derivative, and second derivative at zero as $f(x)$ does. Just as $p_1(x)$ is the best linear approximation to $f(x)$ near zero, $p_2(x)$ is the best

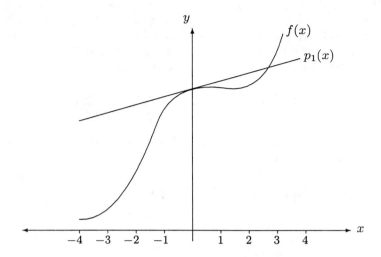

FIGURE B.1. The linear approximation $p_1(x) = f(0) + f'(0)x$ to a function $f(x)$.

quadratic approximation to $f(x)$ near zero (i.e., the best approximation by a parabola). In general,

$$p_n(x) = \sum_{k=0}^{n} \frac{f^{(k)}(0)}{k!} x^k \qquad (\text{B.15})$$

has the same value and derivatives as f at zero, up to the nth derivative, and is the best approximation to f, near zero, by an nth degree polynomial. That is, the nth-order Taylor polynomial approximation to the function $f(x)$ is

$$f(x) \approx \sum_{k=0}^{n} \frac{f^{(k)}(0)}{k!} x^k. \qquad (\text{B.16})$$

If $f^{(k)}(0)$ exists for all k, then taking the limit $n \to \infty$ one might hope that (B.14) always holds, but this is not the case; not all functions are analytic. However, the cases where it fails do not arise for the functions considered in this book, so whenever we need to rewrite a function in terms of its Taylor series expansion (B.14), we will feel free to do so.

Another useful fact is that if $f(x) = \sum_{k=0}^{\infty} a_k x^k$ converges in an open interval containing zero, then $f'(x) = \sum_{k=1}^{\infty} k a_k x^{k-1}$ in the same open interval. In other words, we can differentiate the series term by term. This fact implies that if a series $\sum a_k x^k$ converges for all x in some open interval containing zero, then it is an analytic function on that interval and is its own Taylor series expansion at zero.

We conclude by giving some examples useful for bioinformatics, and we also give the generalization of (B.14) to other points in the domain besides zero. The first example, already given in (B.13), is the geometric sum formula, which we can now see in a new light: The right-hand side is the Taylor expansion of $\frac{1}{1-x}$. Differentiating term by term gives

$$f'(x) = \sum_{k=1}^{\infty} \frac{f^{(k)}(0)}{k!} kx^{k-1} = \sum_{k=1}^{\infty} \frac{f^{(k)}(0)}{(k-1)!} x^{k-1}. \tag{B.17}$$

Using this, we can easily get a new Taylor series from the old one. For example, from (B.13) we get

$$\frac{1}{(1-x)^2} = \sum_{k=1}^{\infty} kx^{k-1}, \text{ for } -1 < x < 1, \tag{B.18}$$

and

$$\frac{2}{(1-x)^3} = \sum_{k=2}^{\infty} k(k-1)x^{k-2}, \text{ for } -1 < x < 1. \tag{B.19}$$

Another basic Taylor series is the one for the exponential function e^x, namely

$$e^x = 1 + x + \frac{x^2}{2} + \frac{x^3}{3!} + \cdots = \sum_{k=0}^{\infty} \frac{x^k}{k!}, \tag{B.20}$$

for all real numbers x. If $|x|$ is small, this implies that

$$e^x \approx 1 + x, \tag{B.21}$$

or, more accurately,

$$e^x \approx 1 + x + \frac{1}{2}x^2. \tag{B.22}$$

Another important Taylor series is that for the (natural) logarithm. If $-1 < x \le 1$, this is

$$\log(1+x) = x - \frac{x^2}{2} + \frac{x^3}{3} - \frac{x^4}{4} + \cdots = \sum_{k=1}^{\infty} (-1)^{k+1} \frac{x^k}{k}. \tag{B.23}$$

If $|x|$ is small, this implies that

$$\log(1+x) = x + o(x) \text{ at zero,} \tag{B.24}$$

or more accurately,

$$\log(1+x) = x - \frac{x^2}{2} + o(x^2) \text{ at zero.} \tag{B.25}$$

As a final example of a Taylor series expansion, we revisit the binomial theorem. From the fact that $(1+x)^\alpha$ is analytic at zero, we get the general version of the binomial theorem:

Theorem (General Binomial Theorem). For $-1 < x < 1$ and α *any real number,*

$$(1+x)^\alpha = \sum_{k=0}^{\infty} \binom{\alpha}{k} x^k. \tag{B.26}$$

A function can be analytic at points other than zero, as the following more general definition shows.

Definition. Let $f(x)$ be a function defined on an open interval containing a. Then f is analytic at a if (1) $f^{(k)}(a)$ exists for all k and (2) for all x in some open interval containing a (this interval being possibly smaller than the domain of f), we have

$$f(x) = \sum_{k=0}^{\infty} \frac{f^{(k)}(a)}{k!} (x-a)^k. \tag{B.27}$$

The nth order Taylor approximation following from this, generalizing equation (B.16), is

$$f(x) \approx \sum_{k=0}^{n} \frac{f^{(k)}(a)}{k!} (x-a)^k, \text{ for } x \text{ close to } a. \tag{B.28}$$

As two particularly important cases, the first-order, or linear, approximation is

$$f(x) \approx f(a) + (x-a)f'(a), \text{ for } x \text{ close to } a, \tag{B.29}$$

and the second-order approximation is

$$f(x) \approx f(a) + (x-a)f'(a) + \frac{(x-a)^2}{2} f''(a), \text{ for } x \text{ close to } a. \tag{B.30}$$

We have thus far defined "analytic" at a point. We will call a function $f(x)$ analytic in an interval if it is analytic at every point in the interval. If $f(x)$ is analytic at every point in its domain we will say that $f(x)$ is an *analytic function.*

B.13 Uniqueness of Taylor series

One of the things that makes Taylor series so useful is the fact that they satisfy certain uniqueness properties. In particular, if $f(x) = \sum_{n=0}^{\infty} a_n x^n$ and $g(x) = \sum_{n=0}^{\infty} b_n x^n$ and $f(x) = g(x)$ for all x in some open interval where both f and g converge, then $a_k = b_k$ for all k. This is because if

$f(x) = g(x)$ for all x in an open interval, then this interval must contain zero, and then $f^{(k)}(0) = g^{(k)}(0)$ for all k; thus

$$0 = f^{(k)}(0) - g^{(k)}(0) = k!(a_k - b_k),$$

which implies $a_k = b_k$ for all k.

This is a convenient fact. From it we see that a random variable with positive integer values is characterized by its pgf in the following sense. Let Y_1 and Y_2 be discrete random variables taking positive integer values and suppose that $\mathbb{p}_1(t)$ and $\mathbb{p}_2(t)$ are their respective pgfs. If $\mathbb{p}_1(t) = \mathbb{p}_2(t)$ on an open interval containing 1, then the probability distributions of Y_1 and Y_2 are identical. This useful observation is used, often implicitly, many times in this book. In the next section we will see that this is true even if we drop the assumption that the random variable takes only positive integer values.

B.14   Laurent Series

A Laurent series is a generalization of Taylor series, being of the form

$$\sum_{k=-\infty}^{\infty} a_k x^k.$$

If a random variable can take negative as well as positive integer values, its pgf is a Laurent series. A Laurent series is not defined at $x = 0$ if $a_k \neq 0$ for some $k < 0$. However, if $\sum a_k$ converges, the series is defined at $x = 1$. When it converges in an open interval containing 1, as it does in all cases we consider in this book, then similar properties hold for these series as did for Taylor series. In particular, a probability distribution is still characterized by its pgf.

B.15   Numerical Solutions of Equations

Many equations arising in practice cannot be solved exactly, and the best that is possible is to find arbitrarily accurate numerical solutions. For example, the calculation of λ from equation (9.4) will normally require these methods. Here we outline Newton's method for the approximate numerical solution of some equation.

Suppose that we wish to find the solution of the equation $f(x) = 0$. Let a be a first estimate of this solution. The equation (B.29) shows that an approximation x_1 to the required solution is given by

$$x_1 \approx a - \frac{f(a)}{f'(a)}.$$

We now take x_1 as a new estimate of the solution and find an improved estimate x_2 from the equation

$$x_2 \approx x_1 - \frac{f(x_1)}{f'(x_1)}.$$

This procedure can be repeated until an estimate of the solution having any desired degree of accuracy is found. In some cases the process can lead to a diverging rather than a converging sequence of values, so some care must be exercised in using it.

B.16   Statistical Differentials

Suppose that X_1 is a random variable with mean μ and small variance (so that X_1 is unlikely to differ much from μ). Equation (B.29) then shows that if f is analytic,

$$f(X_1) - f(\mu) \approx (X_1 - \mu)f'(\mu). \qquad (B.31)$$

Let $X_2 = f(X_1)$. Then taking expectations throughout in the approximation (B.31),

$$\text{mean of } X_2 \approx f(\mu). \qquad (B.32)$$

Squaring in equation (B.31) and then taking expectations leads to the first-order approximation

$$\text{variance of } X_2 \approx (f'(\mu))^2 \text{ variance of } X_1. \qquad (B.33)$$

The approximations made in deriving this formula imply that it must be used with some caution.

B.17   The Gamma Function

The *gamma function* is an analytic function defined for positive u by the integral

$$\Gamma(u) = \int_0^\infty e^{-t}t^{u-1}dt. \qquad (B.34)$$

The gamma function extends the factorial function to all positive numbers, as can be seen from the facts that

$$\Gamma(u+1) = u\Gamma(u), \text{ for } all \text{ positive real numbers } u, \qquad (B.35)$$

and

$$\Gamma(1) = 1, \qquad (B.36)$$

which imply

$$\Gamma(n) = (n-1)! \quad \text{when } n \text{ is a positive integer.} \qquad (B.37)$$

When u is positive but not an integer, equation (B.35) can be used to find $\Gamma(u)$, provided that values of the gamma function for values of u in (say) the interval $(1,2)$ are available. These values are listed in Table 6.1 of Abramowitz and Stegun (1972). A useful value of the gamma function is

$$\Gamma\left(\frac{1}{2}\right) = \sqrt{\pi}. \qquad (B.38)$$

We need two further facts about the derivative of the gamma function. First,

$$\left(\frac{d\,\Gamma(u)}{du}\right)_{u=1} = -\gamma, \qquad (B.39)$$

where γ is Euler's constant defined in (B.6). Second,

$$\left(\frac{d^2 \log \Gamma(u)}{d\,u^2}\right)_{u=1} = \frac{\pi^2}{6}. \qquad (B.40)$$

The fact that $\Gamma(n) = (n-1)!$ when n is a positive integer allows a rapid evaluation of the integral

$$\int_0^\infty \lambda x^n\, e^{-\lambda x}dx, \qquad (B.41)$$

which arises in connection with finding the moments of the exponential distribution (1.59). The change of variable $u = \lambda x$ leads to the calculation

$$\int_0^\infty \lambda^{-n}u^n\, e^{-u}du = n!\lambda^{-n}, \qquad (B.42)$$

for the nth moment of this distribution.

Fractional moments are also sometimes of interest: Thus if X is a random variable having the exponential distribution (1.59),

$$E(X^{1/2}) = \int_0^\infty \lambda^{-1/2}u^{1/2}e^{-u}du$$

$$= \frac{\Gamma(3/2)}{\sqrt{\lambda}} = \frac{\sqrt{\pi}}{2\sqrt{\lambda}} \qquad (B.43)$$

from (B.35) and (B.38).

B.18   Proofs by Induction

One of the most useful methods of proving some proposition $P(n)$ about a positive integer n is a proof by induction. In a proof of this type we (i) first prove that $P(n)$ is true for the particular case $n = 1$, and then (ii) show that the truth of $P(n)$ for the case $n = m$ implies its truth for the case $n = m + 1$. Using (ii) for the case $m = 1$, (i) and (ii) together imply the truth of $P(n)$ for the case $n = 2$. Then using this result and (ii) for the case $m = 2$ implies the truth of $P(n)$ for the case $n = 3$, etc. Thus (i) and (ii) together imply that $P(n)$ is true for all positive integers.

A classic simple example of such a proof is as follows. Define $S(n)$ by $S(n) = 1 + 2 + 3 + \cdots + n$ and then define the proposition $P(n)$ as $S(n) = n(n+1)/2$. This proposition is true for the case $n = 1$, since $1 = 1(1+1)/2$. Suppose that $P(n)$ is true for $n = m$, so that $1 + 2 + \cdots + m = m(m+1)/2$. Then $1 + 2 + \cdots + m + (m + 1) = (1 + 2 + \cdots + m) + (m + 1) = m(m + 1)/2 + (m + 1) = (m + 1)(m + 2)/2$. This shows that $P(n)$ is also true for the case $n = m + 1$.

B.19   Linear Algebra and Matrices

In this section we quote without proof several results from the theory of linear algebra that are relevant to material in the text. It is assumed that the reader is familiar with the basics of linear algebra, including the concepts of matrices and their eigenvalues and eigenvectors, of vectors, of matrix and vector products, and of the transposition operation (denoted here by "$'$"). The vector v is assumed to be a column vector, so that the vector v' is taken as a row vector.

Let P be an $s \times s$ matrix. This matrix has s *eigenvalues*, possibly not all distinct, denoted by $\lambda_1, \lambda_2, \ldots, \lambda_s$, found as the s solutions to the determinantal equation (in λ)

$$|P - \lambda I| = 0. \tag{B.44}$$

It is possible that there are solutions of equation (B.44) having multiplicity exceeding 1: For our purposes we assume that all eigenvalues of P are distinct.

Corresponding to the eigenvalue λ_j is a *right eigenvector* r_j and a *left eigenvector* ℓ'_j, both defined only up to a multiplicative constant, for which

$$P r_j = \lambda_j r_j, \quad \ell'_j P = \ell'_j \lambda_j. \tag{B.45}$$

We now assume that these eigenvectors are normalized so that

$$\ell'_j r_j = 1 \text{ for all values of } j, \quad j = 1, 2, \ldots, s. \tag{B.46}$$

444 Appendix B. Mathematical Formulae and Results

Although this normalization does not uniquely determine $\boldsymbol{\ell}'_j$ and $\boldsymbol{r}_j$, it is sufficient for our purposes. With this normalization, P can be expressed in terms of its eigenvalues and eigenvectors in a *spectral expansion* as

$$P = \sum_{j=1}^{s} \lambda_j \boldsymbol{r}_j \boldsymbol{\ell}'_j. \tag{B.47}$$

If λ_j is an eigenvalue of P with corresponding right eigenvector $\boldsymbol{r}_j$ and left eigenvector $\boldsymbol{\ell}'_j$, then for any positive integer n, $\lambda_j{}^n$ is an eigenvalue of P^n, with right and left eigenvectors $\boldsymbol{r}_j$ and $\boldsymbol{\ell}'_j$, respectively. This implies that the spectral expansion of P^n is

$$P^n = \sum_{j=1}^{s} \lambda_j{}^n \boldsymbol{r}_j \boldsymbol{\ell}'_j. \tag{B.48}$$

This result has relevance to us when P is the transition matrix of a Markov chain, since the spectral expansion of P^n gives useful information about n-step transition probabilities of this chain.

The theory is more complex when some eigenvalues of P have multiplicity greater than 1, but in all cases of interest to us, a spectral expansion of the form (B.48) exists and is used often in Markov chain theory in the text.

Appendix C

Computational Aspects of the Binomial and Generalized Geometric Distribution Functions

The expression for the generalized geometric distribution function $F_Y(y)$ given in equation (6.3) can be written as

$$\frac{(1-p)^{k+1}}{k!} \left(\frac{k!}{0!} + \frac{(k+1)!p}{1!} + \frac{(k+2)!p^2}{2!} + \cdots + \frac{y!p^{y-k}}{(y-k)!} \right)$$

$$= \frac{(1-p)^{k+1}}{k!} \cdot \frac{d^k}{dp^k} \left(1 + p + p^2 + \cdots + p^y \right), \quad y = k, k+1, k+2, \ldots.$$

Thus we can write $F_Y(y)$ as

$$F_Y(y) = \frac{(1-p)^{k+1}}{k!} \cdot \frac{d^k}{dp^k} \cdot \frac{1-p^{y+1}}{1-p}, \quad y = k, k+1, k+2, \ldots. \quad \text{(C.1)}$$

Other expressions for $F_Y(y)$ are possible.

For the geometric distribution case $k = 0$ the distribution function $F_Y(y)$ of Y is also given by equation (C.1) if we interpret the zeroth derivative of a function as the function itself.

It is also useful to calculate $\text{Prob}(Y \leq y - 1)$. Replacing y by $y - 1$ in equation (C.1), we find immediately that

$$\text{Prob}(Y \leq y - 1) = \frac{(1-p)^{k+1}}{k!} \cdot \frac{d^k}{dp^k} \cdot \frac{1-p^y}{1-p}, \quad y = k+1, k+2, k+3, \ldots. \quad \text{(C.2)}$$

The expression on the right-hand side can be calculated in one of two convenient ways. The first is by differentiation. The second uses a recurrence relation, as follows.

We denote the right-hand side in equation (C.2) by $Q_k(p, y)$. If we write $z = (1 - p^y)/(1 - p)$, we obtain

$$\frac{k! Q_k(p, y)}{(1-p)^{k+1}} = \frac{d^k z}{dp^k}. \qquad (C.3)$$

Differentiating both sides of this equation with respect to p, we get

$$\frac{(k+1)! Q_k(p, y)}{(1-p)^{k+2}} + \frac{k!}{(1-p)^{k+1}} \cdot \frac{d}{dp} Q_k(p, y) = \frac{d^{k+1} z}{dp^{k+1}}. \qquad (C.4)$$

Replacing k by $k + 1$ in (C.3), we get

$$\frac{(k+1)! Q_{k+1}(p, y)}{(1-p)^{k+2}} = \frac{d^{k+1} z}{dp^{k+1}}. \qquad (C.5)$$

The right-hand sides in equations (C.4) and (C.5) are equal. Equating the left-hand sides, we arrive at the recurrence relation

$$Q_{k+1}(p, y) = Q_k(p, y) + \frac{1-p}{k+1} \cdot \frac{d}{dp} Q_k(p, y). \qquad (C.6)$$

Given the initial function $Q_0(p, y) = 1 - p^y$, we can easily compute $Q_1(p, y)$, $Q_2(p, y), \dots$ in succession, using this recurrence relation. This provides a second computational approach to finding $\text{Prob}(Y \leq y - 1)$.

The recurrence relation (C.6) can be solved explicitly. We find (Wilf (1999)) that

$$Q_k(p, y) = \sum_{j=k+1}^{y} \binom{y}{j} (1-p)^j p^{y-j}. \qquad (C.7)$$

This is a sum of *binomial* probabilities. The discussion in Section 1.3.5 relating the distribution functions of the binomial and the generalized geometric show why this is so. The event that $y - 1$ or fewer trials occur before failure $k + 1$ is exactly the same as the event that there are $k + 1$ or more failures in the first y trials. $Q_k(p, y)$ is defined as the probability of the former event, and so it is also the probability of the latter event. This is what equation (C.7) states. This observation allows us to write down an expression for the distribution function of the binomial distribution (1.8).

Arratia et al. (1986) in effect show that for the case where y is large and p is small, a close approximation for the probability (C.2) is

$$\text{Prob}(Y \leq y - 1) \approx \frac{1 - y^k p^{y-k} (1-p)^k}{k!}. \qquad (C.8)$$

This is exact for $k = 0$ and when $k = 1$ differs from the exact value by an amount p^y, which, when p is small and y is large, is very small. For larger values of k the approximation (C.8) is also very accurate, although the accuracy decreases slightly as k increases. The approximation (C.8) is useful in many applications, and often saves us making an exact but tedious numerical calculation.

Appendix D
BLAST: Sums of Normalized Scores

Our aim in this appendix is to use the joint density function (9.33) for the case $r = 2$, namely

$$f(s_1, s_2) = \exp\left(-(s_1 + s_2) - e^{-s_2}\right), \qquad \text{(D.1)}$$

to derive the "$r = 2$" case of equation (9.34), namely

$$f(t) = \frac{e^{-t}}{2} \int_0^{+\infty} \exp(-e^{(y-t)/2})\, dy, \qquad \text{(D.2)}$$

for the density function $T_2 = S_1 + S_2$, the sum of the highest two normalized scores in a BLAST calculation.

To do this we introduce the dummy variable Y, defined by $Y = S_1 - S_2$. The domain of the two random variables S_1 and S_2 in the (S_1, S_2) plane is $S_1 \geq S_2$, and this transforms into the domain $Y \geq 0$ in the (T_2, Y) plane. The Jacobian J of the transformation from (S_1, S_2) to (T_2, Y) is 2, and we can write $s_1 + s_2 = t$ and $s_2 = (t - y)/2$. Equations (2.154) and (D.1) then show that the joint density function of T_2 and Y is

$$f(t, y) = \frac{1}{2} \exp(-t - e^{(y-t)/2}). \qquad \text{(D.3)}$$

The density function of T_2 is found by integrating out y over the domain $y \geq 0$ in this joint density function, and this leads immediately to equation (D.2).

References

Abramowitz, M. and I. Stegun (1972). *Handbook of Mathematical Functions*. Dover, New York.

Adams, M.D. et al. (2000). The genome sequence of *Drosophila Melanogaster*. *Science* **287**, 2185-2195.

Altschul S.F. (1991). Amino acid substitution matrices from an information theoretic perspective. *J. Mol. Biol.* **219**, 555–565.

Altschul, S.F. and W. Gish (1996). Local alignment statistics. *Methods in Enzymology* **266**, 460–480.

Altschul, S.F., T.L. Madden, A.A. Schaffer, J. Zhang, Z. Zhang, W.Q. Miller, and D.J. Lipman (1997). Gapped BLAST and PSI-BLAST: a new generation of protein database search programs. *Nucleic Acids Research* **25**, 3389–3402.

Andrews, D.K.W. (2000). Inconsistency of the bootstrap when the parameter is on the boundary of the parameter space. *Econometrica* **68**(2), 399–406.

Arratia, R., L. Gordon, and M.S. Waterman (1986). An extreme value theory for sequence matching. *Ann. Statist.* **14**, 971–983.

Arratia, R., E.S. Lander, S. Tavaré, and M.S. Waterman (1991). Genomic mapping by anchoring random clones: a mathematical analysis. *Genomics* **11**, 806–827.

Bailey, T.L. and M. Gribskov (1998). PROSITE: a dictionary of sites and patterns in proteins. *J. Comp. Biol.* **5**, 211–221.

Benner, S.A., M.A. Cohen, and G.H. Gonnet (1994). Amino acid substitution during functionally constrained divergent evolution of protein sequences. *Protein Engineering* **7**, 1323–1332.

Bernaola-Galván, P., I. Grosse, P. Carpena, J.L. Oliver, R. Román-Roldán, and H.E. Stanley (2000). Finding borders between coding and non-coding DNA regions by an entropic segmentation method *Phys. Rev. Lett.* **85**, 1342–1345.

Biaudet, V., M. El Karoui, and A. Gruss (1998). Codon usage can explain GT-rich islands surrounding Chi sites on the *Escherichia coli* genome. *Mol. Microbiol.* **29**, 666–699.

Blaisdell, B.E. (1985). A method of estimating from two aligned present-day DNA sequences their ancestral composition and subsequent rates of substitution, possibly different in the two lineages, corrected for multiple and parallel substitutions at the same site. *J. Mol. Evol.* **22**, 69–81.

Blom, G. (1982). On the mean number of random digits until a given sequence occurs. *J. Appl. Prob.* **19**, 136–143.

Blom, G. and D. Thorburn (1982). How many random digits are required until given sequences are obtained? *J. Appl. Prob.* **19**, 518–531.

Burge, C. (1997). Identification of complete gene structures in human genomic DNA. Ph.D. thesis, Stanford University, Stanford, CA.

Burge, C. and S. Karlin (1997). Prediction of complete gene structures in human genomic DNA. *J. Mol. Biol.* **268**, 78–94.

Burge, C. (1998). Modeling Dependencies in Pre-mRNA Splicing Signals, Section II.8 in *Computational Methods in Molecular Biology*, S. Salzberg, D. Searls, and S. Kasif, eds., Elsevier Science.

Bussemaker, H.J., H. Li and E.D. Siggia (2000). Building a dictionary for genomes: identification of presumptive regulatory rate by statistical analysis *Proc. Nat. Acad. Sci.* **97**, 10096–10100.

C. Elegans Sequencing Consortium (1998). Genome sequence of the nematode *C. elegans*: a platform for investigating biology. *Science* **282**, 2012–2018.

Chang, J.T. (1996a). Inconsistency of evolutionary tree topology reconstruction methods when substitution rates vary across characters. *Math. Biosci.* **134**, 189–215.

Chang, J.T. (1996b). Reconstruction of Markov chain models on evolutionary trees: identifiability and consistency. *Math. Biosci.* **137**, 51–73.

Chedin, F., P. Noirot, V. Biaudet, and S. Ehrlich (1998). A five-nucleotide sequence protects DNA from exonucleolytic degradation by AddAB, the RecBCD analogue of *Bacillus subtilis*. *Mol. Microbiol.* **31**, 1369–1377.

Chernick, M. (1999). *Bootstrap Methods: A Practitioner's Guide*. Wiley, New York.

Cheung, V.G., J.P. Gregg, K.J. Gogolin-Ewens, J. Bandong, C.A. Stanley, L. Baker, M.J. Higgins, N.J. Nowak, T.B. Shows, W.J. Ewens, S.F. Nelson and R.S. Spielman (1998). Linkage-disequilibrium mapping without genotyping. *Nature Genetics* **18**, 225-230.

Cowan, R. (1991). Expected frequencies of DNA using Whittle's formula. *J. Appl. Prob.* **28**, 886–892.

Daudin, J.J. and S. Mercier (2000). Distribution exacte du score local d'une suite de variables indépendent et identiquement distribuées. *C. R. Acad. Sci. Paris* **329**, 815–820.

Davison, A.C. and D.V. Hinkley (1997). *Bootstrap Methods and their Applications*. Cambridge University Press, Cambridge.

Dayhoff, M.O., R.M. Schwartz, and B.C. Orcutt (1978). A model of evolutionary change in proteins, pp. 3435–358 in *Atlas of Protein Sequence and Structure*, **5**, Supplement 3.

Dixon, M.T. and D.M. Hillis (1993). Ribosomal RNA secondary structure: compensatory mutations and implications for phylogenetic analysis. *Mol. Biol. Evo.* **10**, 256–267.

Dembo, A., S. Karlin, and O. Zeitouni (1994a). Critical phenomena for sequence matching with scoring. *Ann. Prob.* **22**, 1993–2021.

Dembo, A., S. Karlin, and O. Zeitouni (1994b). Limit distribution of maximal non-aligned two-sequence segmental score. *Ann. Prob.* **22**, 2022–2039.

Dudoit, S., Y.H. Yang, M.J. Callow, and T.P. Speed (2000). Statistical methods for identifying differentially expressed genes in replicated cDNA microarray experiments. Stanford University technical report 578, Department of Biochemistry.
www.stat.berkeley.edu/users/terry/Zarray/Html/matt.html.

Dunham, I., N. Shimizu, B.A. Roe, S. Chissoe et al. (1999). The DNA sequence of human chromosome 22. *Nature* **402**, 489–495.

Durbin, R., S. Eddy, A. Krogh, and G. Mitchison (1998). *Biological Sequence Analysis*. Cambridge University Press, Cambridge.

Edwards, A.W.F. (1992). *Likelihood*. Johns Hopkins Press, Baltimore, MD.

Edwards, A.W.F. and L.L. Cavalli-Sforza (1964). Reconstruction of phylogenetic trees. *Phenetic and Phylogenetic Classification*, V.H. Heywood and J. McNeil, eds., Systematics Association Publication 6, 67–76.

Efron, B. (1982). *The Jackknife, the Bootstrap and Other Resampling Plans*. SIAM Applied Mathematics Publication. Philadelphia, PA.

Efron, B. and R.J. Tibshirani (1993). *An Introduction to the Bootstrap*. Chapman and Hall, New York.

Eisen, J.A. (1998). Phylogenomics: improving functional predictions for uncharacterized genes by evolutionary analysis. *Genome Research* **8**, 163–167.

Ewens, W.J., R.C. Griffiths, S.N. Ethier, S.A. Wilcox, and J.A. Marshall Graves (1992). Statistical analysis of *in situ* hybridization data: derivation of the z_{max} test. *Genomics* **12**, 675–582.

Feller, W. (1968). *An Introduction to Probability Theory and its Applications*. Volume I, 3rd edition, Wiley, New York.

Felsenstein, J. (1978a). The number of evolutionary trees. *Systematic Biology* **27**, 27–33.

Felsenstein, J. (1978b). Cases in which parsimony or compatibility methods will be positively misleading. *Syst. Zool.* **27**, 401–410.

Felsenstein, J. (1981). Evolutionary trees from DNA sequences: A maximum likelihood approach. *J. Mol. Evol.* **17**, 368–376.

Felsenstein, J. (1983). Statistical inference of phylogenies. *J. Roy. Statist. Soc. A.* **146**, 246–272.

Felsenstein, J. (1985). Confidence limits on phylogenies: an approach using the bootstrap. *Evolution* **39**, 783–791.

Felsenstein, J., and H. Kishino (1993). Is there something wrong with the bootstrap on phylogenies? A reply to Hillis and Bull. *Syst. Biol.* **2**, 193–200.

Felsenstein, J. (1980–2000). Phylogeny Inference Package. Web resource. (http://evolution.genetics.washington.edu/phylip.html).

Fitch, W.M. (1971). Toward defining the course of evolution: minimum change for a specified tree topology. *Systematic Zoology* **20**, 406–416.

Fukami, K. and Y. Tateno (1989). On the uniqueness of the maximum likelihood method for estimating molecular trees: uniqueness of the likelihood point. *J. Mol. Evol.* **28**, 460–464.

Gojobori, T., K. Ishii, and M. Nei (1982). Estimation of the average number of nucleotide substitutions when the rate of substitution varies with nucleotide. *Mol. Biol. Evol.* **13**, 93–104.

Goldman, N. (1993). Statistical tests of models of DNA substitution. *J. Mol. Evol.* **36**, 182–198.

Goldman, N. and S. Whelan (2000). Statistical tests of gamma-distributed rate heterogeneity in models of sequence evolution in phylogenetics. *Mol. Biol. Evol.* **17**, 975–978.

Golub, T.R., D.K. Slonim, P. Tamayo, C. Huard, M. Gaasenbeek, J.P. Mesirov, H. Coller, M.L. Loh, J.R. Downing, M.A. Caligiuri, C.D. Bloomfield, E.S. Lander (1999). Molecular classification of cancer: class discovery and class prediction by gene expression monitoring. *Science* **286(5439)**, 531-537.

Grant, G.R., V.G. Cheung, E. Manduchi and W.J. Ewens (1999). Significance testing for discrete identity-by-descent mapping. *Ann. Hum. Genet.* **63**, 441-454.

Hall, P. (1992). *The Bootstrap and Edgeworth Expansion.* Springer–Verlag, New York.

Hasegawa, M., H. Kishino, and T. Yano (1985). Dating of the human–ape splitting by a molecular clock of mitochondrial DNA. *J. Mol. Evol.* **22**, 160–174.

Hattori, M. et al. (2000). The DNA sequence of human chromosome 21. *Nature* **405**, 311–319.

Henikoff, S. and J.G. Henikoff (1992). Amino acid substitution matrices from protein blocks. *Proc. Nat. Acad. Sci.* **89**, 10915–10919.

Hillis, D.M. and J.-J. Bull (1993). An empirical test of bootstrapping as a method for assessing confidence in phylogenetic analysis. *Syst. Biol.* **2**, 182–192.

Holmes, S.P. (1999). Phylogenies: an overview. pp. 81–118 in *Statistics in Genetics*, M.E. Halloran and S. Geisser, eds., Springer–Verlag, New York.

Huelsenbeck, J.P. and K. Crandall (1997). Phylogeny estimation and hypothesis testing using maximum likelihood. *Ann. Rev. Ecol. Syst.* **28**, 437–466.

Huelsenbeck, J.P., D.M. Hillis, and R. Nielsen (1996). A likelihood ratio test of monophyly. *Syst. Biol.* **45**, 546–558.

Huelsenbeck, J.P. and B. Rannala (1997). Phylogenetic methods come of age: testing hypotheses in an evolutionary context. *Science* **276**, 227–232.

Huelsenbeck, J.P. and R. Nielsen (1999). Variation in the pattern of nucleotide substitution across sites. *J. Mol. Evol.* **48**, 86–93.

International Human Genome Sequencing Consortium (2001). Initial sequencing and analysis of the human genome. *Nature* **409**, 860–921.

Jonassen, I., J.F. Collins, and D.G. Higgins (1995). Finding flexible patterns in unaligned protein sequences. *Protein Science* **4**, 1587–1595.

Jukes, T.H. and C.R. Cantor, (1969). Evolution of protein molecules. pp. 21–132 in *Mammalian Protein Metabolism*, H.N. Munro (ed.), Academic Press, New York.

Karlin, S. (1994). Statistical studies of biomolecular sequences: score-based methods. *Phil. Trans. R. Soc. Lond. B* **344**, 391–402.

Karlin S. and S.F. Altschul (1990). Methods for assessing the statistical significance of molecular sequence features by using general scoring schemes. *Proc. Nat. Acad. Sci.* **87**, 3364–2268.

Karlin S. and S.F. Altschul (1993). Applications and statistics for multiple high-scoring segments in molecular sequences. *Proc. Nat. Acad. Sci.* **90**, 5873–5877.

Karlin, S. and V. Brendel (1992). Chance and statistical significance in protein and DNA sequence analysis. *Science* **257**, 39–49.

Karlin, S. and V. Brendel (1996). New directions in education: computational tools for the molecular biologist and biological sources for the mathematician. pp. 408–427 in *Education in a Research University*, K.J. Arrow et al. eds., Stanford University Press, Stanford, California.

Karlin, S., C. Burge and A.M. Campbell (1992). Statistical analyses of counts and distributions of restriction sites in DNA sequences. *Nucl. Acids Res.* **20**, 1363–1370.

Karlin S. and A. Dembo (1992). Limit distributions of maximal segmental score among Markov-dependent partial sums. *Adv. Appl. Prob.* **24**, 113–140.

Karlin, S., F. Ghandour, F. Ost, S. Tavaré, and L.J. Korn (1983). New approaches for computer analysis of nucleic acid sequences. *Proc. Nat. Acad. Sci.* **80**, 5660–5664.

Karlin, S. and C. Macken (1991a). Assessment of inhomogeneities in an *E. coli* physical map. *Nucleic Acids Research* **19**, 4241–4246.

Karlin, S. and C.A. Macken (1991b). Some statistical problems in the assessment of inhomogeneities of DNA sequence data. *J. Amer. Stat. Assoc.* **86**, 27–35.

Karlin, S. and H. Taylor (1975). *A First Course in Stochastic Processes.* Academic Press, New York.

Karlin, S. and H. Taylor (1981). *A Second Course in Stochastic Processes.* Academic Press, New York.

Kelly, C. and G. Churchill (1996). Biases in amino acid replacement matrices and alignment scores due to rate heterogeneity. *J. Comp. Biol.* **3**, 307–318.

Kimura, M. (1980). A simple method for estimating evolutionary rate in a finite population due to mutational production of neutral and nearly neutral base substitution through comparative studies of nucleotide sequences. *J. Molec. Biol.* **16**, 111–120.

Krogh, A., M. Brown, I.S. Mian, K. Sölander, and K. Hausler (1994). Hidden Markov Models in Computational Biology: applications to protein modeling. *J. Molec. Biol.* **235**, 1501–1531.

Kuhner, M. and J. Felsenstein (1996). Simulation comparison of phylogeny algorithms under equal and unequal evolutionary rates. *J. Biol. Evol.* **11**, 459–468.

Kullback, S. (1978). *Information Theory and Statistics*. Dover, New York.

Lake, J.A. (1987). Reconstructing evolutionary trees from DNA and protein sequences: evolutionary parsimony. *Mol. Biol. Evol.* **4**, 167–191.

Lander, E.S. and M.S. Waterman (1988). Genomic mapping by fingerprinting random clones: a mathematical analysis. *Genomics* **2**, 231–239.

Lawrence, C.E, S.F. Altschul, M.S. Boguski, J.S. Liu, A.F. Neuwald, and J.C. Wootton (1993). Detecting subtle sequence signals: a Gibbs sampling strategy for multiple alignment. *Science* **262**, 208–214.

Lehmann, E.L. (1991). *Theory of Point Estimation*. Wadsworth, Pacific Grove, CA.

Lehmann, E.L. (1986). *Testing Statistical Hypotheses*. Wiley, New York.

Leung, M.Y., G.M. Marsh, and T.P. Speed (1996). Over and underrepresentation of short DNA words in Herpes virus genomes. *J. Comp. Biol.* **3**, 345–360.

Lewis, P.O. (1998). Maximum likelihood as an alternative to parsimony for inferring phylogeny using nucleotide sequence data. pp. 132–163 in *Molecular Systematics of Plants II*, P. Soltis, D. Soltis, J. Doyle eds., Kluwer, Dordrecht.

Liò, P. and N. Goldman (1998). Models of molecular evolution and phylogeny. *Genome Research* **8**, 1233–1244.

Manduchi E., G.R. Grant, S.E. McKenzie, G.C. Overton, S. Surrey, and C.J. Stoeckert (2000). Generation of patterns from gene expression data by assigning confidence to differentially expressed genes. *Bioinformatics* **16**, 685–698. See also http://www.cbil.upenn.edu/PaGE

Manly, B.F.J. (1997). *Randomization, Bootstrap and Monte Carlo Methods in Biology*. Chapman and Hall, London.

Mott, R. and R. Tribe (1999). Approximate statistics of gapped alignments. *J. Comp. Biol.* **6**, 91–112.

Müller, T. and M. Vingron (2000). Modeling amino acid replacement. To appear in *Journal of Computational Biology*.

Muse, S. (1995). Evolutionary analysis when nucleotides do not evolve independently. pp. 115–124 in *Current Topics on Molecular Evolution*, M. Nei and N. Takahata eds., Penn. State Univ. Press, College Park, PA.

National Bureau of Standards (1953). *Probability Tables for the Analysis of Extreme-Value Data*. Applied Mathematics Series, Washington.

Neuwald, A.F. and P. Green (1994). Detecting patterns in protein sequences. *J. Mol. Biol.* **239**, 698–712.

Norris, J.R. (1997). *Markov Chains*. Cambridge University Press, Cambridge.

Ota, R., P. Waddell, and H. Kishino (1999). Statistical distribution for testing the resolved tree against star tree. pp. 15–20 in *Proceedings of the Annual Joint Conference of the Japanese Biometrics and Applied Statistics Societies*. Sinfonica, Minato-ku, Japan.

Ota, R., P. Waddell, M. Hasagawa, H. Shimodaira, and H. Kishino (2000). Appropriate likelihood ratio tests and marginal distributions for evolutionary tree models with constraints on parameters. *Mol. Biol. Evol.* **17**, 783–803.

Pearson, W.R. (1998). Empirical statistical estimates for sequence similarity searches. *J. Mol. Biol.* **276**, 71–84.

Press, H.W., S.A. Teukolsky, W.T. Vetterling, B.P. Flannery (1992). pp. 275–286 in *Numerical recipes in C* (2nd edition), Cambridge University

Press, Cambridge.

Rabiner, L.R. (1989). A tutorial on hidden Markov models and selected applications in speech recognition. *Proc. IEEE.* **77**, 257–286.

Reinert, G., S. Schbath, and M.S. Waterman (2000). Probabilistic and statistical properties of words: an overview. *J. Comp. Biol.* **7**, 1–46.

Rigoutsos, I. and A. Floratos (1998). Motif discovery without alignment or enumeration. In *Proc. RECOMB98*, S. Istrail, P. Pevzner, and M.S. Waterman eds., 211–227.

Robin, S. and J.J. Daudin (1999). Exact distribution of word occurrences in a random sequence of letters. *J. Appl. Prob.* **36**, 179–193.

Robinson, A.B. and L.R. Robinson (1991). Distribution of glutamine and asparagine residues and their near neighbors in peptides and proteins. *Proc. Nat. Acad. Sci.* **88**, 8880–8884.

Rogers, J.S. and D.L. Swofford (1999). Multiple local maxima for likelihoods of phylogenetic trees: a simulation study. *Mol. Biol. Evol.* **16**, 1079–1085.

Saitou, N. and M. Nei (1987). The neighbor-joining method: a new method for reconstructing phylogenetic trees. *Mol. Biol. Evol.* **4**, 406–425.

Schadt, E.E., J.S. Sinsheimer, and K. Lange (1998). Computational advances in maximum likelihood methods for molecular phylogeny. *Genome Research* **8**, 222–233.

Schbath, S. (1997). Coverage processes in physical mapping by anchoring random clones. *J. Comp. Biol.* **4**, 61–82.

Schbath, S., N. Bossard, and S. Tavaré (2000). The effect of nonhomogeneous clone length distribution on the progress of an STS mapping project. *J. Comp. Biol.* **7**, 47–57.

Schenker, N. (1985). Qualms about bootstrap confidence intervals. *J. Amer. Stat. Assoc.* **80**, 360–361.

Schöniger, M. and A. von Haeseler (1994). A stochastic model for the evolution of autocorrelated DNA sequences. *Mol. Phylogeny Evol.* **3**, 240–247.

Schöniger, M. and A. von Haeseler (1995). A stochastic model for the evolution of autocorrelated DNA sequences. *Syst. Biol.* **44**, 533–547.

Self, S.G. and K.-L. Liang (1987). Asymptotic properties of maximum likelihood estimators and likelihood ratio tests under non-standard conditions. *J. Amer. Stat. Assoc.* **82**, 605–610.

Shaffer, J.S. (1995). Multiple hypothesis testing. *Annu. Rev. Psych.* **46**, 561–584.

Siegmund, D. and B. Yakir (2000). Approximate p-values for local sequence alignments. *Ann. Stat.* **28**, 657-680.

Smith, T.F. and M.S. Waterman (1981). The identification of common molecular subsequences. *J. Mol. Biol.* **147**, 195–197.

Sokal, R.R. and C.D. Michener (1958). A statistical method for evaluating systematic relationships. *University of Kansas Scientific Bulletin* **28**, 1409–1438.

Sprent, P. (1998). *Data Driven Statistical Methods.* Chapman and Hall, London.

Stanhope, M.J., M.R. Smith, V.G. Waddell, C.A. Porter, M.S. Shivji, and M. Goodman (1996). Mammalian Evolution and the Interphotoreceptor Retinoid Binding Protein (IRBP) Gene: Convincing Evidence for Several Superordinal Clades. *J. Mol. Evol.* **43**, 83–92.

Steel, M.A. (1994). The maximum likelihood point for a phylogenetic tree is not unique. *Syst. Biol.* **43**, 560–564.

Steel, M.A., L.A. Székely, and M.D. Hendy (1994). Reconstructing trees when sequence sites evolve at variable rates. *J. Comp. Biol.* **1**, 153–163.

Studier, J.A. and K. Keppler (1988). A note on the neighbor-joining algorithm of Saitou and Nei. *Mol. Biol. Evol.* **5**, 729–731.

Swofford, D.L., G.J. Olsen, P.J. Waddell, and D.M. Hillis (1996). Phylogenetic inference. pp. 407–514 in *Molecular Systematics*, ed. by D.M. Hillis, C. Moritz and B.K. Mable. Sinauer, Sunderland MA.

Takahata, N. and M. Kimura (1981). A model of evolutionary base substitutions and its application with special reference to rapid change of pseudogenes. *Genetics* **98**, 641–657.

Tavaré, S. (1986). Some probabilistic and statistical problems in the analysis of DNA sequences. *Lectures on Mathematics in the Life Sciences* **17**, 57–86.

Tavaré, S. and B.W. Giddings (1989). Some statistical aspects of the primary structure of nucleotide sequences. pp. 117–132 in *Mathematical Methods for DNA Sequences*, ed. by M.S. Waterman. CRC Press, Boca Raton, Florida.

Thompson, J.D., D.G. Higgins, T.J. Gibson (1994). CLUSTAL W: improving the sensitivity of progressive multiple sequence alignment through sequence weighting, position-specific gap penalties and weight matrix choice. *Nucleic Acids Res.* **22**, 4673–4680.

Tourasse, N.J. and W.-H. Li (1999). Performance of the relative-rate test under nonstationary models of nucleotide substitution. *Mol. Biol. Evol.* **16**, 1068–1078.

Trends Guide to Bioinformatics (1998). M. Patterson and M. Handel eds., Elsevier Science, New York.

Tuffley, C. and M. Steel (1997). Links between maximum likelihood and maximum parsimony under a simple model of site substitution. *Bull. Math. Biol.* **59**, 581–607.

Venter, J.C. et al. (2001). The sequence of the human genome. *Science* **291**, 1304–1351.

Waterman, M.S. (1995). *Introduction to Computational Biology*. Chapman and Hall, New York.

Westfall, P.H. and S.S. Young (1993). *Resampling-based Multiple Testing: Examples and Methods for P-value Adjustment*. Wiley, New York.

Whelan, S. and N. Goldman (1999). Distribution of statistics used for the comparison of models of sequence evolution in phylogenetics. *Mol. Biol. Evol.* **16**, 1292–1299.

Wilf, H. (1999). Personal communication.

Wilks, S.S. (1962). *Mathematical Statistics*. Wiley, New York.

Yang, Z. (1993). Maximum-likelihood estimation of phylogeny from DNA sequences when substitution rates differ over sites. *Mol. Biol. Evol.* **10**,

1396–1401.

Yang, Z. (1994). Estimating the pattern of nucleotide substitution. *J. Mol. Evol.* **39**, 105–111.

Yang, Z. (1996a). Among-site rate variation and its impact on phylogenetic analysis. *Trends Ecol. Evol.* **11**, 367–372.

Yang, Z. (1996b). Maximum-likelihood models for combined analyses of multiple sequence data. *J. Mol. Evol.* **42**, 587–596.

Yang, Z. (1997a). *Phylogenetic analysis by maximum likelihood (PAML), ver. 3.* Department of Biology, University College London.

Yang, Z. (1997b). PAML: A program package for phylogenetic analysis by maximum likelihood. *Comm. Appl. Biosci.* **13**, 555–556.

Zharkikh, A. and W.-H. Li (1992a). Statistical properties of bootstrap estimation of phylogenetic variability from nucleotide sequences. I. Four taxa with a molecular clock. *Mol. Biol. Evol.* **9**, 1119–1147.

Zharkikh, A. and W.-H. Li (1992b). Statistical properties of bootstrap estimation of phylogenetic variability from nucleotide sequences. II. Four taxa without a molecular clock. *J. Mol. Evol.* **35**, 356–365.

Zhu, J., J.S. Liu, and C.E. Lawrence (1998). Bayesian adaptive sequence alignment algorithms. *Bioinformatics* **14**, 25–39.

Author Index

Index

DATE DUE

180390			
3/14/04			
ILL (IAH)			
16113375			
03/01/06			